Periodic Table of the Elements

1 H 1.0																	2 He 4.0
3 Li 6.9	4 Be 9.0											5 B 10.8	6 C 12.0	7 N 14.0	8 O 16.0	9 F 19.0	10 Ne 20.2
11 Na 23.0	12 Mg 24.3											13 A1 27.0	14 Si 28.1	15 P 31.0	16 S 32.1	17 C1 35.5	18 Ar 39.9
19 K 39.1	20 Ca 40.1	21 Sc 45.0	22 Ti 47.9	23 V 50.9	24 Cr 52.0	25 Mn 54.9	26 Fe 55.8	27 Co 58.9	28 Ni 58.7	29 Cu 63.5	30 Zn 65.4	31 Ga 69.7	32 Ge 72.6	33 As 74.9	34 Se 79.0	35 Br 79.9	36 Kr 83.8
37 Rb 85.5	38 Sr 87.6	39 Y 88.9	40 Zr 91.2	41 Nb 92.9	42 Mo 95.9	43 Tc (97)	44 Ru 101.1	45 Rh 102.9	46 Pd 106.4	47 Ag 107.9	48 Cd 112.4	49 In 114.8	50 Sn 118.7	51 Sb 121.8	52 Te 127.6	53 I 126.9	54 Xe 131.3
55 Cs 132.9	56 Ba 137.3	see below 57-71	72 Hf 178.5	73 Ta 180.9	74 W 183.9	75 Re 186.2	76 Os 190.2	77 Ir 192.2	78 Pt 195.1	79 Au 197.0	80 Hg 200.6	81 Tl 204.4	82 Pb 207.2	83 Bi 209.0	84 Po 210	85 At (210)	86 Rn (222)
87 Fr (223)	88 Ra (226)	see below 89-103															

57 La 138.9	58 Ce 140.1	59 Pr 140.9	60 Nd 144.2	61 Pm (145)	62 Sm 150.4	63 Eu 152.0	64 Gd 157.3	65 Tb 158.9	66 Dy 162.5	67 Ho 164.9	68 Er 167.3	69 Tm 168.9	70 Yb 173.0	71 Lu 175.0

89 Ac (227)	90 Th 232.0	91 Pa (231)	92 U 238.0	93 Np (237)	94 Pu (244)	95 Am (243)	96 Cm (247)	97 Bk (247)	98 Cf (251)	99 Es (254)	100 Fm (257)	101 Md (256)	102 No (254)	103 Lw (257)

Botany: A Human Concern

Botany: A Human Concern

David Rayle and Lee Wedberg

San Diego State University

HOUGHTON MIFFLIN COMPANY BOSTON

Atlanta Dallas Geneva, Illinois Hopewell, New Jersey Palo Alto London

Cover photograph by Albert Gregory

Chapter openers by Albert Gregory

Chapter 1 opener courtesy of Turtox

Line art by James Loates

Photographs by the authors

Printed in the U.S.A.
Library of Congress Catalog Card Number: 74–15591
ISBN: 0–395–20112–8

Contents

Preface

We began this project for several reasons, but perhaps a prime consideration was our belief in the need for a relevant, readable plant biology text of an intermediate length between the classical and encyclopedic nature of some and the succinct but incomplete nature of others. In short, our aim was to construct a text that was particularly appropriate for a single semester or quarter botany or plant biology course; a text that could be easily read cover to cover by the introductory student.

Throughout this book we have tried to present topics solidly, but with more than casual reference to contemporary problems. Thus, we believe the potential science major will have the necessary background for more advanced courses while the nonmajor will not only see the scientific basis for understanding many of our environmental problems, but in addition will appreciate the value of basic research in attacking those problems.

While we believe the subject material is most easily presented in toto and in the order presented here, other variations might better suit certain individual instructional styles and/or the requirements of the quarter system. For example, some might prefer to read Chapters 4 and 5 before Chapter 3 in order to pursue more vigorously the details of plant breeding. Several chapters are divided into two parts. In these cases we feel that instructors offering a course based on the quarter system may prefer to abbreviate or eliminate the "Part Two" sections of some of the chapters, rather than the single-unit chapters. For some classes greater "molecular" emphasis would be achieved by use of the appendices, particularly Appendix Two, which summarizes respiration.

As with all textbooks, we have treated some subjects in greater detail than others, and perhaps abbreviated those subjects in which your particular interest lies. We welcome comments on content as well as on other matters that the reader may deem appropriate. Indeed, many comments made by students at San Diego State, who tested early versions of the text, have significantly changed the nature of the material in this edition. Your response will ensure that future editions will be even better.

While we have undoubtedly made some errors in interpretation as well as factual matters, we are indebted to those who have reduced their frequency. In particular, we should like to thank Joan Mentze, and Ken Johnson for their assistance. We should also like to acknowledge John E. Benneth of the American Forest Institute, the Weyerhaeuser Company, Dave Burwell of the Rosboro Lumber Company, Al Merrill of Louisiana-Pacific

Corporation, and the Crown Zellerbach Corporation. Their generosity and kindnesses made possible the sections on forestry and paper-making.

For their help as reviewers at various stages of the manuscript, we would like to thank Donald S. Emmeluth, Fulton-Montgomery Community College, Johnstown, N.Y.; Howard M. Lenhoff, University of California at Irvine; Irving McNulty, The University of Utah; Ivan G. Palmblad, Utah State University; Livija Raudzens, University of Massachusetts at Amherst; and James L. Seago, State University of New York at Oswego.

David Rayle
Lee Wedberg

1 Productivity and Consumption: Themes in the Study of Life

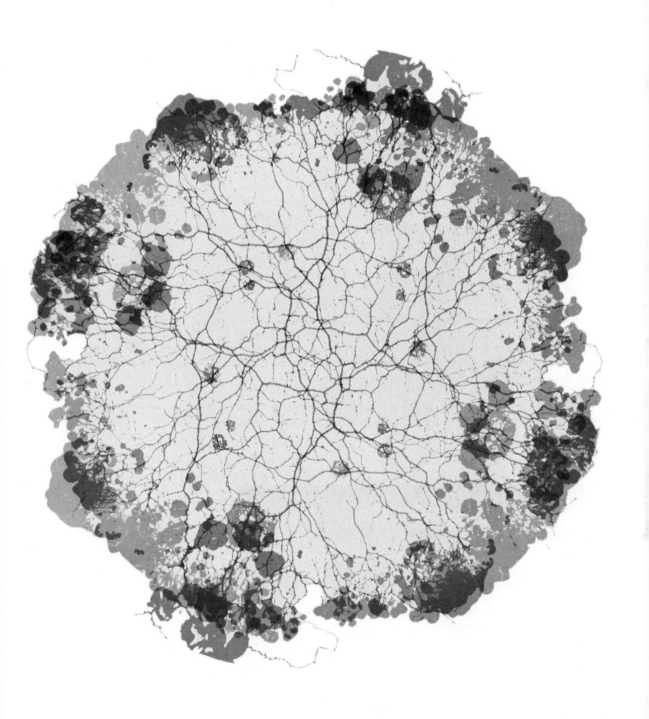

Daily we witness changes in the physical environment which in some way will alter the style and quality of our life. Many accept these changes as inevitable. If, for example, the Los Angeles basin is to be made safe from flooding, then the Los Angeles River must be contained during heavy rainfall. If our expanding population is to have adequate shelter, then dwellings must be provided. Adequate food must be produced in a minimum of space. And clearly, if we are to maintain any kind of sanity in the midst of all this, recreational facilities must be made available. In fact, to a majority of us these sorts of changes are undoubtedly functionally satisfactory, if not aesthetically pleasing (Figure 1–1).

Unfortunately, these changes in our environment have much broader implications than most of us realize, affecting the quality of the air we breathe, the water we drink, and the food we eat. Our environment is a keenly balanced, immensely complicated life-system. As you make your way through this book, you will come to realize that plants are not just pleasant objects to look at and convenient sources of construction material. Rather, they are sophisticated organisms that together with animals form a keenly balanced system which is responsible for life as we know

FIGURE 1–1.

A flood-control project protects the nearby residential, business, and industrial areas from occasional flooding. However, the aesthetic and recreational values of a natural stream are lost.

it. Perhaps we should go even further and say that without plants there would be no life. Aesthetics aside for a moment, you should realize not only that the physical environment is important for maintaining plant life but also that plants themselves play an integral part in maintaining our environment.

Toward Understanding Biological Energy

Perhaps because of the currently revived interest in environmental quality (and the associated conflict with the even more newly recognized energy crisis), some of you might be under the impression that the very important role played by plants is a relatively new concept or discovery. Quite the contrary. Two hundred years ago Joseph Priestley discovered one of the essential features of animal-plant interactions. He found that plants and animals together were parts of a balanced interacting system which kept the air in good condition, as Figure 1–2 and the following excerpts from his original (1772) publication show:

> I flatter myself that I have accidentally hit upon a method of restoring air which has been injured by the burning of candles, and that I have discovered at least one of the restoratives which nature employs for this purpose. It is vegetation . . .
> . . . One might have imagined that, since common air is necessary to vegetable, as well as to animal life, both plants and animals had affected it in the same manner, and I own I had that expectation, when I first put a sprig of mint into a glass jar, standing inverted in a vessel of water; but when it continued growing there for some months, I found that the air would neither extinguish a candle, nor was it at all inconvenient to a mouse, which I put into it.

> Priestley, 1772

Other researchers have amplified Priestley's work. For example, only seven years later Ingenhousz carried Priestley's experiments a little further and discovered that green plants require light in order to "restore" air and that the nongreen parts of plants are unable to do this. Later it was learned that the difference between good and bad air was that bad air had more carbon dioxide (CO_2) and less oxygen (O_2). In 1804 Théodore de Saussure discovered that the carbon (C) in plants came from the CO_2 in the air. And so gradually the idea developed that plants were able to remove CO_2 from the air, use the carbon for building food material, and supply O_2 to the air. This process is called *photosynthesis* and is summarized by the expression:

$$6\,CO_2 \;+\; 6\,H_2O \;\xrightarrow[\text{green plants}]{\text{light}}\; C_6H_{12}O_6 \;+\; 6\,O_2$$

carbon   water   carbohydrates oxygen
dioxide

The food materials manufactured in photosynthesis are represented here as carbohydrates ($C_6H_{12}O_6$), although other foods are also produced (see Chapters 2 and 3). Both plants and animals break down ready-made food materials and produce CO_2. This oxygen-requiring process is called *respiration* and is summarized by the expression:

$$C_6H_{12}O_6 \;+\; 6\,O_2 \;\xrightarrow{\text{all organisms}}\; 6\,CO_2 \;+\; H_2O$$

carbohydrates   oxygen   carbon   water
dioxide

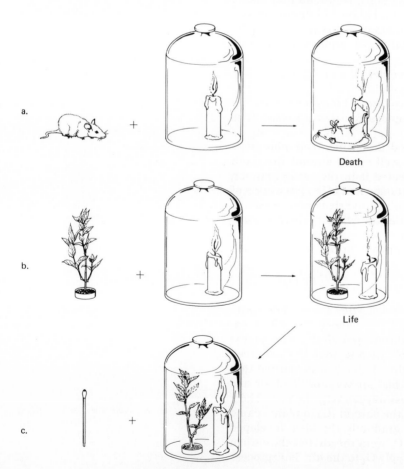

a.

b.

Death

Life

c.

FIGURE 1–2.

Priestley's first experiments with air provided a basis for further experimentation and the eventual discovery of respiration and photosynthesis. (a) Priestley uncovered one of the great relationships in chemistry and in life with a mouse and a candle . . . (b) and a sprig of mint and a candle. Why did the mint live but the mouse die? (c) Why could the candle be made to burn again after the mint had been in the jar for several days? Could a mouse have breathed the air "renewed" by the mint?

Adapted from *Biological Science: An Inquiry into Life,* by the Biological Sciences Curriculum Study, 1963, Harcourt Brace Jovanovich, Inc., Fig. 4.7, p. 77. By permission of the Biological Sciences Curriculum Study.

We are able to understand and appreciate why these dual processes of photosynthesis and respiration are of vital importance in our modern society if we first consider the food-supplying capacity of plants as *productivity*. Biological productivity refers to all those processes that build larger biological molecules from smaller ones and therefore require an input of *energy*. This input is necessary because larger, more complex molecules contain more chemical energy than smaller, simpler molecules. This can be understood readily if one realizes that the chemical bonds that hold atoms together contain energy, and larger molecules have more bonds than smaller ones. As chemical bonds are broken, making smaller molecules out of larger ones, energy is released from the molecule and can be used to perform work. We call this process *consumption*. This concept as it applies to photosynthesis and respiration is summarized in Figure 1–3.

The relationship between productivity and consumption on the one hand and energy flow on the other can scarcely be over-emphasized. It is one of the most important principles in all of biology, and so it is restated once again. Plants are *producers* because they make complex living material out of simple non-living materials (CO_2, H_2O). Plants use some of these complex molecules for their own life processes, but they produce far more than they consume. Animals are strictly *consumers* because they utilize the energy in premade food material by breaking it down to simple nonliving materials. Therefore, biologically speaking, the reason why an animal eats a plant is to obtain energy. Of course, the immediate motivation is hunger, but we are now speaking of the more fundamental biological basis for this behavior. Life's processes require energy, and animals obtain energy by consuming food materials produced directly or indirectly by plants. Plants, of course, also need a direct input of energy, and this energy is derived via the photosynthetic

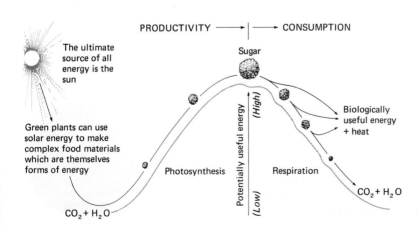

PRODUCTIVITY ⟶ | ⟶ CONSUMPTION

The ultimate source of all energy is the sun

Sugar

Green plants can use solar energy to make complex food materials which are themselves forms of energy

Photosynthesis

Respiration

Potentially useful energy (High) (Low)

Biologically useful energy + heat

$CO_2 + H_2O$

$CO_2 + H_2O$

FIGURE 1–3.

Energy capture and conversion—crossroads in the economy of life. Adapted and reproduced by permission of the publishers. From *Life: An Introduction to Biology,* shorter edition, by George Gaylord Simpson and William S. Beck, © 1957, 1965, 1969 by Harcourt Brace Jovanovich, Inc. Fig. 4.1, p. 70.

Element	Protons (atomic number)	Protons + Neutrons (atomic weight)	Electrons
Hydrogen (H)	1	1	1
Carbon (C)	6	12	6
Nitrogen (N)	7	14	7
Oxygen (O)	8	16	8
Calcium (Ca)	20	40	20

TABLE 1–1.   SOME BIOLOGICALLY IMPORTANT ELEMENTS AND THEIR ELEMENTARY PARTICLES

process which converts light energy to chemical energy in the form of complex food molecules.

On the basis of the foregoing brief discussion, it is clear that there is indeed an important relationship between energy and food. In order to probe this relationship further and to acquire a foundation for later chapters, it is necessary for us to acquire a basic background and vocabulary in chemical principles.

Elements, Atoms, and Particles

The universe and all that it encompasses, living creatures most certainly included, is composed of 92 naturally occurring *elements*. Each element is designated by a chemical symbol which is often the first one or two letters in its English or Latin name (see front endpaper). Some of the most common elements found in the human body include carbon (C), oxygen (O), hydrogen (H), nitrogen (N), and calcium (Ca). Each element is composed of tiny particles called *atoms*. An atom is the smallest complete unit of an element.

Under certain conditions most atoms are able to attach to other atoms, and such combinations of two or more atoms are called *compounds*. The forces that hold the atoms of a compound together are called *bonds*. The controlled formation and breakage of bonds within the cell and the subsequent fate of the energy involved is perhaps the most basic component of productivity, consumption, and the processes of life in general. In order to understand the nature of a chemical bond, we must first understand the basic internal structure of atoms.

For our purpose here we will consider atoms to be composed of three elementary particles: protons, neutrons and electrons. Protons, which have a positive electrical charge (+), and neutrons, which have no charge, are located in the center of an atom where they form what is known as the *atomic nucleus*. Electrons have a negative electrical charge (−) and form a whirling but organized cloud outside the atomic nucleus.

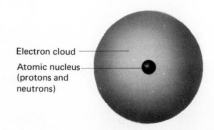

Electron cloud

Atomic nucleus (protons and neutrons)

Carbon-14 (^{14}C) has 2 more neutrons than the stable form of carbon (^{12}C) and thus is radioactive.

Tritium (^{3}H) is also a common "isotope" of hydrogen.

All elements with atomic numbers greater than 82 are radioactive.

TABLE 1-2. RADIOACTIVITY: IMBALANCE IN THE ATOMIC NUCLEUS
Sometimes an imbalance of protons to neutrons in an atomic nucleus causes the nucleus to be unstable. Such a nucleus may spontaneously break down, releasing nuclear particles and energy. An atom which undergoes this behavior is called radioactive.

If we arbitrarily assign a unit weight of 1 to protons and neutrons (it is known their mass is very nearly the same), we find that by comparison the weight of an electron is negligible. Therefore the mass of an atom is concentrated in its nucleus and for practical purposes is the sum of the number of protons and neutrons contained within that particular atomic nucleus (Table 1-1). For most of the elements found in living organisms, the number of protons is equal to the number of neutrons found in the nucleus, but there are some interesting exceptions (Table 1-2).

A second important point to note in Table 1-1 is that for all the elements listed, the number of protons is equal to the number of electrons. This means that each atom is considered to be electrically neutral as a whole. Electrons, of course, do not exist as random clouds around an atomic nucleus but instead travel in discrete paths at fixed distances (also called energy levels) from the nucleus which we call shells. The first shell (the one closest to the nucleus) has one orbital and can hold a maximum of 2 electrons. The second shell has four orbitals and can hold a maximum of 8 electrons. As we move progressively farther from the nucleus, we find that each shell has progressively more orbitals (Figure 1-4).

FIGURE 1-4.

The concept of electron shells. Note that inner shells must be filled before electrons can take up positions in the outer shells. As you will see, this fact has important consequences with respect to chemical bond formation.

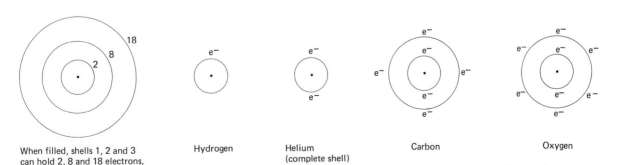

When filled, shells 1, 2 and 3 can hold 2, 8 and 18 electrons, respectively.

Hydrogen

Helium (complete shell)

Carbon

Oxygen

Toward Understanding Chemical Bonds

An atom is electrically and chemically most stable when it has a complete outer shell. For example, hydrogen has a single electron which orbits in the first shell, and thus to be most stable it would "prefer" to have one more electron in order to complete the first shell, or alternatively to lose 1 electron and have none. Helium (2 electrons, 2 protons, and 2 neutrons), on the other hand, has a complete shell and is thus stable and inert (see Figure 1–4).

In order to see how this tendency to complete an outer shell leads to the sharing of electrons between two or more atoms and thus a chemical bond, let us consider the result of a collision between a hydrogen atom and a fluorine atom. In Figure 1–5 we can see that fluorine (F) has an outer shell (shell 2) with 7 electrons, and thus requires one more in order to make a complete set and achieve maximum stability. Hydrogen, on the other hand, can achieve maximum stability by either gaining or losing 1 electron. Therefore, if the two atoms (H and F) were to collide and transfer 1 electron, both atoms would be chemically more stable. This does happen and the resultant compound, hydrogen fluoride (HF), is a gas. That is, in HF the single electron from hydrogen can be shared with fluorine, and one of the electrons from fluorine can be shared with hydrogen. The relative equalness or unequalness of this sharing determines whether we call the particular bond formed between two atoms *ionic* or *covalent*. In the case of HF the sharing of electrons is quite unequal, so that most of the time fluorine has 8 electrons and hydrogen has none.

In biology we are generally more concerned with covalent bonds rather than with ionic bonds such as those formed in HF. In a covalent bond the electrons are more equally shared between two atoms. Consider, for example, what might happen when two hydrogen atoms collide. Each will try to capture an electron from the other in order to achieve maximum stability. Each, however, holds on to its electron equally well, and a kind of tug of war takes place. The net result of this tension is a mutual attraction that keeps the two atoms in contact, through equal sharing of electrons. As a result of each hydrogen atom sharing its electron, both behave as if they had a complete outer shell, and therefore H_2 is a relatively stable molecule (Figure 1–6).

More on the Covalent Bond

In a covalent bond more than one pair of electrons may be involved and shared. For example, oxygen (O) has 6 electrons

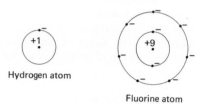

Hydrogen atom

Fluorine atom

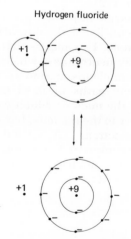

Hydrogen fluoride

FIGURE 1–5.

An ionic bond is characterized by unequal sharing of electrons. A small fraction of the hydrogen fluoride molecules will have the configuration on the top, but in an ionic bond electron (e^-) sharing is unequal, and in the case of hydrogen fluoride the molecular configuration on bottom is the more common. That is, fluorine has an outer shell with 8 e^- and hydrogen has none.

in its outermost shell and requires 2 more for maximum stability. Therefore, when two oxygen atoms collide, stability can be achieved by sharing two pairs of electrons:

$$\ddot{O}{:} + {\overset{xx}{\underset{xx}{\times}}}\ddot{O} \rightarrow \ddot{O}{:}{\overset{xx}{\underset{xx}{\times}}}O$$

oxygen

Electron sharing may also occur between more than two kinds of atoms. For example:

$$\ddot{O}{:} + \overset{x}{H} + \overset{x}{H} \rightarrow H\overset{x}{\underset{\cdot\cdot}{\ddot{O}}}H$$

water

With this background information we are now prepared to deal with the element carbon, the major atom in all biological compounds. Carbon (6 protons, 6 neutrons, and 6 electrons) has an outer shell with only 4 electrons (see Figure 1–4) and therefore could either give up 4 electrons or take on 4 more in order to fill its outer shell. In fact, carbon does neither; instead it always shares its 4 electrons with other atoms in order to achieve stability. For example, carbon can react with four hydrogen atoms to form the molecule called methane:

$$\dot{\underset{\cdot}{C}}{\cdot} + 4\,\overset{x}{H} \rightarrow H\overset{\overset{H}{\cdot x}}{\underset{\underset{H}{x\cdot}}{\times C\cdot}}H$$

methane

In the above diagram each pair of shared electrons (x·) represents one covalent bond. Thus we can see that carbon is tetravalent, that is, it can form four covalent bonds. Rather than always indicating covalent bonds with dot-and-x diagrams as we have done above, chemists and biologists prefer to represent each covalent bond with a straight line and show methane as:

$$\begin{array}{c} H \\ | \\ H-C-H \\ | \\ H \end{array}$$

Sometimes we even abbreviate the expression by omitting the lines. In this way we can write CH_4.

Carbon is unusual not only because it shares 4 electrons but also because it can share these electrons with a variety of biologically important atoms including itself; thus a great variety of *compounds* are possible. For example:

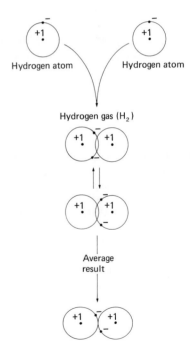

FIGURE 1–6.

A covalent bond involves more or less equal sharing of electrons among the atoms making up the compound.

$$\cdot\overset{\cdot}{C}\cdot + 2\,\overset{x}{\underset{x}{\overset{\cdot}{O}}}^{x} \rightarrow \overset{x}{\underset{x}{\overset{\cdot}{O}}}^{x}\!:\!C\!:\!\overset{x}{\underset{x}{\overset{\cdot}{O}}}^{x}$$

or

$$O{=}C{=}O$$

or

$$CO_2$$

carbon dioxide

also,

$$2\cdot\overset{\cdot}{C}\cdot + 6\,\overset{x}{H} \rightarrow \begin{matrix} H & H \\ H\!:\!\overset{\cdot}{\underset{\cdot}{C}}\!:\!\overset{\cdot}{\underset{\cdot}{C}}\!:\!H \\ H & H \end{matrix}$$

or

$$\begin{matrix} & H & H & \\ & | & | & \\ H\!-\!\!\!&C\!-\!\!\!&C\!\!\!&-\!H \\ & | & | & \\ & H & H & \end{matrix}$$

or

$$C_2H_6$$

ethane

And continuing this reasoning, we can see that almost limitless variations on the structure of carbon compounds are possible.

ethane	ethylene	ethyl alcohol	acetaldehyde	acetic acid
C_2H_6	C_2H_4	C_2H_5OH	C_2H_4O	CH_3COOH

Each of the above compounds is composed of only two carbon atoms, but the properties of the compounds are different. For example, ethane and ethylene are both gases, but the latter acts as a very potent plant-growth regulator while the former does not. Ethyl alcohol is a liquid at room temperature and is also the alcohol found in alcoholic beverages. Acetic acid on the other hand is the chief ingredient of vinegar. Acetaldehyde is sometimes found in small amounts in inexpensive alcoholic beverages and is one of the components responsible for hangovers. So, as you can see, the type of atoms in a compound and their numbers must determine the properties of that compound, and second, with carbon we have the greatest number of variations possible.

Chemical Bonds and Energy

As we have seen, bonding forces hold atoms together with a certain tenacity or strength, which is ultimately determined by the physical properties of the atom itself. The chemical bond, then, is said to represent *chemical energy* or bond energy. This energy may be defined as the amount of work necessary to break a chemical bond. We can consider chemical *compounds* as "packages" of stored chemical energy. When we break chemical bonds, energy is released and work can be done, but in order to make complex molecules, energy must be put into the system. Hence, biological life in its most basic terms involves the capturing of energy with its subsequent conversion into bond energy and the controlled liberation of that energy in order to do work: the twin themes of all life, respiration and photosynthesis.

Energy and Food: Where Do They Go?

Let us return now to the biological role of chemical bonds in a community of organisms, namely, food productivity by plants and the fate of that productivity. We can actually measure the productivity of a plant by measuring the amount of organic matter it produces as it grows and then by calculating the energy contained therein. Just as the food materials that we eat can be measured in energy units (calories), so too can a plant's growth be measured and converted into energy units.

Perhaps it would be useful to look now at the productivity of various kinds of vegetation in order to see how such data can be quantified (Table 1–3). Our data are summarized as weight of plant material (kilograms) produced each year on a small area (square meter) of land or water. (See Appendix One for discussion of metric units.) A number of important points are shown by these data. First, there are substantial differences in the productivity of different kinds of vegetation, raising the question of whether the differences are due to the plants themselves or to the environment in which they live. Second, in the sea, a part of our world that is relatively less familiar to most of us, some areas are highly productive and others much less so, just as are some terrestrial areas. Finally, most agricultural crops are about as productive as the natural vegetation growing in the same area. We will discuss some of the implications of these points in subsequent chapters; however, for the moment we wish only to make you aware of the existence of such figures and their diversity.

Continuing with our major point, let us ask, What ultimately happens to this organic material, this potential energy generated by photosynthesis? The answer is a bit complex because

Community or Region	Mean Annual Productivity $(kg/m^2:$ dry weight)
Aquatic Communities	
Freshwater phytoplankton—Denmark	0.95–1.5
Marine phytoplankton—Denmark	0.26–0.43
Marine phytoplankton, open ocean	0.1
Marine algae, upwelling coastal areas	0.6
Temperate coastal seaweeds	0.9–4.0
Sewage treatment—California	5.6
Subtropical salt marsh—Georgia	3.7
Temperate marsh—Minnesota	2.5
Fertile spring—Florida	2.7
Terrestrial Communities: Uncultivated	
Oak-pine forest—New York	1.20
Deciduous forest—England	0.89
Coniferous forest—England	1.60
Tropical forest—West Indies	6.00
Tropical forest—Ivory Coast	1.34
Temperate grassland—New Zealand	3.20
Terrestrial Communities: Cultivated	
Temperate agriculture; corn, rye—Holland	3.9
Corn, United States average	2.5–4.0
Temperate agriculture, root crop; sugar beet—Holland	1.8
Tropical agriculture; sugar cane—Hawaii	7.2–8.3

TABLE 1–3. ESTIMATES OF PRODUCTIVITY OF VARIOUS SELECTED COMMUNITIES

Source: D. F. Westlake. 1963. Comparisons of plant productivity. *Biological Reviews of the Cambridge Philosophical Society* 38:385–425.
Also K. H. Mann. 1973. Seaweeds: Their productivity and strategy for growth. *Science* 182:975–981.

FIGURE 1–7.

Balance of nature is suggested in this photograph showing moist air coming off the Pacific Ocean and over Torrey Pines State Park, and rainwater returning to the sea via the marsh. Man's activities are evident in the railroad, highway, and distant city.

there are really two answers, one for the organic materials and another for the energy. The raw materials stay here on earth and are recycled. Some plants die, some are eaten by animals which die or are themselves eaten, and so on. Eventually all the material of which plants and animals are composed decays or in some other way returns to the air, water, or soil where it again may be utilized by plants. Thus, the flow of raw materials is cyclic, and the system is in balance (Figure 1–7).

In contrast to the flow of raw materials, the flow of biologically useful energy is not cyclic but one-way. That is, energy can enter the biological world at only one point: photosynthesis. The entire biological world is supported by that one step, the photosynthetic production of food materials that contain chemical bond energy. Without it there would be no population problem—there would be no population—no plants, no animals, only rocks, air (with no free oxygen), and water.

At this point someone may question the noncyclic flow of useful biological energy on the basis of the first law of thermodynamics: Energy is neither created nor destroyed. The catch here is the term *biological energy*. Total energy remains constant, but biologically useful energy continually dissipates. What happens to it? Where does it go? The answer is that at every step in the cyclic flow of materials, some biologically useful energy is lost or converted to a nonusable form such as heat. The body heat maintained by humans and other warm-blooded animals is an example of that energy loss, but the same principle applies in plants and cold-blooded organisms. Consequently, the metabolic machinery which gives life is not 100% efficient. In fact, it is substantially less. We will have more to say about energy flow and the specific nature of biologically useful energy later in the text. The important point here is that essentially all energy in biological systems is eventually lost as heat (Figure 1–8).

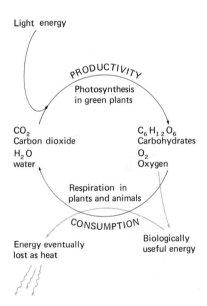

Pollution

The concepts of productivity, cyclic flow of matter, and non-cyclic flow of energy are of immense concern if we are to survive on this spaceship earth. The biological world is carefully balanced in regard to these processes, but this balance may not always be maintained. What happens when consumption exceeds productivity, when the population of the world exceeds the food available? How is productivity affected by the changes wrought by man? Undoubtedly you have been exposed to some of these questions before. Is there anyone who hasn't heard the word *pollution* or the phrase *pollution and the environ-*

FIGURE 1–8.

In the living world the flow of raw materials is cyclic, while the flow of biologically useful energy is one way.

ment? After some thought it should be apparent that environmentalism consists of the basic concepts we have been discussing; pollution, energy flow, and productivity are intimately linked. We cannot really understand one without considering all three. But a problem arises because we don't know the effects of many pollutants and thus can't accurately assess their true effects on productivity and recycling.

Pollution is infinitely more than beer cans in our rivers and trash along our roadsides. A true pollutant is really any object, chemical, or other factor that changes the natural balance of our environment, affecting productivity and thus life as we know it. Oftentimes the pollutant is subtle; perhaps it is even a necessary natural product such as heat. Our civilization generates more heat now than did past generations. But what does this mean?

Some say that the increased heat production may cause the temperature of our atmosphere to increase slightly but significantly. Increased atmospheric temperature may have any number of unforeseen consequences, some of them not particularly desirable. On the other hand, additional heat generated here on earth might simply increase the rate at which it is dissipated into outer space, and we on earth wouldn't see any appreciable difference at all. We just don't know.

Another fear often voiced is that we are introducing too much carbon dioxide into the atmosphere by burning fossil fuels. These fossil fuels—coal, oil and petroleum, and natural gas—were formed from plant material chiefly during the geologic time called the Pennsylvanian period, about 250 million years ago. The concern is that man is returning such large quantities of carbon dioxide to the atmosphere that a change in climate or other adverse consequences may result by very quickly burning these fossil fuels (Figure 1–9). Possibly our increased

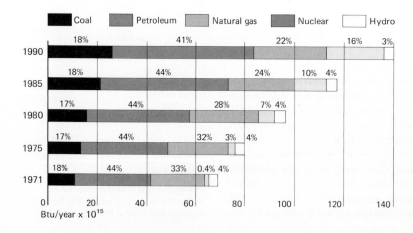

FIGURE 1–9.

Energy consumption in the United States is summarized here according to energy source. The quantities of the various kinds of fuels are converted to a standard unit of energy called Btu (British thermal unit) for purposes of comparison. From "Energy conservation," by G.A. Lincoln, in *Science* 180:156, Fig. 2, April 13, 1973. Copyright 1973 by American Association for the Advancement of Science.

CO_2 production will be a problem, but under average daylight conditions the factor that normally limits the rate of photosynthesis is carbon dioxide. That is, the rate of photosynthesis increases directly in proportion to and as a consequence of increasing concentrations of CO_2, thereby perhaps maintaining a constant level of CO_2, at least for the present.

These two fears illustrate the point that some pollutants are subtler than others and that we presently lack the information and technology to assess properly their impact. This results in considerable controversy and uncertainty. On the one hand, some people point out that the stakes are very high and that we should immediately take effective steps to limit population and industrial growth. On the other hand, it is true that most of us enjoy substantial tangible benefits from expanding industrial technology, and until significant damage of some kind has been clearly demonstrated, there are those who will defend this technological expansion.

Of course, there are other pollutants that directly influence plant productivity and energy flow. The only question is, how serious are their effects? Let's just consider three: detergents, smog, and DDT.

Day after day we run an ocean of water containing detergents down the drain and think that that's the end of it, but of course it isn't. In addition to CO_2 and H_2O all plants require a balance of mineral nutrients for growth and reproduction, and detergents are rich in certain minerals that stimulate the growth of algae and other organisms. Thus, when we pull the plug, we enrich our rivers and lakes with nutrients. This fertilizing can either promote or inhibit algal growth which alters the productivity of our water sources (see Chapter 8 for full discussion). Some detergents have direct effects on other biological systems as well. Protozoa (the simplest animals) and the embryos of echinoderms (the starfish and its relatives), for example, cannot be cultured in glassware that has been washed in detergents, no matter how many times it is rinsed. The residual detergent clinging to the glass kills them. Clearly, washday miracles may have a substantial impact on the biological productivity of our waterways, and we should all be aware of the potential consequences.

What does smog do to plants? The question is not simple because each kind of plant has its own tolerance level, and the effect may not be the same in all cases. Nevertheless, we know that smog enters the leaves of spinach, beets, oats, and a number of other plants through small cellular pores; it then damages or even kills the cells adjacent to these pores and inside the leaf. When that happens, the pores close and cease to function. Since these same pores represent the route by which CO_2 enters the plant, smog may indirectly impair photosynthetic productivity. In addition, ozone (O_3), a particularly damaging component of smog, is known to interfere directly with photosynthesis.

The use of the insecticide DDT (dichlorodiphenyltrichloro-ethane) has recently been banned in the United States for all but emergency uses. Nevertheless, substantial quantities have already been introduced into our soil and water, and DDT now within our environment is deteriorating only very slowly. Some DDT lands on the insects at which it is aimed, some temporarily on plants, and some on soil. But essentially all DDT eventually ends up in water, and nearly all waters run to the sea. Some of the effects of DDT on animals are well publicized (see Chapter 8), and there is also at least one direct effect on plants. DDT in water impairs photosynthesis and also the cell division (growth) of some of the algae growing there. What will be the impact of the DDT and other persistent poisons already released into our environment when they finally find their way to the sea? We simply don't know.

Gathering Some Facts

Let us pause now to review what has been discussed so far. First of all, earth must be regarded as a closed system with regard to the cyclic flow of matter in plants and animals and as an open system with regard to energy. The ultimate source of all energy used in plants and animals is the sun, but the immediate direct source is food, that is, chemical energy incorporated into the food materials produced by plants. This transition of CO_2 and water to food materials in plants through photosynthesis is called productivity. The transition from plant material to animal and back to CO_2 and minerals is called consumption. The ultimate fate of all biological energy, then, is heat which is lost to outer space.

The second point was pollution. The aim was to establish very early in this book an awareness of the impact of environment on basic life processes. The strategy was to mention only a few examples of pollution in order to make the following points. First, many very important, even critical, questions regarding the environmental impact of many common pollutants remain unanswered, and much work is needed now. Second, potentially dangerous changes in our environment may be brought about by seemingly innocent and exceedingly modest pollutants. The example here was detergents, which we all wash down the drain several times a day and thereby potentially impair the growth of at least some algae. The growth of other algae is stimulated by certain ingredients in detergents. Immediately two problems arise. First, since algae are the photosynthetic organisms of the oceans, consider for a moment the impact on food supply or on CO_2 in the atmosphere if algal photosynthesis on a worldwide basis was to decline by 20%. The second problem relates to the complexity of biological communities. Stimulating the growth of some algae while inhibiting the growth of others could lead to simplification of the otherwise complex aquatic communities, and simplification leads to instability. Clearly, we must understand and accurately measure the rates of basic plant processes vital to all life before we can accurately assess the impact of pollution on them.

Up to now the discussion has tried to set the biological context and to emphasize the role of plants in two of the major problems facing the world today, namely, food supply and pollution. Now let us direct our attention to the characteristics and structure of living plants; we will return later to world problems and the relationship of plant structure and function to them.

Life and the Living

To understand plant biology one must understand plant life. Therefore, we first turn to the question, What is life? Life is not a thing but a quality or state of being. There are a number of characteristics of life, and they characterize all living things even though there is no substance or material that is life. These characteristics, listed below, will be discussed in this and subsequent chapters.

1. Reproduction, Replication. An organism giving rise to offspring of its own kind is scarcely a new phenomenon. We know there are a number of organisms that consist of no more than a single cell, the latter being the ultimate building block of all organisms. Bacteria which cause a number of diseases are examples, and we have all heard of and probably seen the amoeba, a protozoan. Certain plants, too, are microscopic and consist of only a single cell. Reproduction for such unicellular organisms usually means simply that the one cell divides into two. Think carefully for a moment what this means. Before a living cell can divide and produce two viable daughter cells, all the various parts within the original cell must duplicate. This is necessary so that both of the daughter cells will have complete sets of all the machinery essential for life.

Thus, reproduction proceeds at several levels of organization. On one level are large plants, such as trees and weeds, most of which reproduce by seeds. On a second level are individual, microscopic cells that reproduce by simple division. The third level of organization is subcellular and is represented by the replication of cellular material. At all levels of organization, reproduction and replication are key characteristics of life.

2. Growth. Growth may be defined as an irreversible increase in size or bulk. Often it is correlated with increase in numbers, as in a population. The essential meaning, however, is increase in size.

3. Specialization. Specialization is the modification of different parts for different functions. That is, a living plant is not simply a homogeneous mixture of living "stuff." Rather, higher plants have roots, stems, leaves, and floral parts. Each of these organs is composed of tissues which perform their respective functions of protection, support, food manufacture (photosynthesis), food storage, transport, and so on. Likewise, specialized parts of cells, called organelles, are associated with particular vital functions. All of these are examples of specialization (Figure 1–10).

4. Metabolism. Metabolism is the sum total of all the biochemistry going on inside an organism. It includes energy conversion, a feature essential to all other aspects of life.

5. Movement and Transport. The sunflower faces the sun as it moves across the sky. A similar phenomenon can also be ob-

FIGURE 1–10.

Specialization is found at many levels of organization. a. Alfalfa plant showing specialized organs. b. Scanning electron micrograph showing the internal anatomy of the cotton leaf. Magnification 1900×.
(b) Photomicrograph from *Probing Plant Structure,* by J. H. Troughton and L. A. Donaldson, 1972, McGraw-Hill Book Company, Plate 40. Courtesy of the Department of Scientific and Industrial Research, Lower Hutt, New Zealand.

a.

b.

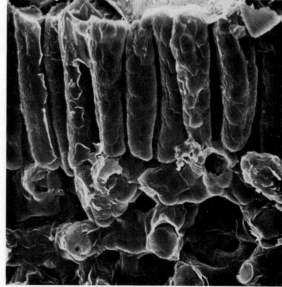

served in a number of other plants (Figure 1–11). Within an individual plant, the food is transported from leaves to other parts in the plant. Within a single cell materials move in and out and from organelle to organelle. At all levels of organization there is movement and transport.

6. Sensitivity and Responsiveness. The sensitive plant, *Mimosa,* quickly retracts its leaves when touched (Figure 1–12). Plant shoots grow upward and toward the light, whereas their roots grow down into the soil. Tendrils spiral snugly around trellis-work. These are all examples of plants' perceiving their environment and responding to it.

7. Complex Organization. The foregoing paragraphs suggest the extremely complex *organization* of plants at all levels. These various levels of organization include the molecular structure of cell parts, the arrangement of cells into tissues and tissues into organs, and the arrangement of organs in the entire plant body. Plants have some of the finest, most precise relationships between structure and function ever contrived. Things are as they are because that's the way they work best in their particular niche, or did at the time they were established. Over 3 billion years of natural selection have been expended in perfecting them, and they are truly remarkable in the complexity of their organization.

Perhaps you are wondering why *all* these characteristics are necessary for life. Wouldn't one or a few of them be adequate? The answer is that all are required because they require each other. Thus, sustained *reproduction* requires *growth,* for without growth reproduction would give rise to increasingly smaller products until some limit was reached, and reproduction would cease. *Growth,* of course, requires *metabolism* because metabolism is the only means by which raw materials can be converted into biologically useful compounds and energy.

FIGURE 1–11.

Orientation of growth with respect to light is a common phenomenon in plants. In this figure young oat seedlings can be seen before (left) and after (right) exposure to unidirectional light. The mechanism responsible for this behavior is discussed in Chapter 6.

Metabolism requires *movement and transport* because raw materials or food for metabolism must be obtained from the environment and brought to the cell's biochemical machinery, and its products distributed throughout the plant body. Clearly, *movement and transport* if they are to be purposeful and directed rather than random and fruitless require some means to perceive certain factors in the environment: *sensitivity.* Finally, all of this could only be possible by means of a very complex apparatus. A simple machine just couldn't handle the job. And so it is understandable that *specialization* into a very *complex organization* is the only course that is open to living organisms.

Careful analysis of the economy of life as outlined heretofore draws our attention to two critical points, namely, reproduction and metabolism. If there is one essential characteristic that is somehow the real essence of life above and beyond the others, it would be the exactness of biological reproduction. We have already indicated, however, that energy is required for any and all life processes, and so we recognize reproduction and metabolism, possibly the greatest "inventions" in the history of the earth, as the two critical points for more detailed discussion.

FIGURE 1–12.

Mimosa, the sensitive plant, responds to touch in a second or two, as shown here before (left) and after (right) touching.

All biological processes depend on cellular or subcellular events, and so it is with reproduction and metabolism. Cells must reproduce before there can be organisms. It can be no other way because organisms are made up of one or more cells. Likewise, the many processes that come under the general heading of metabolism require an exceedingly high level of precision in structure which can only be realized on the microscopic and submicroscopic level of organization. Atom by atom, molecule by molecule, the cellular machinery must be assembled without any major flaws. Any significant error in construction means decreased efficiency in function, and such are the characteristics of the losers in the struggle for existence. But struggle for existence means ability to reproduce. Thus, reproduction and metabolism are the two characteristics which, more than any others, mean life, and they both occur in cells. We must have a closer look at the parts of cells and how they are put together before we can really appreciate life in general and plant life in particular.

Bigger than a Molecule, Smaller than a Breadbox

Higher plants are made up of roots, stems, leaves, and floral parts. These in turn have smaller subunits of structure and function such as the veins and areas between veins in leaves, exterior (bark) and interior areas of stems, and so on. As we look deeper and deeper into the details of how plants are put together, we might ask, what is the ultimate unit of structure and function in plants? The answer to that question required the work of many talented scientists, each one building upon the achievements of others; however, perhaps the most important advance leading to the discovery of cells was the invention of the microscope.

The invention and early development of the microscope (and telescope, too) came during the late sixteenth and early seventeenth centuries, the same period when Queen Elizabeth I was ruling England and William Shakespeare was writing plays for the Globe Theatre on the Thames River in London. The earliest microscopes were simply oval glass lenses mounted in wooden frames with pins for holding the object to be viewed on the other side of the microscope.

A Dutchman, Anton van Leeuwenhoek (1632–1723), described living microorganisms, even bacteria and spermatozoa, using nothing more than this simple device (Figure 1–13). When two or more lenses are mounted in sequence so that an object is viewed through all of them, we have a compound microscope. With this tool Robert Hooke (1635–1703), an English

FIGURE 1–13.

The type of microscope used by Leeuwenhoek in the seventeenth century had a single lens. The specimen to be examined was mounted on the tip of the wire so that fine focusing could be accomplished by gently squeezing the wire with the thumb and forefinger as the microscope was held to the eye.

physicist, in 1665 described a thin section of cork and in his description established the terms *cell* and *wall* as they are used today in plant biology:

> I took a good clear piece of Cork, and with a Pen-knife sharpen'd as keen as a Razor, I cut a piece of it off, and thereby left the surface of it exceeding smooth ... But judging from the lightness and yielding quality of the Cork, that certainly the texture could not be so curious, but that possibly, if I could use some further diligence, I might find it to be discernable with a *Microscope*, I with the same sharp Pen-knife, cut off from the former smooth surface an exceeding thin piece of it, and placing it on a black object Plate, because it was itself a white body, and casting the light on it with a deep *plano-convex Glass*, I could exceeding plainly perceive it to be all perforated and porous, much like a Honey-comb ...
>
> ... for the *Interstitia*, or walls (as I may so call them) or partitions of those pores were neer as thin in proportion to their pores, as those thin films of Wax in a Honey-comb (which enclose and constitute the *Sexangular cells*) are to theirs.
>
> *Hooke, 1665*

The foregoing shows how new and better techniques have often made possible giant steps forward in our understanding of life. Such techniques, often developed by one or very few

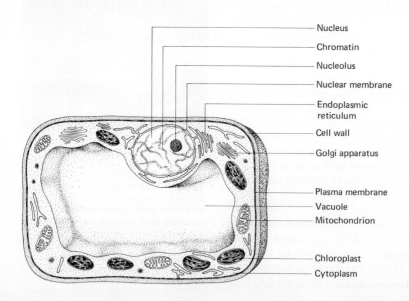

- Nucleus
- Chromatin
- Nucleolus
- Nuclear membrane
- Endoplasmic reticulum
- Cell wall
- Golgi apparatus
- Plasma membrane
- Vacuole
- Mitochondrion
- Chloroplast
- Cytoplasm

FIGURE 1–14.

Diagram of a generalized cell from a higher plant. Organelles are described in the text.
From *Contemporary Biology: Concepts and Implications,* by Mary E. Clark, 1973, W. B. Saunders Company, Fig. 3.23, p. 82.

individuals, have expanded the knowledge and understanding of whole populations in a revolutionary way. Thinking and meditating alone are often not enough, and on this point we must take exception to Kahlil Gibran's statement: "No man can reveal to you aught but that which already lies half asleep in the dawning of your knowledge. . . . For the vision of one man lends not its wings to another man."

Many years passed, and biologists were seeing cells in a variety of organisms and developing better microscopes, better strains, and other preparation techniques. Finally, in 1838, the German botanist Matthias J. Schleiden (1804–1881) concluded that in fact all plants were made of and derived from cells. It is interesting to note that this tremendously important finding was made by a man originally trained as a lawyer. Only after failing as a lawyer and in a subsequent suicide attempt did Schleiden turn to plant science and achieve fame. A year later Theodor Schwann (1810–1882) concluded that animals were also composed of cells and their derivatives. Since those early days when cells were first seen and then recognized as common to all organisms, we have come to recognize the cell as the smallest unit of structure and function that maintains all the characteristics of life. Let us have a closer look at cells and some of the things they can do.

The Complete Cell

As you view a "typical" plant cell (Figure 1–14), one of the most obvious features is the outer covering, the *cell wall*. The wall may have specialized layers that we call primary and secondary walls. Animal cells do not have a cell wall, and thus we have already noticed one of the major differences between plant and animal cells. The cell wall is a very important functional constituent of plant cells since, among other things, its physical characteristics directly regulate the growth of some cells, and in others it forms the rigid support for the plant body as a whole.

The primary cell wall is made up almost entirely of sugar units linked together in long chains. Perhaps the most important and certainly the most abundant molecule within the wall is *cellulose*. When we view the microscopic structure of the wall (Figure 1–15), many threadlike strands of cellulose are apparent. These strands are interwoven in a manner that gives the wall an amazing amount of strength. Of course, there are other less conspicuous components that also contribute to the properties of the cell wall, but as yet we do not fully understand how these interact with cellulose.

In addition to this relatively thin primary cell wall, certain kinds of cells produce an additional wall of considerable thick-

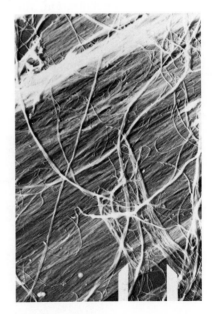

FIGURE 1–15.

The cell wall of a cotton fiber. The interwoven strands of cellulose give great strength to the wall but do not prevent the flow of materials into or out of the cell.
U.S.D.A. photo by Southern Regional Research Center, New Orleans, Louisiana.

ness internal to the primary wall. We call this the secondary wall. Botanically, wood is chiefly secondary cell walls.

Just inside the wall we find another organelle termed the cell membrane or plasma membrane (see Figure 1–16). This organelle has the important job of regulating the passage of salts and other molecules into or out of the cell. The wall is much too porous to function in this regard. Thus, the plasma membrane separates the cell's internal contents from the external environment. This selective control of the materials entering or leaving a cell is obviously of great significance because it must maintain a rather fixed and specific environment for the business of life in an external environment subject to rapid change.

Let us briefly examine just how the plasma membrane performs this regulating function; in doing so we will gain some insight into an important biological process. Plant membranes (and animal membranes, too) are constructed in such a way that they are extremely permeable to water and are much less permeable to dissolved solutes, for example, salts. Another rather general principle in biology is that when a membrane separates two solutions that differ in solute (dissolved molecules other than water) concentration, there is an intrinsic tendency for water to move through the membrane toward the solution containing the greater solute concentration. Looking at it in another way, there is a tendency for water to move in a direction so as to equalize the solute concentration on each side of the membrane. The term for movement of water in this manner is *osmosis* (Figure 1–17).

FIGURE 1–16. THE CELL MEMBRANE.

a. In this photomicrograph a stain was used to darken just the plasma membrane of onion stem cells. The cells can be stained in various ways to illustrate particular features. b. When cells are broken open by mechanical means, the plasma membrane fractures, and the pieces can be isolated. Note that the membrane fragments appear to fold back on themselves, forming small spheres (vesicles). Magnification in both photomicrographs is 35,000×.

Photomicrographs by W. Van Der Woude, University of California, Riverside.

a.

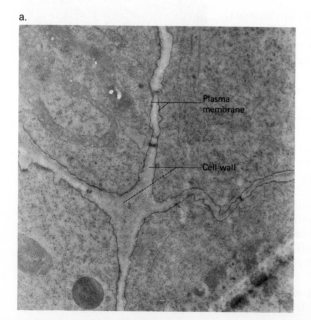

Plasma membrane

Cell wall

b.

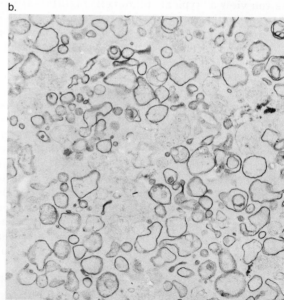

a. *Diffusion.* If we place a sugar cube in the bottom of a glass of water and allow it to stand undisturbed, the sugar eventually dissolves and becomes evenly distributed throughout the water. This uniform distribution is ultimately due to the random movement of the sugar molecules in water and is called *diffusion.*

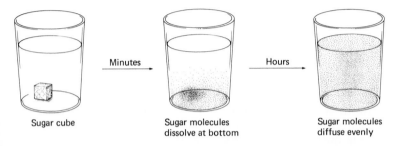

| Sugar cube | Sugar molecules dissolve at bottom | Sugar molecules diffuse evenly |

b. *Osmosis.* Assume that we have a selectively permeable membrane that is permeable to water but not sugar. This membrane is initially separating two fluid compartments of equal height. Compartment 1 contains a water solution with sugar; compartment 2 contains pure water.

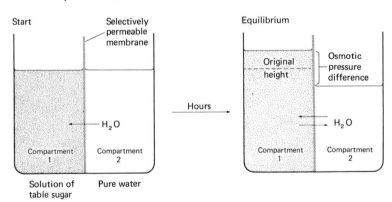

With time there will be a net movement of water from compartment 2 into compartment 1. This movement occurs because there is a difference in the concentration of water in the two compartments; the sugar molecules in compartment 1 will eventually produce a pressure difference (called osmotic pressure) between the two compartments. Eventually this pressure difference becomes equal to the tendency for water molecules to move across the membrane, and an equilibrium state is reached in which there is no net flow of water into either compartment. The amount of osmotic pressure developed depends directly on the concentration of sugar molecules in compartment 1. The actual movement of water across a membrane in response to a concentration difference is termed *osmosis.*

FIGURE 1–17.

Diffusion and *osmosis* are important concepts in understanding the activities of cells.
From *Contemporary Biology: Concepts and Implications,* by Mary E. Clark, 1973, W. B. Saunders Company, Figs. 3.15, 3.16, pp. 71, 72

Let us consider what happens when a plant cell is bathed in a solution of pure water. We know that the cell contains various solute molecules (salts, sugars, and so forth); therefore, according to the principles of osmosis, water will flow from the more dilute solution (water) through the membrane to the area of higher solute concentration (the cell). What does this mean? Well, for one thing, it means the volume of the cell will increase, and this in itself could be disastrous if uncontrolled. Just consider what would happen if you blew up a balloon, and blew and blew. What is to prevent the cell from expanding to the point at which it will pop just like a balloon? Answer: The cell wall. Plant cells are restrained from continuous swelling by the cell wall. We can easily visualize this if we imagine blowing up a balloon in a box. Obviously, as we continue to push air into the balloon, its sides will come in contact with the box. Eventually air pressure builds up within the balloon in the box until no more air can be forced into the balloon. And so it is with plant cells (Figure 1–18a). The osmotic entry of water proceeds until further flow is prevented by the physical pressure exerted by the wall (wall pressure).

In nature most cells are fully turgid. They contain as much water as is osmotically possible, and thus the membrane is pushing firmly against the wall, making the entire unit firm like an inflated tire. This condition is important because it is responsible for the rigidity of certain plant organs—leaves and young stems, for example. When conditions such as the lack of water and extreme heat cause water to be lost from plant cells, this physical pressure no longer extends the walls, and the cells become limp (Figure 1–18b). When this occurs in nature, we call it wilting. In principle these conditions are illustrated by a tire with 30 lb pressure and by one that is flat.

Thus far we have talked only about the transport of water across the membrane; we said the membrane was almost impermeable to solutes. This statement can now be modified slightly. Solutes such as salts and sugars certainly don't move as rapidly through a cell membrane as water does; nevertheless many solutes, salts in particular, are necessary for growth and must be obtained from the outside environment. Therefore, they must pass through the membrane. Sometimes this occurs by diffusion, i.e., the salts simply move passively from areas of high concentration to areas of lower concentration. The mechanism, of course, is based on the random motion of particles in solution. Because diffusion requires no input of energy, this type of movement into and out of cells is called *passive transport* (Figure 1–19). For many kinds of molecules and ions, however, special physiological pumps are built into the membrane itself, and these pumps can preferentially bring some materials into the cell and exclude others. This sort of selective work by the

a. Turgidity: cell in pure water

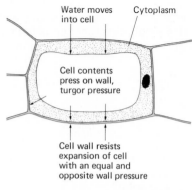

b. Plasmolysis: cell in water plus solute

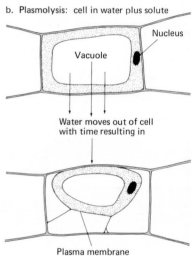

FIGURE 1–18.

a. *Turgidity.* As water moves into a cell, the cell contents are pushed against the wall by a pressure called turgor pressure. The wall resists this pressure with an equal and opposite force called wall pressure. The turgidity (that is, rigidity) of the cell is maintained by these equal but opposite forces. b. *Plasmolysis.* If a cell is placed in a solution containing more solute than found in the cell sap, water diffuses out of the cell, and the cytoplasm shrinks away from the cell wall (top, before plasmolysis; bottom, after plasmolysis).

cell membrane requires energy, and so it is called *active transport* (see Figure 1–19). It is important that you understand the distinction between passive (simple diffusion, no energy required) and active (selective movement, energy required) transport.

After this discussion about the significance of the plant wall and the membrane, let's move on and probe more deeply into the cell. Once past the plasma membrane, we find the cell is composed of a rather viscous (thick) fluid in which are embedded smaller units of cellular function called organelles. This internal organization is far from static; rather it is a bit like a rapid transit system within the cell. The internal viscous fluid and organelles suspended therein constantly move about. This streaming, as it is called, is poorly understood physiologically, but it could contribute to intracellular transport and also to more rapid exchange of materials with adjacent cells or the environment.

Let us now introduce some of the more conspicuous subcellular organelles seen in Figure 1–14 and briefly explain their function. In subsequent chapters we will explain each of these organelles in more detail.

In a mature plant cell the most conspicuous subcellular structure, occupying most of the volume, is the *vacuole*. It is sur-

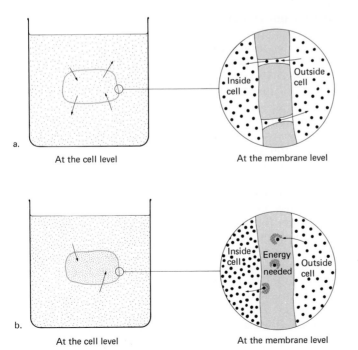

a.

At the cell level   At the membrane level

b.

At the cell level   At the membrane level

FIGURE 1–19.

a. *Passive transport.* No cellular energy required; molecules diffuse through the membrane according to concentration gradients. b. *Active transport.* Cellular energy required; molecules may be transported against a concentration gradient if energy is expended.

rounded by a membrane similar in appearance to the plasma membrane and contains a watery solution of many different solutes. Animal cells usually lack such an extensive vacuole but may contain several smaller ones. The function of the vacuole is not entirely clear, but it may regulate water uptake and thus the turgidity of the cell in the way we have already seen. It may also serve as a reserve storage area for various substances.

The cytoplasm is located between the vacuole and the plasma membrane. All the organelles that form the true machinery of the living process are located here. Let us first discuss *mitochondria* (singular, mitochondrion). This organelle, bounded by a double membrane, is primarily responsible for breaking down complex food molecules progressively to smaller units and for ultimately releasing CO_2, water, and energy. This process, as noted previously, is part of cellular *respiration* (see Appendix Three). It is important to realize that all organisms respire—yes, even plants do. One often loses sight of this because plant-oriented texts tend to emphasize the photosynthetic activities of plants. The point is that plants not only make food material but also break it down. Fortunately, the production apparatus operates more rapidly than the degradation machinery, and therefore plants have a net food-producing ability, which we previously called productivity. Respiration within the mitochondria is important because the breakdown of complex molecules releases energy which can be used to do cellular work (Figure 1–20).

Chloroplasts are organelles that are generally larger than mitochondria and have an entirely different function. It is within the chloroplast that photosynthesis takes place. The detailed structure and function of chloroplasts will be discussed in Chapter 2.

There is also a conspicuous network of membranes within plant cells. This system, called the *endoplasmic reticulum* (ER), is apparently involved in intracellular and possibly in extracellular transport. For example, some materials are known to move out of the nucleus (see below) to the cytoplasm by means of the ER. In addition, some ER has small granules called *ribosomes* attached to it. Ribosomes carry out the important function of protein synthesis (see Chapter 5).

The last organelle to be discussed at this time is the *nucleus*. This relatively large organelle is the control center of the cell, the information-containing organelle. Important? That would be a gross understatement. Complicated? Very much so, and the more we learn about it, the more complex it seems to become.

The nucleus is bounded by a special membrane, the *nuclear envelope*, which contains rather large pores (Figure 1–21).

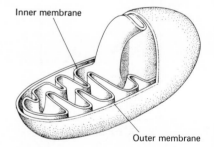

Inner membrane

Outer membrane

FIGURE 1–20.

Within the mitochondrion much of the cell's adenosine triphosphate (ATP) is formed as a result of the breakdown of food materials. When ATP is broken down it releases energy which is used to do cellular work.

From *Contemporary Biology: Concepts and Implications,* by Mary E. Clark, 1973, W. B. Saunders Company, Fig. 3.18 (part a), p. 74.

Pneumococcus	$0.1 \times 0.2\ \mu$
Ostrich egg	75 mm
Human egg	$100\ \mu$
Axon (nerve cell)	Several feet long
"Typical" plant or animal cell	C. 10–$20\ \mu$

TABLE 1–4. THE SIZES OF CELLS
The size of a cell can vary considerably from organism to organism. We have listed some of the extremes as well as the size one can expect for the average plant or animal cell. The micron (μ) is equal to 1/1000 mm.

These pores may function in transporting materials into and out of the nucleus. Within it we find a dense aggregation of material called the *nucleolus* (actually there may be several in each nucleus), and the *chromosomes*. The function of the nucleolus is not known for certain; however, it seems to serve as the site where the material that transports information into and out of the nucleus can aggregate. Chromosomes are long, threadlike objects composed primarily of DNA (deoxyribonucleic acid) and protein. They are the actual information carriers within cells, and you will hear much about them in future chapters.

One final word about cells has to do with size. You may ask how large are these things which perform such wonders. Table 1-4 gives some averages and extremes. Most cells are in the general range of 10–20 microns in diameter, but they come in different sizes and shapes, depending on their location and function (see Appendix One for metric conversions).

The upper size limits for cells probably are set by their surface

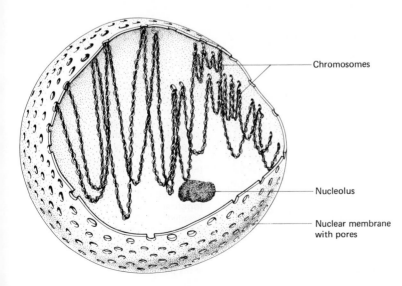

— Chromosomes

— Nucleolus

— Nuclear membrane with pores

FIGURE 1–21.

The detailed structure of a nondividing nucleus is suggested in this model by Dr. D. E. Comings. The nuclear membrane is perforated by pores, to the inside of which are attached the chromosomes.
Courtesy of Dr. David E. Comings, City of Hope Medical Center, Duarte, California.

to volume ratio and physiological needs. Perhaps this can be most easily understood by considering a simple cube whose edge is 1 cm long. If the cubic centimeter was sliced at 1 mm intervals along each of the three planes, the same volume now has 60 cm² surface area instead of 6 cm² (Figure 1–22). As this applies to cells, larger cell size means lower surface to volume ratio and therefore relatively less area through which exchange of nutrients and wastes can take place as compared with smaller cells. This restricted exchange of materials presumably limits physiological activities, and so smaller cells have the advantage in competition. The lower size limit of cells probably is set by space required for the physiological machinery necessary to carry on the mode of life to which a cell is destined.

The largest single cells (bird and reptile eggs) come complete with stored food and a protective shell, and their way of life is to persist as an independent unit for some extended time with little exchange of materials with the outside world. Physiologically active cells, such as those in a photosynthetic leaf, are on the order of 10–20 microns in diameter. Such a size is large enough to contain chloroplasts, mitochondria, a nucleus, and a complete set of organelles but at the same time is small enough to provide a favorable surface to volume ratio.

One of the main purposes of Chapter 1 has been to make the point that life consists of processes, not just substances; it is things happening, not just things. Life is action—lots of it, all the time. That sort of action wherever it occurs requires energy. All biological energy conversion takes place within cells, and the discussions about cell structure were intended to show you where this action takes place. We will continue by describing in more detail some of the action itself. Although these discussions could begin in a number of other areas, we will start by examining photosynthesis and its control.

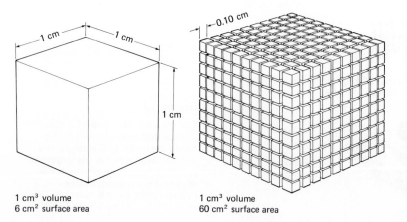

1 cm³ volume
6 cm² surface area

1 cm³ volume
60 cm² surface area

FIGURE 1–22.

The surface to volume ratio of a cell is important because it directly regulates the transport of materials from the environment into the cell. A cube 1 cm on a side has 1 cm³ volume and 6 cm² surface area. If the cube is sliced at 1 mm intervals along all three planes, we have 1000 small cubes 1 mm on a side, which together still have 1 cm³ volume. Adapted from *Botany: An Ecological Approach,* by William A. Jensen and Frank B. Salisbury. © 1972 by Wadsworth Publishing Company, Inc., Belmont, California 94002. Reprinted by permission of the publisher, Fig. 24.2, p. 417.

Selected Readings

Asimov, I. 1962. *The World of Carbon*. Garden City, N.Y.: Doubleday. Semipopular paperback on the chemistry and properties of carbon.

Broecker, W. S. 1970. Man's oxygen reserves. *Science* 168:1537–1538. Reports that fears about man's oxygen supply are not founded in fact.

Cole, L. C. 1969. Thermal pollution. *BioScience* 19:989–992. Calculation of the earth heat budget and predictions.

Fowler, W. A. September 1956. The origin of the elements. *Scientific American* 195(3):82–91. A good place to begin further reading.

Holter, H. September 1961. How things get into cells. *Scientific American* 205(3):167–180.

Kennedy, D. 1965. *The Living Cell: Readings from the Scientific American*. San Francisco: W. H. Freeman. Designed for nonbiologists, this series is a very good introduction to the properties of cells.

ReVelle, C., and P. ReVelle. 1974. *Sourcebook on the Environment. The Scientific Perspective*. Boston: Houghton Mifflin. Good source of information on many pollutants and potential pollutants.

Scientific American. September 1971. Vol. 225, no. 3. (Offprint nos. 661–671.) Entire issue deals with energy and the biosphere.

Storer, J. H. 1966. *The Web of Life*. New York: New American Library. A short introduction to ecology in paperback.

Westberg, K., N. Cohen, and K. W. Wilson. 1971. Carbon monoxide: Its role in photochemical smog formation. *Science* 171:1013–1015. The role of carbon monoxide in ozone production.

2 Photosynthesis and the Green Machine

Perhaps the single most important concept that the beginning student in botany should grasp is the fundamental role that plants play in supporting all other forms of life. As Chapter 1 emphasized, life as we know it could not exist without plants. In this chapter we will examine in detail two major reasons why this fact is so. We will see that plants are responsible for supplying a large part of our atmospheric oxygen (an obvious requirement for the life of air-breathing animals) and, second, and perhaps more significantly, how plants provide directly or indirectly for almost all the organic matter (carbohydrates, fats, and protein) on which animals depend for daily maintenance and growth.

Reviewing briefly, you will recall that the key to the latter concept is not just that we consume some plant material as

FIGURE 2–1.

The organisms (consumers) making up a food chain will vary from region to region. Nevertheless, all food chains must begin with plants (producers). A phenomenon common to all food chains is the progressive decrease in food energy going from producer to consumer, which is reflected in proportionally fewer organisms the higher they are in the food chain.

food but also that the nonplant material we eat is itself ultimately derived from the plant material animals consume. We can illustrate this dependence on plant material by the use of what is known as a food chain or food web (Figure 2–1). By necessity, every food chain must begin with organisms capable of making their own foods, in other words, green plants. Organisms that consume plants directly are called *primary consumers* or *herbivores*. Next in the chain are the *secondary consumers*, the meat eaters or *carnivores*, which feed on herbivores. Carnivores may in turn be eaten by other carnivores and so on. Clearly, all the members of any food chain depend on one another's existence, and all are directly or indirectly dependent on the bottommost level—plants, and the process photosynthesis.

Our first task in this chapter will be to understand something about the biochemistry of photosynthesis. Second, we will consider some of the environmental factors that limit photosynthesis in the field and discuss how man has learned to maximize photosynthesis under certain special conditions. Third, we will begin to address ourselves to the timely concept of how we can produce more food for the rapidly growing world population and to the question of whether or not our need for food will outstrip our ability to produce it. One interesting and fruitful way to begin is with a little history.

Historical Perspectives

The early Greeks believed that plants obtained their organic material directly from the soil. Simplifying slightly, we can say the concept developed that somehow a plant's roots consumed a portion of the soil and that this contributed exclusively to the increase in mass of a growing plant. The "soil eater" concept was perhaps further fostered by the observation that soil additives (fertilizers) increased crop production. This notion remained firmly entrenched until the seventeenth century when a simple but elegant experiment performed by Jean Baptiste van Helmont was revealed.

BY EXPERIMENT, THAT ALL VEGETABLE MATTER IS
TOTALLY AND MATERIALLY OF WATER ALONE

That all vegetable matter immediately and materially arises from the element of water alone I learned from this experiment. I took an earthenware pot, placed in it 200 lb of earth dried in an oven, soaked this with water, and placed in it a willow shoot weighing 5 lb. After five years had passed, the tree grown therefrom weighed 169 lb and about 3 oz. But the earthenware pot was constantly wet only with rain or (when necessary) distilled water; and it was ample in size and imbedded in the ground; and, to prevent dust flying around from mixing with the earth, the rim of the pot was kept covered with an iron plate coated with tin and pierced with many holes. I did not compute the weight of the deciduous leaves of the four autumns. Finally, I again dried the earth of the pot, and it was found to be the same 200 lb minus about 2 oz. Therefore, 164 lb of wood, bark, and root had arisen from the water alone.[1]

Jean Baptiste van Helmont

It would be difficult indeed to disagree with van Helmont's conclusion that plants are not soil eaters; however, his explanation of how they gained their added mass, via water, although logical considering the scientific sophistication of the period, is now known to be basically incorrect. The key knowledge van Helmont lacked was that the atmosphere contains gases and that one of these is of prime importance in photosynthesis. A more complete understanding of the process of photosynthesis gradually evolved as the primary products produced by plants were studied and as a greater understanding of the molecules which make up our environment was gained.

Joseph Priestley (1733–1804) performed experiments (see Chapter 1) which demonstrated plants were able to restore air (replace oxygen) that had become "noxious in consequence of animals either living and breathing or dying and putrifying in it." Jan Ingenhousz (1730–1799) was able to confirm Priestley's work, and furthermore he was able to show that this restoration or purification of the air took place only if the green parts of plants were supplied with light. One of the important controls in this experiment consisted of showing that sunlight itself did not have the ability to "mend" air in the absence of plant life. Ingenhousz proposed that plants were not just purifying air but that in sunlight they were absorbing carbon dioxide (CO_2), utilizing the carbon (C) for nourishment, and expelling oxygen (O_2). Théodore de Saussure (1767–1845) further confirmed these notions and was able to show that the bulk of the dry weight of a plant comes from the carbon taken in as CO_2. He further recognized that water plays an important but subtle role in the process. Summarizing these important findings, we are able to write the overall reaction for photosynthesis (see also Chapter 1).

[1]From Jean Baptiste van Helmont, "By Experiment, that All Vegetable Matter Is Totally and Materially of Water Alone," translated by Naphtali Lewis in Mordecai L. Gabriel and Seymour Fogel, eds., *Great Experiments in Biology,* © 1955, p. 155. Reprinted by permission of Prentice-Hall, Inc., Englewood Cliffs, New Jersey.

$$6\,CO_2 \; + \; 6\,H_2O \; \xrightarrow[\substack{\text{green} \\ \text{plants}}]{\text{light}} \; C_6H_{12}O_6 \;\; + \;\; 6\,O_2$$

carbon dioxide　　water　　green plants　　organic material (carbohydrates)　　oxygen

While the results of these experiments represent important steps in our understanding of photosynthesis, they immediately suggest numerous questions. For example, we might first ask, what are the intermediate steps in the conversion of CO_2 to carbohydrates? Later we will consider the precise role of light and the limiting factors in photosynthesis.

CO₂ to Carbohydrates—
A Most Important Step for Mankind

If CO_2 is to give rise to food molecules, it must enter the cell and either (1) be combined with some organic molecule already present or (2) become bonded to another CO_2 molecule, thus forming within the cell a 2-carbon complex. In either case the product formed would then enter into further reactions leading to the production of more complex organic molecules. Put another way, the first question we can consider is whether the first stable produce of photosynthesis is a 2-carbon molecule or something larger.

Before proceeding, however, you should be aware of the fundamental problems involved in seeking a solution to this question. Carbon is one of the most abundant elements on the earth and is found in all organic molecules. Therefore, the critical problem involved identifying which particular organic molecules were formed directly from CO_2 during photosynthesis. In order to do that, the carbon in CO_2 had to be somehow identifiable. This part of the problem was solved shortly after World War II when radioactive carbon dioxide ($^{14}CO_2$) became available to research scientists. At the University of California at Berkeley a group led by Melvin Calvin and Andrew Benson pioneered much of the early research in this area. Briefly, the Berkeley group supplied $^{14}CO_2$ to actively photosynthesizing cells, allowed them to utilize $^{14}CO_2$ for a few minutes, then killed the cells, and identified the products (organic molecules) that contained the radioactive carbon derived from radioactive CO_2 (Figure 2–2). Their early results showed that when cells were harvested after only a few minutes in $^{14}CO_2$, many compounds were radioactive.

Although this may have been interesting, it hardly solves our particular problem. Clearly, if the *first* stable product was to be identified, the time interval between the feeding of $^{14}CO_2$ and the killing-extraction must be shortened drastically. The real key to the success of Calvin and Benson's work was that they devised techniques so that exposure times of only a few seconds could be easily studied. Under such conditions it was found that a 3-carbon compound (*not* a 2-carbon molecule) identified as 3-phosphoglyceric acid (PGA) predominated. This finding, along with other evidence obtained after years of study, established that PGA *was* the first stable product of photosynthesis. Later Calvin and Benson established that PGA was formed by adding CO_2 to a 5-carbon molecule, forming an unstable 6-carbon intermediate which breaks down immediately to give two molecules of PGA.

$$
\begin{array}{ccc}
\text{CH}_2\text{O}\,\textcircled{P} & & \text{CH}_2\text{O}\,\textcircled{P} \\
| & & | \\
\text{C}=\!\text{O} & & \text{HO}-\!\text{C}-\!\text{H} \\
| & & | \\
\text{H}-\!\text{C}-\!\text{OH} & \text{HO}-\!\text{C}-\!^{14}\text{COOH} & ^{14}\text{COOH} \\
| & | & \\
\text{H}-\!\text{C}-\!\text{OH} & \text{C}=\!\text{O} & \text{COOH} \\
| & | & | \\
\text{CH}_2\text{O}\,\textcircled{P} & \text{H}-\!\text{C}-\!\text{OH} & \text{H}-\!\text{C}-\!\text{OH} \\
& | & | \\
& \text{CH}_2\text{O}\,\textcircled{P} & \text{CH}_2\text{O}\,\textcircled{P}
\end{array}
$$

ribulose 1,5-diphosphate →($^{14}CO_2$)→ intermediate sugar acid →(H_2O)→ phosphoglyceric acid (2 molecules)

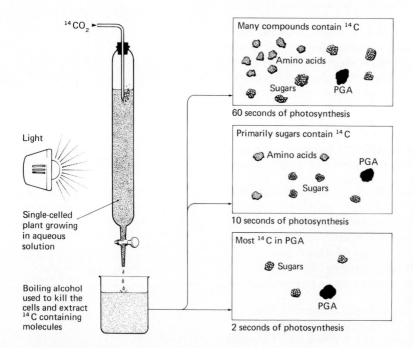

Light

Single-celled plant growing in aqueous solution

Boiling alcohol used to kill the cells and extract ^{14}C containing molecules

$^{14}CO_2$

Many compounds contain ^{14}C
Amino acids
Sugars
PGA
60 seconds of photosynthesis

Primarily sugars contain ^{14}C
Amino acids
PGA
Sugars
10 seconds of photosynthesis

Most ^{14}C in PGA
Sugars
PGA
2 seconds of photosynthesis

FIGURE 2–2.

Shortly after World War II, radioactive CO_2 became available and was supplied to plants for short periods of time to identify the products formed from photosynthesis. In Calvin and Benson's studies, the plant selected was a freshwater alga which could be conveniently manipulated as shown here. The radioactive compounds were identified by a technique known as paper chromatography. When the exposure to $^{14}CO_2$ was brief (i.e., 2 seconds), phosphoglyceric acid (PGA) was found to be the first stable product formed.

It is now known that PGA once formed enters a series of complex reactions that lead to the production of molecules we immediately recognize as "food molecules," sucrose and starch among others. Second, a portion of the PGA produced is recycled through another series of reactions eventually leading to the reformation of the 5-carbon acceptor or intermediate sugar acid, ribulose 1,5-diphosphate (Figure 2–3).

Since the pioneering experiments of Calvin, Benson, and the Berkeley group elucidating the just-described cycle of reactions, we have obtained evidence that at least one other major carbon-incorporating system operates in certain groups of plants, for example, corn, sugar cane, and their relatives. In these plants, in addition to the Calvin-Benson cycle, CO_2 may be accepted by a 3-carbon compound giving rise to a 4-carbon compound. This latter molecule may have a number of fates including donation of the accepted CO_2 molecule to the 5-carbon acceptor compound in the Calvin-Benson cycle. Plants making use of this interesting option are called C-4 plants.

Let us now consider two other rather fundamental questions about photosynthesis. Where in plant cells do the photosynthetic

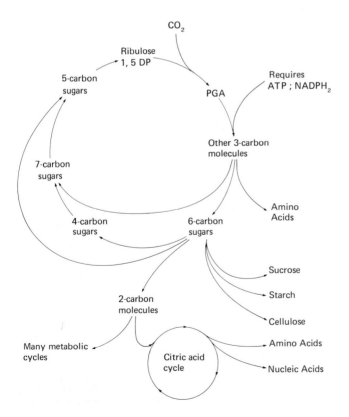

FIGURE 2–3. THE CALVIN-BENSON CYCLE AND BEYOND

Note that CO_2 fixation is accomplished by a cyclic sequence of reactions which can spin off excess sugar molecules. The sugar can subsequently be converted into molecules we ordinarily associate with food or daily maintenance. These reactions are actually much more complex than shown here, and each step requires the presence of a specific catalyst or enzyme. One should also note the requirement for ATP and NADPH₂.

reactions take place, and why is light required? Much of the remainder of this chapter will deal with these two problems. We will first proceed with the intracellular location of the capturing or fixing machinery of CO_2.

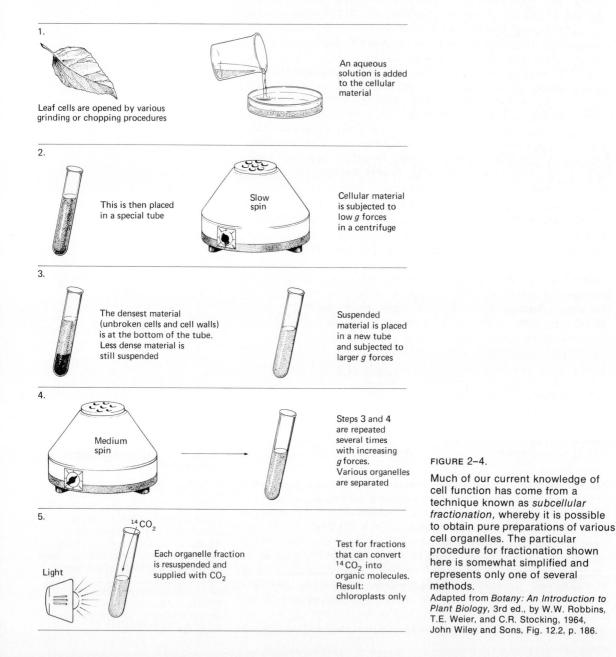

1.

Leaf cells are opened by various grinding or chopping procedures

An aqueous solution is added to the cellular material

2.

This is then placed in a special tube

Slow spin

Cellular material is subjected to low g forces in a centrifuge

3.

The densest material (unbroken cells and cell walls) is at the bottom of the tube. Less dense material is still suspended

Suspended material is placed in a new tube and subjected to larger g forces

4.

Medium spin

Steps 3 and 4 are repeated several times with increasing g forces. Various organelles are separated

5.

$^{14}CO_2$

Light

Each organelle fraction is resuspended and supplied with CO_2

Test for fractions that can convert $^{14}CO_2$ into organic molecules. Result: chloroplasts only

FIGURE 2–4.

Much of our current knowledge of cell function has come from a technique known as *subcellular fractionation,* whereby it is possible to obtain pure preparations of various cell organelles. The particular procedure for fractionation shown here is somewhat simplified and represents only one of several methods.

Adapted from *Botany: An Introduction to Plant Biology,* 3rd ed., by W.W. Robbins, T.E. Weier, and C.R. Stocking, 1964, John Wiley and Sons, Fig. 12.2, p. 186.

The Chloroplast

Plant cells and animal cells are composed of various subcellular units termed organelles. The organelles are located in a viscous matrix, the cytoplasm (see Chapter 1). Does photosynthesis take place in the cytoplasm, in all organelles, or in a particular organelle specialized for photosynthesis? This problem can be approached by using what is now a classic technique in modern biology—*subcellular fractionation*. Plant (or animal) cells can be gently broken open by various and sometimes ingenious means. However, one of the gentlest techniques available is simply chopping plant material into small bits with a razor blade. The material obtained from ruptured cells (cytoplasm and organelles) can then be *fractionated* to yield pure or homogeneous populations of the various subcellular units. Various techniques are used to accomplish fractionation, and one of the most popular methods makes use of the difference in density of the various cellular organelles (Figure 2–4). Briefly, the mixture of cytoplasm and organelles is suspended in an aqueous solution, placed in a tube, and subjected to varying gravitational (g) forces within a centrifuge. This method shows that denser organelles sediment rapidly and at low speeds or g forces, while the smaller particles require stronger g forces. Thus, by repeated centrifugations at greater and greater speeds (equalling greater and greater g forces), it is possible to collect various organelle fractions which differ in density. Once such fractions are obtained, the next step is to test each for its ability to convert CO_2 into organic molecules in the presence of light.

To make a long story short, in higher plants the *chloroplast* contains all the photosynthetic machinery and is unique to plants; all other organelles in plants and animals lack the ability to carry on photosynthesis.

The chloroplast is disc-shaped and surrounded by a double membrane system of some complexity (Figure 2–5). Its most distinguishing characteristic is, however, not its shape or complexity but rather its green color. Plants are green because they have chloroplasts, and chloroplasts are green because they contain a molecule known as *chlorophyll*. Chlorophyll is not evenly distributed throughout the chloroplast but rather is contained in a portion of the membrane system. This membrane system gives rise to the stacklike configurations seen in Figure 2–6. We call such structures *grana*, and the surrounding non-chlorophyll-containing material is called the *stroma*. The CO_2-incorporating apparatus we have just reviewed is contained within the stroma (the nongreen portion) of the chloroplast. You might therefore think that grana stacks and chlorophyll are unimportant in the photosynthetic process and that our problems are solved. This is not so. As pointed out earlier in the chapter, light is essen-

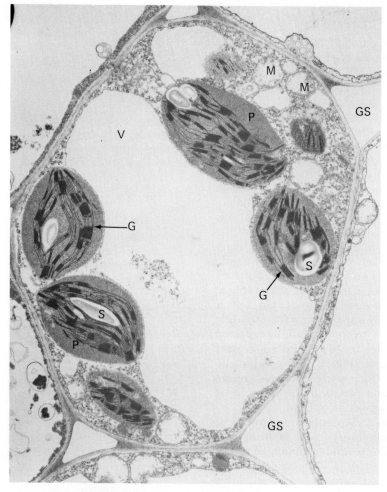

FIGURE 2–5. LEAF MESOPHYLL CELL WITH CHLOROPLASTS

Cellular structures are indicated as follows: vacuole (V), mitochondrion (M), gas-filled cavity between cells (GS), chloroplast (P), starch (S), and granum (G). Note especially the chloroplast membrane system; details are shown in Figure 2–6. Magnification 17,000×.

From *Introduction to the Fine Structure of Plant Cells,* by M. C. Ledbetter and K. Porter, 1970, Springer-Verlag, New York, Inc., Plate 8.1, p. 132.

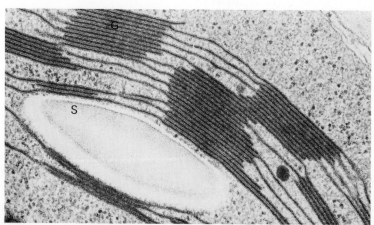

FIGURE 2–6.

This is an electron micrograph of a vertical section through several grana (G) of a leaf cell. Note the stacklike arrangement and interconnecting membranes. Starch (S) is a storage form of fixed carbon and is commonly found in chloroplasts. Magnification 100,000×.

From *Introduction to the Fine Structure of Plant Cells,* by M. C. Ledbetter and K. Porter, 1970, Springer-Verlag, New York, Inc., Fig. 8.2.12, p. 144.

tial for photosynthesis, but we have not yet directly implicated it in the process. Neither have we dealt with the production of O_2. We will do so now with a slight initial detour to examine some of the properties of light itself. Presently you will see that there is indeed an interaction between the stroma and the chlorophyll-containing grana stacks and that light plays an integral part in this interaction.

Light, the Light Reactions of Photosynthesis, and Intraorganelle Communication

Under normal field conditions the only source of light for photosynthesis is sunlight, so we will begin our investigation of this phase of photosynthesis by briefly considering the nature of sunlight itself. Solar radiation isn't simple to comprehend, nor is it easy to explain. Part of the reason for this is that solar radiation is actually a mixture of various kinds of radiation, each type having a different energy value and wavelength (Figure 2–7). It is quite normal for most students to find this concept rather odd, partly, at least, because we are directly acquainted only with a small portion of the solar radiation. That is, our eye-brain complex is put together in such a way that it is only sensitive to the so-called visible spectrum. Within the visible portion of the spectrum, the longest wavelengths we perceive are red in color; the shortest are violet. We also recognize wavelengths of intermediate length, such as orange, yellow, green, and blue.

The visible radiation produced by the sun is a source of energy. In fact, as we will see, it is the ultimate source of virtually all the biological energy on the earth. However, there is a catch to understanding this because, in order for light energy to be useful in any biological process (photosynthesis most certainly included), it must first be absorbed and converted to a form of chemical energy useful to the cell. Molecules and atoms within cells have the ability to absorb units of light energy. These units of energy are called *photons*. But a given molecule or atom will not absorb light of just any wavelength or energy. Rather, light is absorbed only when the photon is of just the right energy

FIGURE 2–7.

It is possible to think of light in terms of a wave motion phenomenon. The length of the wave is indicated by the Greek lambda (λ). Solar radiation consists of a mixture of many different wavelengths; however, plants and the human eye are sensitive to only a small portion of the total radiation reaching the earth —the so-called visible fraction. From Arthur W. Galston, *The Green Plant*, © 1968. By permission of Prentice-Hall, Inc. Englewood Cliffs, N.J. Fig. 1.5, p. 5.

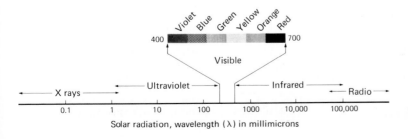

Solar radiation, wavelength (λ) in millimicrons

level *to move an electron* from one energy level to another more energetic level (Figure 2–8). Different molecules vary drastically with respect to what is "just the right energy level." Some will absorb light of the visible spectrum while others require different wavelengths. Since we know there are various wavelengths of sunlight, we might first ask, What portion of the sun's total spectrum is essential for photosynthesis? You are probably already aware of the answer, but let us proceed for the sake of emphasizing an important point. This question is perfectly testable, assuming that we can separate various portions of the sun's spectrum—and such separation is indeed possible. That is, we can expose a plant to x rays or visible rays or any other part of the solar radiation spectrum, and note whether the plant responds by fixing CO_2 into food molecules. The result obtained would show that *visible radiation* was the effective form of light for inducing photosynthesis.

We know that visible radiation is really a mixture of light of various colors, so we can logically ask, What portions of the

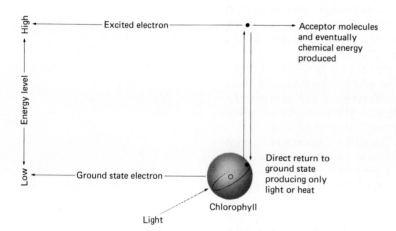

FIGURE 2–8.

When a molecule such as chlorophyll absorbs light energy, an electron is raised from the ground to an excited state. The high-energy electrons so produced can return immediately to the ground state and in so doing give off their extra energy as light or heat, or more importantly some of the excited electrons can be captured by other acceptor molecules in the chloroplast; this leads to the eventual conversion of light energy into chemical energy.

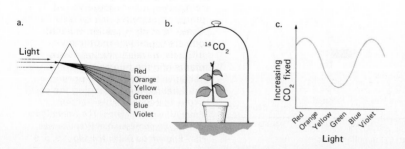

FIGURE 2–9. ACTION SPECTRUM OF PHOTOSYNTHESIS

a. *Preparation.* Our ability to separate visible light with a prism into its component wavelengths makes it possible to determine exactly what type of light is effective in initiating photosynthesis. b. *Test.* Each wavelength can be tested for its effectiveness in causing $^{14}CO_2$ to become incorporated into sugars. c. *Result.* Red and blue are found to be most effective in supporting photosynthesis, while green is relatively inactive.

visible spectrum (e.g., red, orange, green, blue) are most efficient in supporting carbon fixation? In order to answer this question, one would proceed as before. Simply supply various colors
of light to plants and measure the amount of photosynthesis
that occurs in response to each type of light. When we do an experiment of this sort, we call the result an *action spectrum.*
An action spectrum tells us what kinds of light produce a particular response; the methods and results of such an experiment are shown in Figure 2–9. Most carbon fixation takes place
in response to red or blue light, while green is relatively ineffective. Furthermore, since we have just said that for light to
be effective physiologically, it must first be absorbed, there
must be something within the chloroplast that absorbs red and
blue light but is not capable of absorbing significant amounts of
green. In other words, we would expect to find something that
has an *absorption spectrum* resembling the action spectrum
for photosynthesis. When *chlorophyll* is isolated from chloroplasts and its absorption spectrum determined, it is found that
of all the molecules within the chloroplast, the absorption
spectrum of chlorophyll most closely parallels the action spectrum of photosynthesis (Figure 2–10). This, along with other
types of evidence, has clearly implicated chlorophyll (see Figure
2–11 for chemical structure) as the primary receptor of light
within photosynthetic cells. Therefore we have a portion of the
information we desire. The light requirement for photosynthesis
is actually a requirement for red or blue light, and the molecule
responsible for capturing the energy contained in these wavelengths of light is chlorophyll. But we still must relate this phenomenon to the conversion of CO_2 into food reserves.

We saw previously that when a molecule absorbs a photon of
light, an electron is driven to a higher energy level (review
Figure 2–8). In other words, we have an excited or energized
electron with an increment of energy. The situation at this
point, however, is a bit complex because a molecule with an
electron in such a state is unstable. As a result, one of three

FIGURE 2–10. ABSORPTION
SPECTRUM OF CHLOROPHYLL

Many kinds of molecules present in
a leaf cell can absorb visible light,
but chlorophyll is the primary light-
absorbing pigment in photosynthesis.
a. *Preparation.* Chlorophyll and
various other molecules are extracted from plant leaves; the fractions are separated and tested individually. b. *Test.* Various wavelengths of light are directed at the
preparations; the less light that
passes through the solution, the
greater is the absorption. c. *Result.*
The absorption spectrum of chlorophyll most closely matches the action
spectrum of photosynthesis.

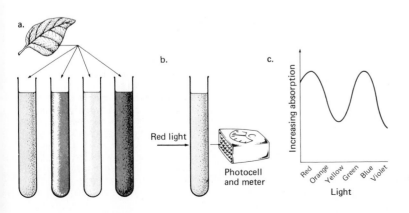

things can happen: (1) the electron can lose its energy as heat and return to its original state, (2) it can re-emit the energy as light, or (3) it can participate in energy-transfer reactions, thus endowing other molecules with increased energy. In a chloroplast possibilities 1 and 2 would be relatively useless and therefore, by the process of elimination, we are left with the notion that the excited chlorophyll molecule must somehow participate in energy-transfer reactions. Such reactions are exceptionally complex, even for a biological process, and are not fully understood. We can, however, present the following somewhat simplified picture.

The chlorophyll molecules in the grana are excited by light. Most of these molecules, however, do not participate in directly transferring electrons to other nonchlorophyll molecules but rather funnel the electrons into other geometrically special (but chemically identical) chlorophyll molecules. These special molecules (perhaps 1 in 200) can then transfer electrons to other molecules. It is now believed that there are actually two different, but interrelated, electron-transfer schemes which we call photosystem I (PS I) and photosystem II (PS II) (see Figure 2–11). First, we will simply describe what is thought to happen

FIGURE 2–11.

a. The "light reactions" of photosynthesis. This diagram, sometimes called the "Z" scheme, shows that light initiates a flow of electrons from chlorophyll molecules. Directly or indirectly this results in the formation of $NADPH_2$, ATP, and O_2. The first two products are required in CO_2-fixing reactions. b. The structure of chlorophyll a. Note the ring complex or head, which contains magnesium (Mg), and the attached tail.

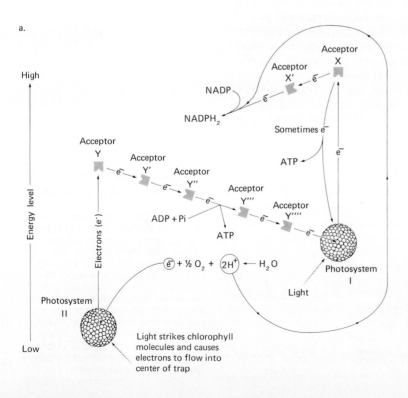

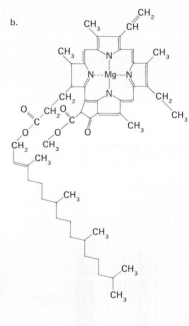

in the entire scheme, and then we will single out certain points that should be emphasized.

Starting first with PS II, we can see that an excited chlorophyll molecule can contribute an energized electron to acceptor molecule Y. This electron then flows from one kind of acceptor to another, and during this process a certain amount of the original energy is lost. The electron eventually finds its way to a chlorophyll molecule which is in PS I. In order for the chlorophyll in PS I to accept this electron, it must have been previously excited by a photon of light and have lost one of its own electrons. The electron lost from PS I can also be transferred from acceptor to acceptor, but finally it ends up in a molecule known as $NADPH_2$ (nicotinamide adenine dinucleotide phosphate). $NADPH_2$ is a stable molecule after it has accepted two electrons, and thus can be regarded as "storing" a portion of the energy originally harvested as light by PS I and PS II. You should also note that in several places indicated in Figure 2–11, molecules of ATP (adenosine triphosphate) were formed. This molecule can be regarded as another stable form of stored energy. The final piece of information that we need to complete this scheme is to understand how the electrons lost from PS II are replaced. We are unsure of the details here, but we do know that somehow water is split, liberating hydrogen ions (H^+), oxygen gas (O_2), and electrons (e^-). The latter particles can then flow into PS II to replace those that have been lost.

Because all this information can be rather staggering at first, even for biologists, it would be good to re-emphasize the most important points. (1) Light initiates the flow of electrons from special chlorophyll molecules to other nonchlorophyll acceptor molecules. (2) As a result of the energetics of electron flow, a portion of the light energy is converted into chemical energy stored as ATP and $NADPH_2$. (3) The entire process can be repeated thousands of times because the electrons lost from the chlorophyll molecules are replaced either from an acceptor (PS I) or from water (PS II). In the latter case, oxygen is a by-product.

Above all else, however, one must realize the importance of the production of ATP and $NADPH_2$; these compounds are necessary for the carbon-fixing process. For convenience, then, we can recognize two phases or sets of reactions in photosynthesis. First, the light reactions, summarized in Figure 2–11, immediately and directly require light energy in order to activate chlorophyll. Second, the dark reactions, summarized in Figure 2–3, require the ATP and $NADPH_2$ produced in the light reaction but do not require light itself. That is, light is needed for photosynthesis because the carbon-fixing process requires ATP and $NADPH_2$, and the only way of generating sufficient ATP and $NADPH_2$ in the stroma is via light-initiated electron-transfer reactions in the grana.

Limiting Factors in Photosynthesis

Now that we have some idea about how and where photosynthesis takes place, we are in a position to investigate some of the factors that can influence the rate of photosynthesis. Such an exercise is far from trivial because the world's food supply *does* depend on the efficiency of photosynthesis, and in theory at least, there seems to be ample room for improvement. For example, it has been calculated that the efficiency of photosynthesis could be around 30% under near-perfect conditions. This means that 30% of the radiant energy falling on a field could be converted into chemical energy, and this would indeed be remarkable for any energy-conversion system. In practice, however, conversion rates of 1% or 2% are most common, although in some plants, under special circumstances, efficiencies of 5% or more have been reported. We will see that the underlying reasons for efficiency rates of only 1% or 2% are not simple, but we will also learn about some novel and interesting ideas for improving the situation.

If we are going to talk about photosynthetic rates and efficiency, we should first know how to measure such rates. One might think it would be simple. CO_2 is used up in photosynthesis, and O_2 is given off. Therefore, with space-age technology it should be possible to buy, or to design and build, a machine capable of monitoring the amount of CO_2 taken up by a plant or the amount of O_2 evolved. Such instruments are available, but unfortunately the problem is slightly more involved than the above might indicate. It must be remembered not only that plants make food via photosynthesis but that they also use up food materials for sources of energy. This latter process, *respiration*, as we have seen, is common to both animals and plants (see Appendix Three); the overall process is just the reverse of photosynthesis:

$$(CH_2O)_n + O_2 \rightarrow CO_2 + H_2O + \text{useful energy}$$

organic   oxygen   carbon   water
material   dioxide

With this in mind, perhaps you can see a way to measure net photosynthesis. We can monitor the net uptake of CO_2 and efflux of O_2, but we must correct these values for respiration. This is simply done by simultaneously measuring the exchange rates of the two gases in another plant kept in total darkness.

If you think about this situation for a moment, it should be obvious that if we have a plant in total darkness and gradually increase the light intensity, we will eventually reach a point at which the rate of photosynthesis equals the rate of respiration. In other words, the plant is at a point at which it is on a no-weight-gain diet—just enough food is made to equal that broken down. We call the situation at that light intensity the *compensation point* (Figure 2–12). This point varies slightly from species to species, but as a sort of rule of thumb, the light intensity required just to exceed the compensation point in most plants is roughly 50–250 foot-candles (about what you would have in a well-lit room). To put this in perspective, it must be realized that this sort of intensity is only a small fraction of the radiation that would normally fall on a field during a cloudless summer day (perhaps up to 10,000 foot-candles). Consequently, during the peak daylight period, photosynthesis more than compensates for respiration (perhaps at a ratio of 10:1). In fact, photosynthetic activity is sufficient to produce such an excess of food material that only a very small fraction of the total is used up during the night hours when only respiration occurs.

Light and Photosynthetic Activity

At this point we can begin to address ourselves directly to the problem of *limiting factors* in photosynthesis. Let's first consider light intensity since we have already dealt with it briefly. Could we increase the photosynthetic rate in the field if the light intensity was increased? With only a few exceptions the answer is an unqualified no. Under field conditions, during peak daylight hours, light intensity does not limit the rate of photosynthesis since further increasing the intensity does not speed up the photosynthetic process. Under very dim conditions or in the dark, light does limit the amount of photosynthesis that can occur, but we are primarily concerned with field conditions where we have relatively high intensities. In fact, many plants are *light saturated* in rather shady locations, as the following discussion should make clear.

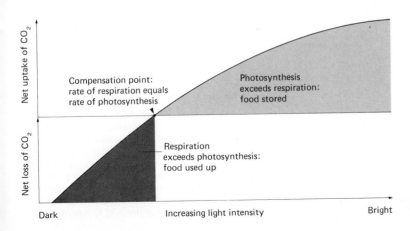

FIGURE 2–12.

It is important to realize that plants carry on respiration as well as photosynthesis. When there is little or no light, there will be a net loss of CO_2 due to respiration. As the light is gradually increased, respiration continues but photosynthesis is enhanced. When there is no net gain or loss of CO_2, the two processes balance one another and this is called the *compensation point*.

You may have noticed that many nurseries list a particular species as either a *shade plant* or a *sun plant.* You now have the information to understand this distinction scientifically. Compensation points and the relative light intensity which saturates the photosynthetic machinery determine whether a plant is a shade or sun plant. Let's consider Figure 2–13. Here you can see that a typical shade plant (e.g., a woodland fern) is light saturated at about 1000 foot-candles. That is, higher intensities of light fail to increase the amount of photosynthesis that occurs when other factors such as CO_2 and temperature are held constant. On the other hand, a sun plant (e.g., soybean) requires perhaps 3000 foot-candles or more to saturate the photosynthetic apparatus. Of course this saturation phenomenon in itself does not endow shade plants with a particular advantage in low-intensity light. Rather, it is this phenomenon coupled with the fact that shade plants have a lower compensation point that allows them to survive under conditions of illumination that would be intolerable to a sun plant. Put another way, it means that less net photosynthesis has to take place in shade plants in order for them to stay ahead of respiration than is necessary in sun plants. Under shady conditions a sun plant would starve to death!

Temperature, CO₂, and Photosynthetic Activity

Photosynthesis can take place over a rather amazing temperature range. For example, certain primitive microscopic plants called blue-green algae can carry on photosynthesis in thermal pools and hot springs at temperatures of $75°C$, while pines and certain other evergreen trees can maintain a low rate of photosynthesis at $6°C$. But these are rather extreme examples; for most plants the range is much narrower with the

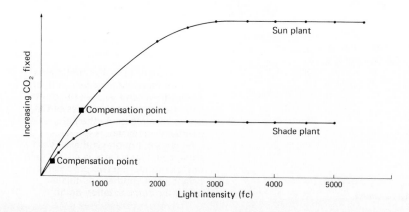

FIGURE 2–13.

As a general rule, plants are classified by the light intensity needed to saturate their photosynthetic machinery. *Shade plants* (e.g., club moss, forest oxalis, impatiens, ivy, moss, woodland fern) saturate at relatively low intensities and are further characterized by the low intensity needed to reach the compensation point. *Sun plants* (e.g., chrysanthemum, corn, daisy, potato, sunflower, tomato) need higher light intensities to reach the compensation point and to saturate.

ideal or optimal temperature somewhere between 20–35°C. This fact suggests circumstances in which temperature could limit photosynthesis. For example, if there is adequate light and CO_2, photosynthesis will occur more rapidly at 25°C than at 0°C. However, again under field conditions during the peak growing period, temperatures are usually *already optimal* for photosynthesis, and therefore temperature is not usually a limiting factor under ideal field conditions.

Now we come to the crux of the matter. On a nice, warm, sunny day, the CO_2 concentration in the atmosphere is generally the factor that limits photosynthesis. This phenomenon is demonstrated experimentally by placing a plant in an illuminated room where one can artificially adjust the CO_2 levels and also continuously monitor photosynthesis. In Figure 2–14 we can see that at a moderate light intensity and temperature, photosynthesis increases as the CO_2 concentration is increased. That is, additional CO_2 effectively increases net photosynthesis until the CO_2 concentration reaches a level of approximately 0.1% by volume. In the atmosphere, CO_2 is only about 0.03% by volume. Therefore, even in relatively dim light, CO_2 will limit photosynthesis, and certainly on a sunny day under field conditions, the limiting factor is almost always the availability of CO_2. Stated in another way, under favorable field conditions there is sufficient light to produce all the ATP and $NADPH_2$ needed for the CO_2-fixing reactions, but the amount of sugar produced is limited by the availability of new carbon in the form of CO_2.

It is worth re-emphasizing that CO_2 is only a very small component of the atmosphere (0.03%). This percentage can be contrasted with the amount of nitrogen (78%) and oxygen (21%).

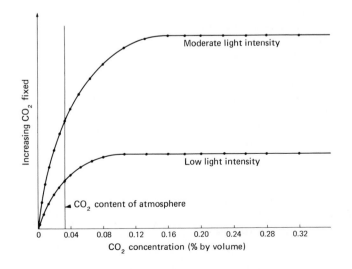

FIGURE 2–14. CO_2 AND PHOTOSYNTHESIS

CO_2 is commonly the limiting factor in photosynthesis as is shown by the fact that photosynthesis increases with increasing concentrations of CO_2. Light becomes the limiting factor when the curves level off. Note that even at low light intensities the normal atmospheric concentration of CO_2 is not sufficient to saturate photosynthesis.

There is even less CO_2 in the atmosphere than the gas argon which we classify as rare! Indeed, with such a small percentage of CO_2 in the atmosphere, there is an amazing quantity of it fixed annually: about 6×10^{13} kg (see Appendix One for metric conversion chart) or approximately one-thirtieth of the total amount available in the atmosphere. And with such a large amount of the total atmospheric CO_2 used annually, it is important to have some rather efficient recycling mechanisms working to replace continually the CO_2 being fixed. Let's have a look at some of these mechanisms.

The Carbon Cycle: Recycling Is Not a New Concept

First, we know that both plants and animals respire, and therefore living organisms do not just act as CO_2 traps but also are CO_2 producers. As you recall, plants are able to photosynthesize more rapidly than they respire so that net uptake of CO_2 is favored. This phenomenon would be serious for the total atmospheric level of CO_2 if it was not for the activities of various plants and animals known as *decomposers*. When a plant or animal dies, the molecules that make up its body gradually decompose. You can well imagine a world where this never occurred—we would be buried in our own wastes. This vital life process is carried on primarily by soil microorganisms. During decomposition, CO_2 is released into the atmosphere as a result of the respiratory activities of the decomposers, and the cycle is complete. Atmospheric CO_2 is fixed into food molecules and eventually is released from these molecules and returns to the atmosphere (Figure 2–15).

In addition to this cycle, we have an extensive reserve of CO_2 in the oceans: roughly 80 times the concentration in the atmosphere. This amount is far from trivial, considering the fact that the oceans occupy nearly three-fourths of the surface of the earth. All these processes—photosynthesis, respiration, and decomposition—also occur in the ocean just as they do on land. In addition, however, other complex cycles exist because a considerable amount of CO_2 in seawater is tied up in animal shells and rocks as calcium carbonate. While all of this is important, the main point you should remember is that the oceans are a vast reservoir of potential CO_2 and that an important interface is formed by the oceans and the atmosphere. It is most likely, therefore, that a kind of equilibrium exists which serves to stabilize the atmospheric level (see Figure 2–15). For example, if the atmospheric concentration falls slightly below the equilibrium value, additional CO_2 will diffuse from the ocean and increase the atmospheric level. If, on the other hand, the atmospheric concentration rises above the equilibrium value, more CO_2 will become dissolved in the water and once again stabilize the level. It would be difficult to overestimate the importance of the oceans in this regulatory role. All food chains are ultimately dependent on photosynthesis, and photosynthesis is commonly limited by CO_2. A drop in the atmospheric level would immediately lower our already strained food base. It should also be pointed out that the CO_2 cycles in freshwater and saltwater bodies are exceedingly complex and are not fully understood. But obviously they *are* dependent on the "good health" of the water and the organisms that live there. Pollution and overexploitation of these complex bodies of water may have far-reaching effects which we cannot fully forecast but which could presumably alter such cycles.

Summarizing Limiting Factors

Before proceeding, it would be helpful to attempt to summarize some of this information concerning the limiting factors in photosynthesis. First, under various conditions, light, temperature, or CO_2 can be the limiting factor. Plants in the dark cannot photosynthesize because light is necessary for ATP and $NADPH_2$ production, and the latter molecules are necessary for

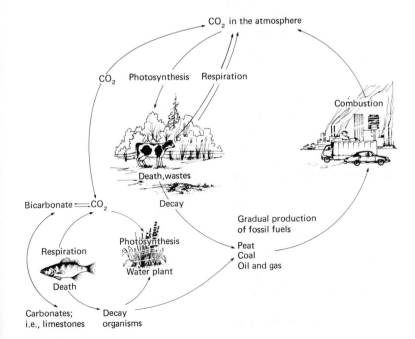

FIGURE 2–15.

Arrows indicate the direction in which carbon dioxide (CO_2) is cycled in the biosphere. Try to visualize the consequences if a particular part of the cycle is eliminated or partially destroyed.

From *Population, Resources, Environment: Issues in Human Ecology*, 2d ed., by Paul R. Ehrlich and Anne H. Ehrlich. W. H. Freeman and Company, Copyright © 1972. Fig. 7.2, p. 198.

the carbon cycle. As the light intensity is increased, it is more difficult to say for certain what factors may be limiting. Sun plants in the deep shade, during the early morning or later afternoon hours, or on cloudy days may indeed be limited by light intensity or CO_2, depending on the particular circumstances. Shade plants, however, are adapted to lower light intensities and are characterized by a lower compensation point and a photosynthetic apparatus that saturates at relatively low light intensities.

Temperature is a limiting factor only when it is extremely hot (leaves wilt and prevent CO_2 entry) or when it is extremely cold (CO_2-fixing reactions are greatly slowed down due to reduced enzyme activity). In general, it would appear that under favorable growing conditions, CO_2 is the factor that limits photosynthesis. It is important for all of us, therefore, to take care not to disrupt the natural factors that maintain the CO_2 level.

Before ending this chapter, it is interesting to speculate on how some of the theoretical information on photosynthesis might be put to practical use. We all know that there is a food shortage. But can this situation be improved by a scientific application of the principles we have just discussed? It is a difficult question, and the answer is yes *and* no! Let us briefly investigate two of the more obvious possibilities, and then in the next chapter pursue some of the subtler aspects of plant productivity.

Getting the Most out of the Green Machine: Light

Superficially, it would seem that the single most obvious way to increase plant productivity would be to illuminate plants artificially during the hours of normal darkness. Although we all realize that this would be economically impossible for large-scale farming operations, it is nevertheless worth considering as a possibility for greenhouse use. Indeed, artificial light does increase the productive capacity of a great number of plants and is used by some commercial greenhouse operators. However, as curious as it may at first seem, artificial illumination during the normal hours of darkness is not useful in growing all plant species. The reason for this phenomenon is that plants and other organisms rely on light-dark cycles to regulate some aspects of their physiology and development. We will discuss this particular aspect of plant physiology in a later chapter, but just to give you some idea of this subtle phenomenon, an example or two seems in order.

Flowering in chrysanthemums normally occurs in the fall

when day length shortens. This particular response is known to be related to an actual measurement of day length—rather than simply to a requirement for the plants to achieve a certain age or ripeness—because one can induce flowering during the summer by artificially shortening the days. Likewise it is also possible to stimulate vegetative growth in chrysanthemums during the fall and winter and to prevent flowering. To do this we simply turn the lights on for about 4 hours during the middle of the night in order to shorten the dark period. Chrysanthemums, then, *require* short days and long nights in order to flower, and so they are called *short day* plants. *Long day* plants require long days and short nights. Carnations, for example, produce more flowers if day length is extended to 14 hours by artificial lighting during November through March. Commercial growers, being able to tell time themselves, have made good use of the plants' ability to do so, and as a consequence we can have flowers all year round.

What all this means is that some plants (as well as some animals) have evolved mechanisms that can somehow measure time (Figure 2–16), and that in the case of flowering the time-measuring mechanism involves light-dark cycles. Therefore, returning to our original question concerning continuous light, it is possible in some plants to reap benefits by prolonging the total light period and thus the total time a plant carries on photosynthesis. *But* constant light may also lead to secondary effects (flowering or not flowering, among others) which may in turn serve to defeat the original purpose which is, of course, an increase in *usable* yield. Indeed, as we proceed, you will see that plants respond to their environments in complex and subtle ways, and it is difficult to alter drastically one particular aspect of a plant's metabolism or development without changing others in sometimes unwanted ways.

Carbon Dioxide and Productivity

We have seen that under field conditions photosynthesis is typically limited by the concentration of carbon dioxide in the air. It would therefore seem obvious that if we could somehow increase the level of CO_2 available to plants, we could increase their overall productivity. The data presented in Table 2–1 appear to support fully this notion. That being the case, why don't we get on with it and supply all our crops with extra CO_2? As you may suspect, the problems involved tend to cast doubt on whether or not even this seemingly practical solution to the problem of increasing productivity would prove worth the effort in the long run.

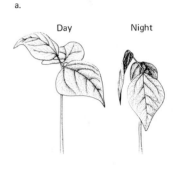

a.

Day   Night

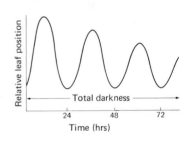

b.

Relative leaf position

Total darkness

24   48   72

Time (hrs)

FIGURE 2–16.
There are many ways to illustrate the phenomenon of time measurement. In one, the flowering response of plants, day-night cycles are important; however, other regularly timed physiological responses (called rhythms) occur in plants which do not necessarily key on a light-dark cycle, for example, *leaf movement*. a. *Observation*. Leaf movement is a phenomenon exhibited by many plants, including the bean. b. *Experiment*. Because bean leaf movements occur in total darkness on a 24-hour cycle, the mechanism which regulates the timing of these movements must be intracellular.

	Lettuce, Fresh Weight[1]	Tomatoes, Fresh Weight[2]
Normal CO_2 (125–500 ppm)	1.4	9.8
CO_2 enriched 3× to 7× (800–2000 ppm)	2.4	14.0

TABLE 2–1. EFFECTS OF CO_2 ENRICHMENT OF PLANT PRODUCTIVITY

[1]Pounds per 10 heads, average 3 varieties.
[2]Pounds marketable fruit per plant, February 15 to June 1.

Source: From S. H. Wittwer and W. Robb. 1964. Carbon dioxide enrichment of greenhouse atmosphere for food crop production. *Economic Botany* 18:34 ff.

To begin with, most units that produce CO_2 (combustion or burning of organic compounds is a common method) also emit impurities which are toxic to plants (Table 2–2). Thus, even in closed systems such as greenhouses, setting up a workable source of CO_2 can be expensive. The second problem involves technology. How does one increase the CO_2 level in field crops? It has been done by bringing it in with long pipes at ground level and by supplying Dry Ice (frozen CO_2) between rows. Some results are encouraging, although, as you can imagine, plowing is extremely tricky. The third problem may be even more serious. Larger and faster-growing plants would require increasing the frequency and amount of fertilization. This means added expense and could lead to increased weed-control problems and an ever-increasing economic spiral. Again we can see that altering a single environmental factor has multiple and sometimes unexpected effects. You may now begin to see how really complex our environment is and how finely tuned are the metabolic processes in living things. But even if all these problems could be worked out, there is still some question about whether or not we should seriously consider this approach. A plant's metabolism oftentimes responds in unexpected ways to increasing CO_2. Generally we find an increase in sugar production but a relative decrease in certain amino acids and organic acids in plants grown in a CO_2-enriched environment. Therefore, the popular appeal or marketability of such plants may be enhanced, but their nutritive value, especially protein content, may decline (see Chapter 3 for a discussion of protein in the diet). And finally, the climatic effects of increasing CO_2 in agricultural areas could be adverse.

In summary, then, we can increase the productivity of the crop plants we already have, and there is undoubtedly an even greater potential present than has been explored. However, as with almost all radically new processes, unexpected problems

Gas	PPM
Ethylene	0.1
Sulfur dioxide	0.5
Ammonia	10.0
Nitrogen oxides	25.0
Hydrogen sulfide	50.0
Carbon monoxide	500.0
Carbon dioxide	20,000.0

TABLE 2–2. HIGHEST LEVELS OF SELECTED GASES WHICH TOMATOES CAN TOLERATE WITHOUT INJURY

Source: From S. H. Wittwer and W. Robb. 1964. Carbon dioxide enrichment of greenhouse atmosphere for food crop production. *Economic Botany* 18:34 ff.

and a certain degree of danger do exist. Yet these examples are but a small part of the Green Revolution. Let's now pursue the entire problem in the depth it deserves.

Selected Readings

Arnon, D. I. November 1960. The role of light in photosynthesis. *Scientific American* 203(5):104–118. Good summary by a distinguished researcher in the field.

Bassham, J. A. June 1962. The path of carbon in photosynthesis. *Scientific American* 206(6):88–100. (Offprint no. 122.) Easy-to-read account of the experiments leading to the elucidation of the Calvin-Benson cycle.

Hatch, M. D., and C. R. Slack. 1966. Photosynthesis by sugar cane leaves. *Biochemical Journal* 101:103–111. Technical article, and one of the first on C-4 photosynthesis.

Levine, R. P. December 1969. The mechanism of photosynthesis. *Scientific American* 221(6):58–70. A good review.

Plass, G. N. July 1959. Carbon dioxide and climate. *Scientific American* 201(1):41–47. (Offprint no. 823.) Readable account of how increasing CO_2 levels may affect climate.

Salisbury, F. B., and C. Ross. 1969. *Plant Physiology*. Belmont, Calif.: Wadsworth. Good source to pursue the details of photosynthesis.

3 The Green Revolution

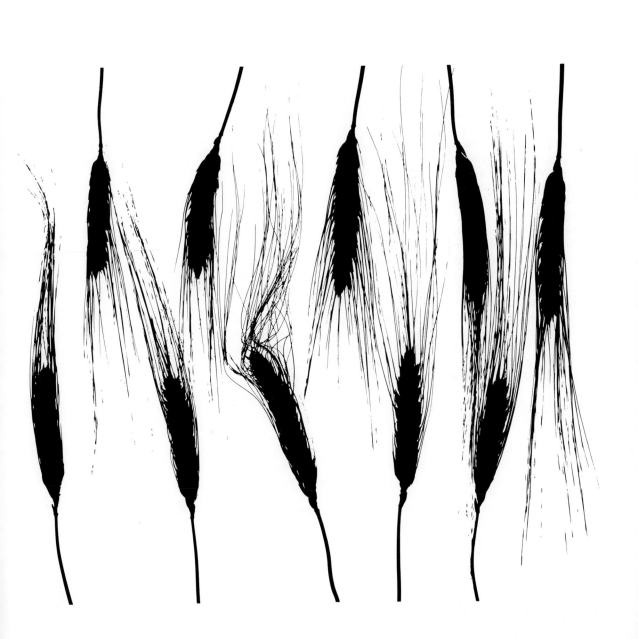

The population of the world is increasing. Beyond the sheer statistics it is something we can feel, something that affects our day-to-day existence. We notice it in our cities where traffic congestion, pollution, and crime threaten to extinguish a lifestyle that has existed for generations. We notice it in the so-called energy crisis, in the chlorinated water we drink, and in the air we breathe. Away from the cities many of our campgrounds are overcrowded, our rivers overfished, and our alpine meadows overloved. In fact, as many of you know, permits are required for cross-country hiking in some of our wilderness areas. Yes, the population is increasing. But the rather pedestrian examples just used, as unpleasant and irritating as they can be on occasion, scarcely give any indication of the other infinitely more serious problems that may be generated by the population explosion: worldwide famine, epidemic disease, and economic and social diaster. One scarcely has to open a newspaper or turn on the radio before encountering a modern-day prophet predicting famine and mass starvation in the near future—perhaps before 1980. But can we really believe such statements? We also hear statements from the less vocal but equally ardent extreme who claim that science and technology will so increase our production of crops and raw materials that we will be able to support an increasing population for an indefinite time. Just where does the truth lie; can we afford the consequences if our guess is incorrect? Perhaps the best most of us can hope to do is assess the information at hand, weigh the alternatives, and then decide what future mankind has in the next century.

World Population

The first thing we need to know is just how fast the world population is growing. Figure 3–1 will give you some feeling for the situation. You'll notice that in the left part of the figure, we can only mark with certainty those figures after 1650 because we do not have sufficient historical information to estimate accurately the population much before this period. Nevertheless, the interval between 1650 and the 1900s spans a sufficient time to appreciate the situation. Our aim is to show not simply that the world's population has increased but rather that the *rate* of population growth has increased. There are several ways to illustrate this phenomenon, but perhaps the most convenient approach is to consider the time necessary to add a half billion people to the earth. The first half billion took thousands of years beginning when man first appeared on the earth until about 1650. The time necessary to add another half billion was two centuries and has continued to grow successively shorter. Now only about seven years are required.

What is the reason for this increased rate of population growth? Formally, the growth rate of any population is estimated by subtracting the deathrate from the birthrate since the deaths obviously represent a subtraction and the births an addition. Because we have experienced an overall increase in world population, clearly the births have exceeded the deaths in most years. In the United States, for example, in 1967 the birthrate was about 17.3 per 1000 people, while the deathrate was approximately 9.6 per 1000 people. This means that in that given year there was a net increase of 7.7 people per 1000 or an increase of 0.77%. This figure is about average for a developed country, but for the world as a whole the increase in this period was about 2%.

Just viewing such figures may give you some feeling for population growth over a short period of time, but they could give you the wrong impression when dealing with longer time intervals. This is because such figures imply that the world population grows at a rate of 20 people per year per 1000 inhabitants (2% of 1000 is 20). But does this mean 20 this year, 20 next year, and 20 the year after? No, the rate is greater than this because the new people added to a population will themselves eventually produce more people. Thus, in the second year we must calculate the growth rate at 2% of 1020 and so on year after year. It's exactly like investing your money in a savings account at compound interest; you earn money not only on the original investment but also on the interest *compounded*

FIGURE 3–1.

World population in the past and present with projections for the future.

From "Human Food Production as a Process in the Biosphere" by Lester R. Brown. Copyright © 1970 by Scientific American, Inc. All rights reserved.

Year (A.D.)	Population (billions)	Number of years to double
1	0.25 (?)	1,650 (?)
1650	0.50	200
1850	1.1	80
1930	2.0	45
1975	4.0	35
2010	8.0	?

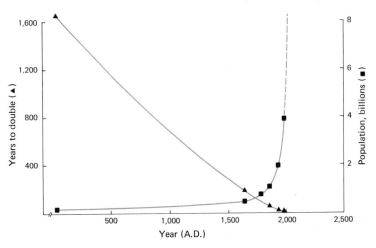

annually. Thus, you can see that the growth of any population, where birthrates exceed deathrates, will continually spiral upward, and the time necessary to double the original number will grow progressively shorter and shorter (Figure 3–2).

Even the optimists among us should agree that this trend cannot continue indefinitely. Clearly, with finite resources of space, both land and water, doubling our present population of 4 billion is a proposition different from doubling 1 billion to 2, or 2 billion to 4. Yet, even this kind of reasoning doesn't present as bleak a picture as actually exists because in such an argument the entire world is treated as a single unit while in reality each country has different resources and degrees of technical skill. Sadly, the population explosion is most severe in those countries where various technical skills are less advanced. This, of course, results in a decreased ability to exploit their particular resources in order to provide adequately for rapid population growth, and so the problem is, in reality, much severer when examined in detail.

The Problem: Food

Just for the moment let us assume that we can somehow cope with the social problems of crowding and pollution, and ask about the world's food supply. How long can it keep pace with our current rate of population growth? On this point the experts differ. For the past 10–15 years, food production has been increasing at about the same rate as the population. This might not sound too ominous at first; perhaps it even sounds hopeful.

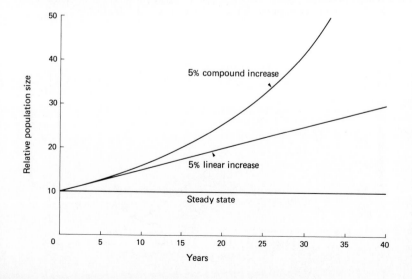

FIGURE 3–2.

Steady state, linear increase, and compound increase. When the birth rate equals the death rate, a population remains the same size year after year, and the existing situation is called a steady state. An increase each year of 5% of the original population gives a linear increase; the number doubles after 20 years and triples at 40 years. A 5% compound increase gives a curve wherein the population doubles after 15 years, triples at 23 years, and quadruples at 29 years.

But is it? In the first place, it is difficult to detect a difference between linear growth rates and compound growth rates over such a short span. We know that the population can continue to grow at the compound rate, but can food production? Second, there are presently about 2 billion people on this planet with insufficient food to meet their daily needs. Perhaps 10–20 million people starve to death each year. If we are simply able to increase our food supply at the current rate, does this not mean that the mass of undernourished and starving people will also increase as our population grows? It does indeed. In the past the less-developed countries have turned to America as the granary of last resort when their own agricultural productivity was insufficient to meet their needs. Foods were made available at low costs—even free—and were supplied from our food surpluses. The federal government was then paying farmers to keep land out of production. Presently, however, farmers are encouraged to produce as much as possible, but even so, American reserves are at their lowest levels in over 20 years.

Obviously we must do more than just keep pace with our current growth rate. We must greatly increase the food available on a per capita basis, and this is precisely what this chapter is about. We will examine in detail many of the options and opportunities available for expanding our present food-producing capacity, and in doing so, you will come to understand just what is meant by the Green Revolution—its potential *and* its limitations.

Of course it would be pointless to talk about food and food production without first considering just what we mean by food. Let's start by defining *food* as any material that provides an organism with energy and building materials for growth and daily maintenance. We have already seen that green plants can subsist on the rather monotonous diet of CO_2 and water. Animals, however, require ready-made food, and furthermore, growth and body function depend on the correct amounts and combinations of various food materials. Not all food sources are equally good in providing a balanced diet. What all of this means is that we can't just increase the production of any food crop; we must also see to it that this increased bulk contains all the nutrients necessary to support man. The next step is to examine these nutritional requirements.

Food and Diet

Approximately 96% of the weight of the human body is made up of just four elements: carbon (C), nitrogen (N), hydrogen (H), and oxygen (O). Therefore, most of our nutritional needs center

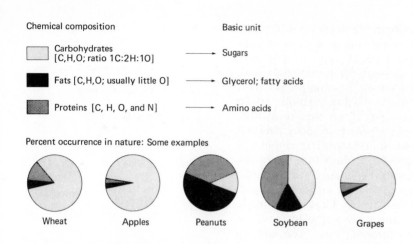

Chemical composition

Basic unit

Carbohydrates
[C,H,O; ratio 1C:2H:1O]   → Sugars

Fats [C,H,O; usually little O]   → Glycerol; fatty acids

Proteins [C, H, O, and N]   → Amino acids

Percent occurrence in nature: Some examples

Wheat   Apples   Peanuts   Soybean   Grapes

FIGURE 3–3.

Our metabolic needs center around obtaining carbon (C) hydrogen (H), oxygen (O), and nitrogen (N). We obtain these by ingesting the basic types of food listed here. Note the different distributions in various plant products (percent occurrence in nature).

around providing these basic elements. We do this by ingesting them in various foods, which can be broadly classified as containing proteins, carbohydrates, and fats (Figure 3–3). Of these, proteins are undoubtedly of the greatest importance. One might actually consider the world's food problem to be one of protein supply—solving it would largely resolve current and anticipated food shortages. This point can be rather forcefully made by pointing out that at the present time a significant portion of the world's population is existing on a protein-deficient diet (Figure 3–4). Protein starvation is especially serious for young children, and if not dealt with early in a child's life, it

FIGURE 3–4.

Hunger does not require a passport. From *Population, Resources, Environment: Issues in Human Ecology*, 2d ed., by Paul R. Ehrlich and Anne H. Ehrlich. W. H. Freeman and Company, Copyright © 1972. Fig. 4.4, p. 83. Data from FAO *Production Yearbook 1968*.

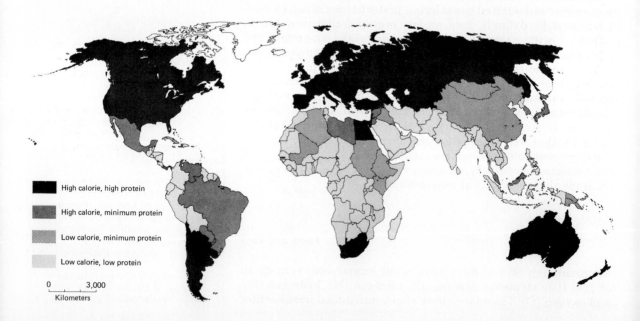

High calorie, high protein

High calorie, minimum protein

Low calorie, minimum protein

Low calorie, low protein

0   3,000
Kilometers

FIGURE 3–5.

Protein deficiency (Kwashiorkor disease) can often be seen in African children following weaning. Note the child in left-center.
Photograph by Richard A. Dewey.

can lead to growth retardation, brain damage, and death (Figure 3–5).

In order to understand properly the nutritional requirement of protein in the diet, we must first briefly review what proteins do in the cell and how they are constructed. Proteins are indispensable components of the body for two reasons. First, they serve as structural material within every cell (in fact, approximately 50% of the dry weight of a cell *is* protein). They make up part of every membrane system and cell wall and are structural components of chromosomes and all other organelles. Second, special proteins called enzymes are necessary to catalyze almost every metabolic reaction that occurs in cells (Figure 3–6). In many ways the functional and structural features of a cell simply reflect the properties of its proteins. Proteins are large molecules made up of smaller units called amino acids, and it is, in fact, these small building blocks that we require in our diets (Figure 3–7). The dietary requirement for amino acids stems from the fact that animals lack the ability to construct all the amino acids necessary for protein production and there-

FIGURE 3–6. AN INTRODUCTION TO THE ENZYME CONCEPT

Enzymes are cellular catalysts that speed up the rate of chemical reactions in cells without themselves being changed in the process. In the absence of enzymes, most biologically important reactions would occur at a very slow rate or not at all. Each cellular reaction is catalyzed by a specific enzyme. This specificity between the enzyme and its substrate leads to the formation of an enzyme-substrate complex. Forces within the enzyme cause chemical changes in the substrate (indicated by the arrows) which result in the formation of new molecules—the products.
From *Contemporary Biology: Concepts and Implications*, by Mary E. Clark, 1973, W. B. Saunders Company, Fig. 3.7, p. 65.

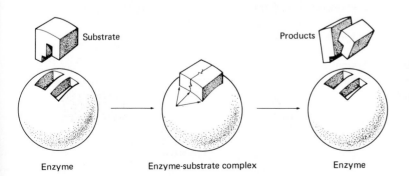

Substrate   Products

Enzyme   Enzyme-substrate complex   Enzyme

fore must obtain these building blocks as part of their diet. We could ingest amino acids as such, but the usual way is to consume some food source which contains protein. We don't make use of preformed protein directly, however, but instead during digestion degrade protein to amino acids in the stomach and small intestine. These compounds then enter our cells and are reassembled back into new protein as needed. If we follow this argument back far enough, it becomes obvious that some organisms had to make initially all the amino acids necessary for functional proteins, and these organisms are plants. Plant cells require proteins for structural as well as enzymatic purposes, but fortunately they can construct all the necessary amino acids from sugars and nitrogen which they obtain via photosynthesis and absorption from the soil, respectively.

In summary, we can say that all cells contain proteins and they are indispensable parts of our cells. Plants and animals construct these large molecules from amino acids. Oftentimes the amino acid source is a previously constructed protein which yields amino acids on digestion. Plants also require protein molecules for cellular processes but, unlike animals, can construct all the necessary building blocks from simpler raw materials.

Within a protein, amino acids can be arranged in different ways, thus giving rise to a large possible number of different proteins, each of which will have a different function. This variety is a result of both primary and secondary structure.

a.

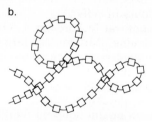

b.

a. *Primary structure.* Assume amino acids represented by numbers 1 to 20. Chains could exist composed of various combinations. b. *Secondary structure.* Proteins can consist of chains of more than 200 amino acids which often exist in a folded and non-symmetrical structure, depending on the primary structure.
c. *Function.* Function is dependent on physical characteristics which in turn depend on primary and secondary structure. Variety in structure means variety in function.

c.

Protein 1 ⟶ Participates in starch synthesis

Protein 2001 ⟶ Part of a cell membrane

FIGURE 3–7. AMINO ACIDS AND PROTEINS

Approximately 20 different kinds of amino acids are needed for protein production in man. These amino acids are required in differing amounts and ratios, but these ratios are most closely met via the ingestion of animal protein. Plant sources are often low in certain necessary amino acids and thus, in theory, could give rise to an inadequate diet. The figure shows how various amino acids are used and how a variety of proteins can exist.

We will now proceed to an examination of various protein sources and see what problems arise when we attempt to optimize the world's supply.

In most cultures man prefers to obtain most of his protein from animal rather than plant sources; in fact, a few cultures exist entirely on a diet of animal protein and fat. The reason for this preference is primarily cultural; however, it is important to realize that there are firm biological reasons as well for choosing animals, rather than plants, as a primary protein source. Most animal protein is readily digested and provides man with all the essential amino acids *in the proper proportion* or ratio. This isn't true of plant protein in general. Thus, it would perhaps be most desirable if we could simply increase the supply of animal protein, that is, raise more cattle, pigs, and sheep. However, the cost of producing animal protein is high, and this fact alone would seem to eliminate this possibility.

Animals obtain their protein from plant sources, but they are *inefficient* converters of plant protein to animal protein. In the transfer of protein from plants to animals to man, we lose a large amount of valuable food due to wastage and the metabolic activity of the animal itself. Cattle, for example, eat only part of the plant. The roots go unused and thus are wasted. Cattle also use up energy (and indirectly protein) in their daily activities—they "walk off" part of their energy intake. And, of course, man only consumes part of the animal as his food source, wasting much potentially useful protein.

The phenomenon alluded to here can be expressed in an even more general way by reference to what we call a *food pyramid*. Understanding the principles of a food pyramid may aid you in understanding the aforementioned argument and, more important, in seeing the general form of the dilemma regarding the world's food shortage. You will recall from Chapters 1 and 2 that plants and animals constitute an interacting system in which food material is produced by plants (primary producers) utilizing solar radiation as the energy source. Absorbed solar energy is converted to chemical energy and stored as food molecules; in this form it is passed on to herbivorous animals which eat plants (primary consumers). The animals then utilize this chemical energy for their daily activities and for growth. Other animals (secondary consumers) may eat the primary consumers, and the food value so obtained represents energy which can be used for the various life processes. Secondary consumers can then be consumed by tertiary consumers and so on.

The essential point that we now want to make about food pyramids is that the transfer of chemical energy from one consumer level to another is an inefficient process. Only about 10% of the total chemical energy present in food material at one level ends up as chemical energy at the next level (Figure 3–8). This is a serious consideration because the energy *input* is regulated entirely by the photosynthetic activity of plants. This means that when we eat cattle, we are decreasing by at least tenfold the total food energy that could be available to us as compared with eliminating the "middleman" and eating plants directly. In short, eating meat is a very inefficient way to harvest protein, while the direct consumption of plant protein is more efficient and of course cheaper. So, some might ask, what's the problem? Why don't we just discontinue raising animals for meat and eat only plant protein?

The problem is really twofold, as we previously explained. Cultural practices have developed which cause most men and women to regard meat as an indispensable part of any diet, and second, the nutritional value of plant protein isn't as satisfactory as animal protein. That is, in plant protein some of the amino acids needed by humans are present in rather low amounts. Protein obtained from seeds (a rich source) is often low in the two sulfur-containing amino acids, cystine and methionine. Cereal grains, another important source, are low in three amino acids: lysine, methionine, and tryptophane. Another complication arises from the fact that plant proteins are generally less digestible than are proteins obtained from animal sources. These problems are not trivial, but we do not wish to imply that the scientific aspects of the problem are unsolvable. On the contrary, our future sources of protein must come from plants; we simply must *select* those that have the most desirable characteristics, and such selection is an integral part of the Green Revolution (see Figure 3–3).

You can now understand the major problem: to produce more food, primarily plant protein, and to select for strains that are nutritionally superior to the ones presently available. Let us now

FIGURE 3–8.

After the rabbit eats one head of lettuce, there must always be another for his next meal. Thus, there must be more plant material than there are herbivores to eat it. Likewise, the hawk needs a ready supply of rabbits in order to survive. So, there must be more herbivores (primary consumers) than there are carnivores (secondary consumers). This concept of decreasing numbers with increasing levels of the food chain is called the food pyramid.

turn to the various options open to us and attempt to analyze their value and the potential results of their exploitation.

OPTION 1: Redistribution and Reduction of Food Losses

We have tried to build carefully a case for the frantic need to produce more and better food, and it is true that we must do this. However, as ironic as it may seem, presently the world produces enough food (calories *and* protein) to feed adequately the entire world population. How can this be, you might ask, when we already know that approximately 2 billion people are underfed and millions are starving? The answer lies in the word *produces*, rather than *delivers*. Unfortunately, vast quantities of food are lost in transit and storage, and secondly, food is unevenly distributed. Sadly, the underdeveloped countries suffer most on both counts. Ehrlich and Ehrlich[1] write that:

> . . . in 1968 rats devoured almost 10 percent of India's grain production, and others think 12 percent is more nearly accurate. It would take a train almost 3,000 miles long to haul the grain India's rats eat in a single year. And yet in 1968 India spent $265 million on importing fertilizers, about 800 times as much as was spent on rat control. The rats in the Philippine provinces in 1952–1954 devoured 90 percent of the rice, 20–80 percent of the maize and more than 50 percent of the sugar cane.

Rats, of course, are not the only problem. Fungal and bacterial diseases, insects, and other organisms contribute heavily to the yearly food losses. Clearly, very real progress could be made in cutting these losses, and it would seem that we should expend the capital and develop the techniques required as soon as possible.

With respect to distribution, the inequities between developed and underdeveloped countries are equally clear. The greatest problem exists with respect to protein production and distribution. South America exports beef; Asia and Africa export shellfish. They export in spite of the fact that they do not have enough protein to feed their own people. The situation has developed to this absurd level largely as a result of economic considerations. Underdeveloped countries must export what is available, that which can be produced without massive capital development and technical skills. Agricultural and seafood products fit this description. Obviously, this situation can be improved only by enhancing the economic position of these countries and by a

[1]Paul R. Ehrlich and Anne H. Ehrlich, *Population, Resources, Environment: Issues in Human Ecology*, 2d ed. (San Francisco: W. H. Freeman and Company, 1972), p. 112.

AREA IN BILLIONS OF ACRES

Continent	Population in 1965 (millions of persons)	Total	Potentially arable	Cultivated*	Acres of cultivated* land per person	Ratio of cultivated* to potentially arable land (percent)
Africa	310	7.46	1.81	0.39	1.3	22
Asia	1,855	6.76	1.55	1.28	0.7	83
Australia and New Zealand	14	2.03	0.38	0.04	2.9	2
Europe	445	1.18	0.43	0.38	0.9	88
North America	255	5.21	1.15	0.59	2.3	51
South America	197	4.33	1.68	0.19	1.0	11
U.S.S.R.	234	5.52	0.88	0.56	2.4	64
Total	3,310	32.49	7.88	3.43		

TABLE 3–1.   PRESENT POPULATION AND CULTIVATED LAND ON EACH CONTINENT, COMPARED WITH POTENTIALLY ARABLE LAND

*Our cultivated area is called by FAO "Arable land and land under permanent crops." It includes land under crops, temporary fallow, temporary meadows, for mowing or pasture, market and kitchen gardens, fruit trees, vines, shrubs, and rubber plantations. Within this definition there are said to be wide variations among reporting countries. The land actually harvested during any particular year is about one-half to two-thirds of the total cultivated land.

Source: From *Population, Resources, Environment: Issues in Human Ecology*, 2nd ed., by Paul R. Ehrlich and Anne H. Ehrlich, Table 5–2. By permission of W. H. Freeman and Company. Copyright © 1972. After the President's Science Advisory Committee. 1967. *The World Food Problem*, vol. II, Table 7–9.

shift from the relatively inefficient production of beef as a protein source to the direct use of more plant protein.

OPTION 2: Increase the Acreage Now Under Cultivation

On paper at least, increasing farm acreage seems to be a very promising option. Approximately 3.5 billion acres are now under cultivation, and this represents about 10% of the land surface of the earth. Another 7–8 billion acres are potentially useful. That is, these remaining acres could be brought under cultivation *if* we had the necessary capital and technology (Table 3–1). Let us first consider the matter of cost and then return to the technological problems. Under the best of circumstances, estimating the cost of opening up new land is a chancy business. Such land must be cleared of existing vegetation, but this isn't all. Roads must be built, the soil must be prepared, people must be relocated, and houses and irrigation projects completed. These costs vary greatly and can be quite high, but for the purpose of our argument, let's take a medium estimate of $400 per acre, as suggested by Ehrlich and Ehrlich,[2] and let's estimate that one acre will feed one person. From such figures they calculate that it would cost about $28 billion a year to open up new lands just to keep pace with the current population growth. This is a high cost indeed, but the more discouraging aspect of the problem is, in fact, not the cost but the chance for technical success. For example, more than half of the potentially arable land is in the tropics. Therefore, in order to assess properly the technical problems, we need to know a great deal about sound agricultural practices in these areas. We don't. We know most about agricultural practices in temperate climates (western Europe and North America) and very little about the difficulties and ramifications of converting tropical forests into agriculturally productive areas. Furthermore, attempts to put what little knowledge we do have into practice have met with little success. The basic reason for this is that very shallow topsoil found in the tropics is adversely affected by removal of the existing vegetation and is thus subject to decreased fertility, erosion, and flooding.

Of course, we also have potentially arable land in the deserts but once again have the serious problem of providing an adequate water supply. This problem is made even more difficult because agriculture greatly increases the amount of water lost from the soil. The obvious solutions including large-scale water projects and desalting plants are wrought with difficulties. Construction and transport costs are high. Furthermore, agreements involving political boundaries and the control of water rights are becoming more and more difficult as our supply of fresh water diminishes. In short, it seems unlikely that we can expand the acreage under cultivation at anywhere near the rate that will be necessary to cope with the population's need.

OPTION 3: Food From the Ocean

It was once thought that the oceans were a virtually limitless source of food. "There is no need to worry, the oceans will save us" was a popular but apparently ill-conceived notion, as can be seen from a few simple calculations. It is estimated that the total annual fish production is about 200–400 million tons. It is generally believed that we can catch about one-half of the annual production or 100–200 million tons each year without depleting the stocks. At the present time the annual catch of fish is about 60 million tons. What does all this mean? Simply that we are already catching about 25%–50% of the fish-producing capacity of the ocean, and judging from the trends toward in-

[2]Ibid., p. 92.

	Total Protein	Plant Protein	Animal Protein	Fish
Grams per person per day	73.0	39.4	32.2	3.9
Percent of total protein	100.0%	53.9%	44.2%	5.4%

TABLE 3–2.   WORLDWIDE PROTEIN DIET

creased fishing, we will soon be harvesting up to 100%. However, substantial increases in the ocean's harvest would not alleviate the world's food problems nearly as much as we might hope. For example, on a worldwide scale, fish contribute about 5% of the total protein diet (Table 3–2), with the expected large variations from country to country. Thus, even doubling the ocean's harvest would contribute only a modest share of the world's needs.

So it is clear that the ocean doesn't contain a limitless supply of food. It may be possible to "squeeze" a little more productivity out of the ocean by placing greater emphasis on the harvest of some of the other organisms that live there: plankton, kelp, snails, and worms. That is, we would concentrate on organisms near the base of the food pyramid. This again appears workable on paper, but it is often difficult to make such foods acceptable to large numbers of people because of cultural practices and taboos. Nevertheless, as most of you know, there are some meat substitutes now being successfully marketed in the United States. Their main attraction is price. "Ham" made from soybean curd costs less than ham made from hogs! What kinds of synthetic meats and dairy products can you find in your local markets? Is it not true that such substitutes must taste, look, and smell exactly like the foods they mimic in order for large groups of people to consider them seriously as substitute protein? Other sources of novel foods may also help but are unlikely to be a panacea.

Perhaps the greatest potential benefits will arise from growing single-celled organisms on petroleum or crude oil. Such organisms could in theory make up much of the current and predicted protein deficit. The problems involved in making this plan a reality include constructing large industrial plants for the culture of these organisms, extracting and purifying the protein, and finding a way acceptable to the general public to present this material—that is, convince them that it is food! Pilot plants are now producing such protein extracts from bacteria, so it is not just a far-fetched dream. Other ideas that are either being considered or under development include extracting leaf protein from forest trees, making cattle feed from wood, herding unorthodox animals such as rodents and even insects, and growing algae on fecal slime.

**OPTION 4: Increase the Yield per Acre, or
the Agricultural Efficiency**

In the long run greater yield may prove to be the most workable method of increasing food production. There are really two separate parts to this proposal: (1) selection and cultivation of high-yielding plant varieties and (2) getting the most out of these plants while they are in the ground. The latter part of this proposal simply means increasing the agricultural input: fertilizers, water, pest-control systems, and better mechanical equipment. In short, it means increased costs and expansion of agriculture-related industries.

The theory of selection and cultivation of high-yielding crop varieties requires more explanation. The most recent advances in agricultural efficiency and our expectation for the immediate future are only possible because of a long history of selection and cultivation which have brought us to our present level of technology. Therefore, let us begin our detailed discussion of Option 4 with the origin of plant cultivation itself. Then we can look at the mechanisms for improving such plants and the prospects for the future.

The Origin of Cultivated Plants

There are no written records concerning the origin of plant cultivation, and therefore we do not know all the details. Nevertheless, through fossil records and other geologic clues it is possible to envision the following scenario. Around the end of the last glacial period (see Chapter 10), the climate was becoming drier and warmer. Much of the once lush countryside was gradually becoming scrub, grassland, and barren. Small streams and lakes were drying up, and food became scarce except along a fairly narrow strip on each side of the larger rivers. The nomadic tribes around the Tigris and Euphrates basins were increasingly confined to the larger waterways in their quest for the necessities of life. The nomadic habit was no longer practical, and tribes settled more or less permanently in favorable sites along the rivers. Under these circumstances (i.e., the sedentary habit, inadequate food supply, and changing climatic conditions) somewhere, sometime, a group or groups began the cultivation of plants for agricultural use. At first the practice was so rudimentary as to be scarcely recognizable, and chance probably played an important part in its development. Perhaps the first step in what is known as the agricultural revolution was simply keeping the foraging wild animals away from a small clearing where wild wheat plants grew. Perhaps the second was clearing other competing plants, such as small shrubs and weeds, out of a field so that more wheat plants could grow. And so, little by little, cultivation techniques developed and agriculture was born.

The agricultural revolution occurred, it is thought, concurrently and independently in western Asia (around Burma and southern China), in the Near East east of the Mediterranean Sea, and in Middle America. We do not know why, but each of these origins for agriculture appears to have involved a localized center of fairly intense development in a temperate or subtropical latitude, which exchanged information and new advances with a much larger area, called a noncenter, in the tropics (Figure 3–9). Ancient man brought a great many plants under cultivation: corn, potatoes, squash, and tomatoes in America; barley, wheat, oats, rice, beans, peas, and lentils in Asia. All of this happened about 6,000–10,000 years ago, depending on the particular crop and geographic area. Interestingly, modern man has not added much to this list in spite of our greatly increased knowledge of botany. Rather, we have concentrated our efforts on improving the yields of these plants. Throughout this long history, cultivated plants have changed slowly but substantially. For example, a series of fossils in the central highlands of Mexico shows us that modern corn developed gradually and in stages from a large grass that produced several seeds in a cluster. Would you call this ancient structure a corncob (Figure 3–10)? In order to understand how these changes came about, we must first review something about the mechanisms through which all lasting biological changes occur.

Evolution: Change Through Time

Evolution is a term used to indicate change through time. We can speak of the evolution of art, architecture, farming, and lan-

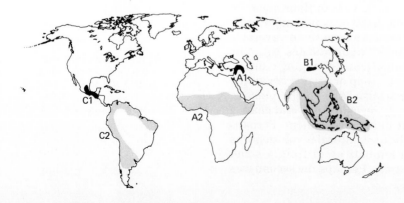

FIGURE 3–9.

Centers and noncenters of agricultural origins. A1, Near East center; A2, African noncenter; B1, North Chinese center; B2, Southwest Asian and South Pacific noncenter; C1, Meso-American center; and C2, South American noncenter.
From "Agricultural origins: Centers and noncenters," by Jack R. Harlan, in *Science* 174(4008):472, Fig. 6, October 29, 1971. Copyright 1971 by American Association for the Advancement of Science.

guage. In plant biology we generally use *evolution* to mean change in the characteristics of some kind of plant through successive generations. Furthermore, this change must be heritable, that is, it is caused by a change in the genes (units that determine hereditary characteristics) in an individual (see Chapter 5). Substantial changes take place gradually, over a period of hundreds or thousands of generations. Occasionally, however, more spectacular and rapid results can be obtained in the laboratory or in the field when environmental conditions are severe.

The mechanism of evolution can be summarized by the following expressions:

Mutation + recombination = variation
Variation + selection = evolution

Mutation is simply the origin of a new type, a change in a gene or genes. Mutations occur spontaneously all the time although usually at a low rate, on the order of one in a million for each gene in every generation. The rate at which new mutations occur can be increased substantially by the use of various mutagenic agents. The mutagenic effect of x rays was first reported in 1927, and since that time a number of other mutagens have been discovered, including other kinds of radiation and certain chemicals. *Recombination* indicates that various combinations

FIGURE 3–10.

Evolution of corn at Tehuacán starts (far left) with a fragmentary cob of wild corn of 5000 B.C. date. Next (left to right) are an early domesticated cob, 4000 B.C.; an early hybrid variety, 3000 B.C.; and an early variety of modern corn, 1000 B.C. Last (far right) is a cob from about the time of Christ.
From "The origins of new world civilization," by Richard S. Macheish, in *Plant Agriculture, Readings from Scientific American*, 1970, W. H. Freeman and Company, p. 18. By permission of the R. S. Peabody Foundation, Andover, Massachusetts.

of genes arise through the sexual process in such a way that no two individuals of sexually reproducing organisms are identical (see Chapter 4). *Variation*, then, is the sum of recombination and mutation: Each of us has a unique collection or combination of genes, which in part is due to *sexual* reproduction—a unique combination of chemically different *gametes* or sex cells uniting (fertilization)—and in part to *mutation*—some of the genes in those gametes have never existed before.

Thus, as you can see, each generation displays to the world a wide range of *variation*. *Selection* acts on that variation, and the result is *evolution*. What does that mean? We assume the obvious, that all seeds produced by all plants cannot germinate and grow into mature adults; there simply isn't enough space. But which seeds are the chosen few? How is it decided which individuals will develop to maturity and which will not? That difference between those who do and those who do not is the result of selection. It is called *natural selection* if it occurs in response to the environment. In modern agriculture, however, man selects for the purpose of maintaining high yield and quality and so the process of evolution can be accelerated and directed.

Wheat: A Case History in Accelerated Evolution

We do not know when man first began propagating and selecting wheat in order to achieve and maintain certain desirable features, but it probably occurred quite early in agricultural history. In the beginning and for many years following, progress must have been slow and haphazard. However, in the last 30 years or so, our knowledge of inheritance and basic biological principles has increased greatly. As a result there have been tremendous advances achieved in the field of plant breeding and in the selection of specific traits. The principles and goals behind this work can perhaps best be illustrated by considering the work of a particular scientist, Norman Borlaug (Figure 3–11). In 1970 Borlaug won the Nobel Peace Prize for his role in developing a new and particularly high yielding strain of wheat. By so doing, he initiated what is now known as the Green Revolution.

Borlaug's original training and interests were in forestry and the control of plant disease. Early in his career he worked in the U.S. Forest Service for a couple of years. In 1944 he and three other agronomists went to Mexico with the same general purpose: to improve the productive yield of certain crop plants. With the financial support of the Rockefeller Foundation and the International Maize and Wheat Improvement Center, Borlaug's plan was to bring into Mexico highly productive wheat varieties or strains from around the world. He then attempted to produce

FIGURE 3–11.

Dr. N. E. Borlaug and his prize-winning wheat at the International Maize and Wheat Improvement Center, Londres, Mexico.
Photograph courtesy of Norman Borlaug.

hybrids (offspring of two different parents) among these various strains, including those that were indigenous (native), in order to bring the most desirable traits from a number of sources together into a single type of wheat. The ultimate goal was to produce a relatively stable and nonvariable variety which would produce high yields in the varied environments found in Mexico.

While all of this was going on, in 1946 a U.S. Department of Agriculture scientist, S. C. Salmon, brought back from Japan seeds of a number of dwarf wheat varieties called collectively Norin wheat, and distributed them to breeders in the United States. Working with these Japanese wheats, a group at Washington State University, under the direction of another Department of Agriculture scientist, O. A. Vogel, succeeded in producing several strains which were very successful in the Northwest. One of these produced a world record of 216 bushels per acre, as compared with 10–12 bushels per acre in most of India, for example. In 1953 Borlaug received seeds of Norin wheat and also some of the hybrid selections which Vogel's group had on hand. Borlaug's first crop of Norin was planted near Mexico City and was a complete failure; the entire crop was wiped out by wheat rust, a fungal disease (Figure 3–12). The following year he planted in a different geographic area, near the west coast of Mexico, where the plants grew well. Next he crossed his new dwarf strain with everything else he had around, and this was the beginning of the revolution, so to speak.

Before the dwarf Japanese wheats became available, all the strains Borlaug worked with were the normal tall varieties on the order of 4 feet high. The grains of this wheat are produced at the tip of rather slender stems, and consequently, a plant

FIGURE 3–12.

Wheat stem rust can spread from one diseased wheat plant throughout an entire field by means of spores released from the surface of the stem or leaf. The spores can also infect barberry.

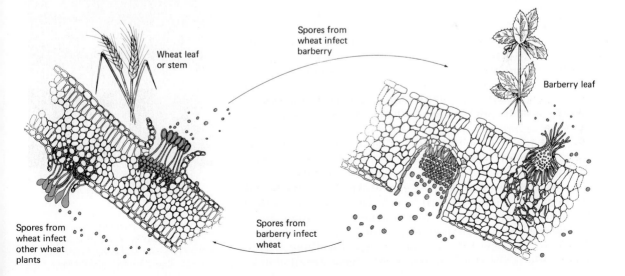

Wheat leaf or stem

Spores from wheat infect barberry

Barberry leaf

Spores from wheat infect other wheat plants

Spores from barberry infect wheat

that produces excessive amounts of grain often topples from the weight. One toppled plant falls on the next, and so even a modest breeze at just the right time can mean disaster for a field of wheat. This was the problem for which the dwarf wheats provided a breakthrough. Borlaug bred the tall with the dwarf strains and produced a semidwarf, large enough to take full advantage of fertilizers and irrigation, and at the same time short and stout enough to stand erect against normal vagaries of the weather (Figure 3–13).

A second major contribution of Borlaug's work was that some of his so-called miracle semidwarf wheats grew well in several climates. He managed this trick by two manipulations, one involving genetics and the other ecology. First, through an extensive breeding program Borlaug was able to bring together the inherited potential for growing well in varied environments. This was his hybrid-production program already discussed. Second, he tested his various hybrid combinations in fields near Mexico City and also near Mexico's northern border. These two stations are about 800 miles and 10 degrees latitude apart and are very different in climate and soil. This was quite important because, as previously stated, wheat is susceptible to wheat rust; and due to its complex life cycle, rust can be a particularly serious problem in lower latitudes where it is fairly warm all year. The ability of new wheats to grow in the lower latitudes meant that they could tolerate the temperature, moisture, and soil condition there and *still* resist the wheat rust. Borlaug's wheat strains are so successful that Mexico has recently exported vast quantities of seeds to other countries including India and Pakistan. In India where the new strains are grown, the yield per acre is five times what it was ten years ago.

Balancing the Scales

FIGURE 3–13.

Dwarf wheat growing in Mexico.
Photograph by Morris Brinkman, San Diego State University, California.

To have a complete and balanced picture, however, you must keep in mind that the newly developed miracle wheats have not solved our problems to the point at which we can afford to be complacent. Far from it. Very serious problems remain not only with food supply in general but also with wheat in particular.

Borlaug himself, in fact, has suggested that his work and that of his fellow scientists "will have been useless unless we strike a proper balance between populations and food resources" (*Newsweek*, November 2, 1970, pp. 50–51). Even the optimists who praise the Green Revolution and Borlaug's work as the answer to the world's most serious problem today admit that we have done nothing more than to buy a little time, to postpone the moment of decision. "The prophets of doom will undeniably be proved right in the long run unless their basic assumptions are nullified by concrete acts, and soon" is the way L. P. Reitz summarized it in 1970. According to L. R. Brown (1970), "If within 15 years we are not well on our way to stabilizing world population growth, Dr. Borlaug's monumental contribution will have been in vain." At this writing, 4 of those 15 years have already passed.

Let us examine some of the specific problems that remain and their ramifications. First, the new wheats may not be as productive in the long run as our experience up to now suggests. So far they have proved remarkably resistant to wheat rust, but their record should not suggest that this status will continue. The genes in the fungus that cause virulence are mutating all the time, and there is no doubt that sooner or later a new strain of rust will appear which can successfully attack the miracle wheats. When that moment comes, the wheat breeders of the world must have a new resistant strain ready or the consequences will be serious indeed.

Other problems of a more humanistic nature arise from the Green Revolution. For example, some people simply don't like the taste or texture of the new wheats and rice and feed it to their farm animals. Perhaps this sounds crazy, but human nature itself is often unpredictable. Probably more serious is the impact that the new wheat strains can have on the economy of a nation. Nearly 70% of the so-called Third World people depend on agriculture for their livelihood, and most of these are small farmers. Unfortunately only wealthier farmers—and nations—can afford the investments in machinery, fertilizers, irrigation systems, and so on which are required to take full advantage of the new wheats (Figure 3–14). This serves to widen the gap between the wealthy few and the many poor, bringing serious economic consequences. Furthermore, the substantial quantities of chemical fertilizers, herbicides, and pesticides that are needed to grow new crops raise other important considerations involving pollution and the environment (see Chapter 8). In fact, it is possible that as a result of the Green Revolution, we might see cases of more calories but a poorer diet and an overall reduction in the quality of life. Indeed, this appears to have occurred in the Philippines where chemicals necessary to grow a new high-yielding rice have polluted streams and killed fish, which formerly were an important source of animal protein for the local people. And finally, the distribution of food is a problem. For example, the Mexican strains of wheat can be grown only on about one-quarter of India's agricultural land, with three-quarters lacking irrigation, and the technological hardware required for the new strains. Even now India finds herself with local areas of great productivity and with warehouses overflowing with surplus wheat, while most of her people remain poorly fed.

The Green Revolution brings mixed blessings. On the one hand we have bought a little time, on the order of 15 years or so, during which most of the peoples of the world could be adequately if not properly fed—provided we could get the food to them. On the other hand we see that serious problems face us, including a substantial economic input required for a successful harvest, and realize that the new miracle wheats may not be able to overcome natural catastrophe. Even if all goes well for the foreseeable future, population growth must be brought under control, or inevitable starvation will involve that many more billions of people. What are the prospects for such control in the time available? Slim, very slim.

One group forecasting a culmination of the current population

FIGURE 3–14.

Unfortunately, only the large-scale farmer can afford the investments required to take full advantage of the new wheats.

From *Seed to Civilization: The Story of Man's Food,* by Charles B. Heiser, 1973, W. H. Freeman and Company, Fig. 5.6 (parts a and d), pp. 82, 83. Courtesy of FAO and USDA.

problem was described by *Time* magazine as "70 eminently respectable members of the prestigious Club of Rome." Their effort was funded by a quarter-million-dollar grant from Volkswagen Foundation, and the actual work was carried out by an international team of scientists at Massachusetts Institute of Technology. In a nutshell, they foresee a decline in natural resources beginning now, industrial output and food per capita reaching a peak around the year 2000 or soon afterward, and the population peaking around 2050 (Figure 3–15). Most of us living today will probably see if these predictions are accurate.

Up to now we have dealt with the question of survival. This is a popular problem discussed in magazines and books. It addresses the question, How many people can the earth support? Think for a minute what that means, and you might agree that it is not really the most meaningful question. If we are to exist with anywhere near the number of people that the earth is capable of supporting, then it is likely that we must all share more or less equally from available resources (see Figure 3–4). Sharing in that way among such a large population would necessarily mean a low standard of living for all and necessarily a decrease in the quality of life. Therefore, perhaps a more meaningful question relates to the optimum world population: How many people can the earth support at a comfortable standard of living? The only realistic alternative based on previous experience is that the haves will continue to have, and the have-nots will continue to have not even hope. And that doesn't seem like a defensible goal from any viewpoint.

Let us turn our attention briefly to this question of the optimum size of world population. Remember that our argument here is that optimum is best, not most, and best means a good standard of living for all. Our first problem is to select criteria. What is a satisfactory standard of living? As the basis for discussion, we should perhaps use our own American standard, although as you will see, this is not a realistic one. We will disregard open spaces and wilderness, because these pleasantries of life would be the first to be eliminated, and will summarize the "essentials" of our lifestyle in three groups: food, other renewable resources, and nonrenewable resources. We have already discussed *food* as a possible limiting factor, but what is the picture when we evaluate the prospects for feeding the whole world in the style to which *we* are accustomed? The arithmetic is complex and involves such things as bringing under cultivation deserts and other nonagricultural acreage, optimizing land now used for grazing animals by growing plant crops to be used directly by man, and so on. But after all this is calculated, the answer is not encouraging. It looks as though the earth could provide 1.2 billion people with a diet comparable to American food standards.

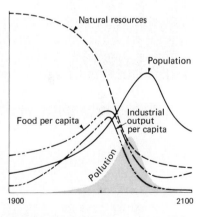

FIGURE 3–15.

Projection for disaster: The population problem.

From "The worst is yet to be?" *Time* magazine, January 24, 1972, p. 32. Reprinted by permission from *Time, The Weekly Newsmagazine;* Copyright Time, Inc., 1972.

Other renewable resources include lumber, paper, and other forest products. Here again the outlook is not hopeful. A population of slightly over a billion could use these resources at the rate at which we do here in America.

Nonrenewable resources include many of our energy sources, such as coal, oil, and gas, plus all the other indexes we use to measure our industrial growth and standard of living. And

FIGURE 3–16.

Overpopulation leads to insignificance in life . . . as in death.
(top) Photograph by Frank Siteman, Stock, Boston. (bottom) Photograph by Peter Menzel, Stock, Boston.

again the world's industrial complex could provide America's standard of living for only about a billion people.

There are nearly 4 billion individuals on earth (Figure 3–16). The Green Revolution offers some hope that most of them and their offspring might be fed for another dozen years or so. But when we compare our own comfortable lifestyles with those of people sleeping on the sidewalks in Calcutta, we realize that there is potentially more to human life than mere subsistence— also potentially less. Obviously our long-range thinking should involve not only survival but also the question of a reasonable standard of living for all mankind. But when we do think along these lines, we are forced to conclude that the earth is simply not capable of providing more than about a third of us here now with what most Americans consider a reasonable standard of living. This is not a pleasant picture, but it does fit the facts as they now exist. We can perhaps revise this prediction as times change and new scientific "revolutions" occur. Each of us *can* provide some useful input into this situation, and indeed we should, for an informed population must be the first step toward a more optimistic scenario.

Selected Readings

Borgstrom, G. 1974. *The Food and People Dilemma*. Belmont, Calif.: Duxburg Press. Data compiled from official publications from the U.N. on the complexity of the food-people problem.

Brown, L. R. 1970. *Seeds of Change*. New York: Praeger Publishers. Describes the development of high-yielding crop strains and some of the associated problems.

De Kruif, P. 1928. *Hunger Fighters*. New York: Harcourt, Brace. Tales of researchers in nutrition.

Ehrlich, P. 1968. *The Population Bomb*. New York: Ballantine. A widely read paperback calling for population control.

Elting, M., and M. Folson. 1967. *The Mysterious Grain*. New York: M. Evans. Describes the hunt for the ancestor of corn.

Meadows, D. H., D. L. Meadows, J. Randers, and W. W. Behrens. 1972. *The Limits to Growth*. New York: Universe Books. Predictions from an M.I.T. team on the present growth trends and possible consequences.

Wharton, C. R., Jr. April 1969. The Green Revolution: Cornucopia or Pandora's box? *Foreign Affairs* 47:464–476. Very readable summary of the social and economic consequences of the Green Revolution.

PART ONE

This chapter and to some extent the next will examine in detail the principles of plant reproduction. For the most part our concern is with *sexual reproduction* which involves the fusion of special male and female sex cells. We will see that *variation* is possible primarily through the sexual process and that the sexual process, in turn, has important implications concerning evolution.

However, before dealing directly with the sexual process, you should first be acquainted with an alternative method of reproduction called *vegetative* or *asexual reproduction*. As a vehicle for illustrating the advantages as well as the disadvantages of the asexual process, we will discuss the development of the seedless navel orange.

Asexual Reproduction and the Navel Orange

Oranges have a long and interesting history, as do most other agriculturally important plants. Citrus trees have been in cultivation so many years that their wild ancestors are not even known and are probably extinct. The earliest recorded reference of any kind to oranges (one of the several kinds of citrus) is in an ancient Chinese book, *Tribute of Yu*, from around 2205 to 2197 B.C., which is the time Emperor Ta Yu reigned. In fact, citrus are thought to be native to southern China, although their earliest cultivation is lost in antiquity. Seeds of one species of *Citrus*,[1] the citron (*C. medica*), were found in some excavated ruins of southern Babylonia which were dated around 4000 B.C. It is curious that the Bible contains no direct reference to any citrus, although other fruits are often mentioned.

Piecing together the available facts, it would appear that oranges were introduced from the Orient to the Mediterranean area and Europe by the Arabians who worked the Genoese overland trade route to India. This probably occurred later than 1200 and soon after the Crusades but perhaps not until the fourteenth century. The climate in much of southern Europe at that time, as well as today, was a bit too cool for oranges, and frost damage

---

[1]Each kind of plant has a scientific name consisting of the genus and species. Various similar species are grouped together as several species within one genus. The genus *Citrus* includes 12 species, among them the lemon, *C. limonia*; lime, *C. aurantifolia*; grapefruit, *C. paradisi*; and, of course, the orange, *C. aurantium*. These names are always italicized or underlined.

was a constant threat during winter. To cope with this problem, special houses called orangeries were built primarily to protect the prized citrus, although other exotic plants soon found their way inside. This was the beginning of the greenhouse.

Oranges quickly became an important part of European diets, and so it was quite understandable that Columbus took seeds of oranges as well as other important plants on his second voyage when he established a colony on Haiti in 1493. Shortly after this, oranges were spread throughout the Americas by the early explorers and settlers.

Of particular interest is the establishment of oranges in Bahia, on Brazil's east coast, in 1549. In Bahia soon after 1800, a so-called bud variation (also called sport or mutation) arose on one particular plant of a seedy variation, wherein certain parts of the flowers aborted; as a result the new variant produced fruit containing *no seeds* whatsoever. The value of this new variation or sport was seen immediately, and so it was propagated *asexually* by budding.

To do this, one slits open the bark of a normal tree and carefully inserts a bud taken from the branch of the variant (Figure 4–1). If the operation is successful, the bud "takes" and begins growing, drawing water and nutrients from the *stock* to which it was grafted. Finally, the stock's own branches are removed, leaving only the branch that developed from the grafted bud. A reasonable number of these asexually reproduced seedless orange trees were growing in Bahia when the first Presbyterian missionary to the state, the Reverend F. E. C. Schneider, wrote to the U.S. Department of Agriculture in 1869 concerning this exceptional fruit. The following year he sent twelve budded trees to Washington, D.C., and that is why we call them Washington navels, although they probably should be known as Bahia navels. These twelve trees were immediately propagated and distributed to several areas in the United States. Notably, in 1873 two trees were sent to and grown in a private residence in Riverside, California, and from those two trees the navel orange industry of California and much of the world developed. Thus, as you can see, the parentage of every navel orange tree in the world goes back through generations of asexual reproduction to a single individual that appeared spontaneously in Brazil around 1800. Indeed, it could be no other way; navel oranges cannot reproduce sexually because they do not produce seeds.

Strawberries: Another Example of Asexual Reproduction

New strains of strawberries are produced by sexual means; however, established strains are maintained through the asexual production of runners (Figure 4–2). Runners are stems

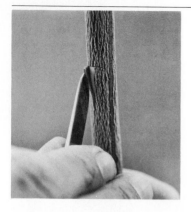

a. First, the bark of the stock is carefully slit in the shape of a T, as deep as the cambium. b. Next, a scion is removed from the tree being propagated . . .

c. . . . and is very quickly slid under the bark of the stock. d. The entire graft is then wrapped.

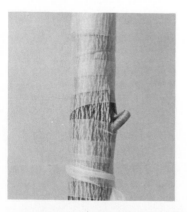

e. Elastic wrapping stretches as the branch grows, and can be left indefinitely. f. The buds may grow several feet the first year.

FIGURE 4-1.

Budding in citrus is accomplished by methods comparable to those used with most other species. The *scion* is the part containing the bud whose aerial part is being propagated, and the *stock* is the older branch with roots to which the scion is being grafted.
Photographs through the courtesy of the UCR Botanic Gardens, Riverside, California.

that grow horizontally along the surface of the soil and sprout roots and sets of leaves at *nodes* (the point on the stem where the leaf is attached) spaced at intervals along their length. Such reproduction is strictly asexual because flowers and fruits, which are the organs associated with sexual reproduction in plants, are not at all required.

Many other commercial plants, and a fair smattering of wild ones, reproduce by a wide variety of asexual means. Such widespread means of reproduction have certain virtues as well as serious drawbacks which we will now examine. Asexual reproduction is, of course, required for plants that are sterile, such as navel oranges. Strawberries, on the other hand, are fertile, but we still rely on asexual means to propagate them because only via asexual techniques can we be sure that all individual plants of a particular strain will be identical. Different leaves on the same twig, different twigs on the same branch, or different branches on the same tree all have the same inheritable traits or genes. When we break off several branches from one plant and make several new plants out of them, as when grafting, the new plants are really parts of the original plant in a sense, although, of course, they are no longer physically attached to each other. Individuals produced in this manner are members of the same *clone* (a group of individuals all descended by asexual reproduction from a common parent or ancestor). So the primary advantage of asexual reproduction is that large numbers of *identical individuals* can be produced fairly quickly. This consequently has obvious advantages. in agriculture, assuming that the original parent is unique in some advantageous way. Sexual reproduction, on the other hand, gives rise to variation, and no two individuals of sexually reproducing organisms are identical.

FIGURE 4-2.

Strawberries reproduce asexually by runners. The parent plant is at the far right, a younger one is in the center, and a new one is just beginning at the tip of the runner on the left.

The chief difficulty with asexual reproduction is that *all* the genes, both those that contribute to normal development *and* those that are potentially harmful, are carried along to the offspring. In Chapter 3 it was pointed out that mutations happen all the time—although at a low rate, to be sure—but sooner or later essentially all genes mutate. Furthermore, most mutations are harmful in the sense that their effect is to decrease the vigor or reproductive potential of the individuals that carry them. The net result, therefore, is that the quality and vigor of an asexually reproducing strain will gradually decline because all the mutations which "accumulate" in successive generations are transmitted to the progeny.

The methods of asexual reproduction described arise from vegetative structures—stems, leaves, and roots but no flowers and fruits. Thus, while the particular tactics exploited by plants may vary from case to case, the strategy is always to produce more cells, all of the same kind. This is because all plant parts are made of cells, and in order to grow more plant parts, we must produce more cells (see Chapter 1). It follows that we must understand how cells reproduce if we are to understand the basis for asexual reproduction.

One Will Get You Two

Cells have alternating periods of growth and replication (division) until they become specialized as part of some kind of mature tissue where they continue to function metabolically until they die. In order for a cell to divide (for one cell to become two cells), it isn't enough simply to split down the middle. Rather, each daughter cell must contain the same information as the parent cell. This means that the particles in the parent cell which contain information must produce two exact copies. This vital information is the hereditary material, or *genes*, which is contained in the nucleus as part of special organelles called *chromosomes* (see Chapter 1). In the jargon of biology, we say that genetic material must be duplicated or replicated. Once each daughter cell has a full information content, it then has the ability to produce or direct the machinery to make itself exactly like its parent, and it of course lacks the ability to do anything wildly different. This very important process of replication and subsequent division of hereditary material is called *mitosis*.

The first step in the mitotic cycle is to replicate all genetic material. The molecular details of this process will be discussed in Chapter 5; however, even without rigorous knowledge of the mechanism, it is easy to demonstrate that replication does occur at an early stage in the mitotic process. The evidence for this comes from our ability to tag or label molecules. For example, the substance thymidine can be manufactured using radioactive hydrogen (^{3}H) as one of the materials. Such molecules of thymidine can be detected by special techniques when they are present in a plant cell, and are therefore said to be tagged or labeled. Thymidine is one of the basic building blocks for the DNA found in chromosomes (see Chapter 5). Therefore the experimental technique used to demonstrate time of replication is first to culture some rapidly growing plant tissue, such as onion root tips, in a solution containing radioactive thymidine (Figure 4–3).

The time during which the cells are grown in the radioactive solution can be varied, and then the chromosomes are examined to see if they have become radioactive. If they have, it means that the cells were making new chromosome material at the time they were in the radioactive solution.

In a way the technique can be viewed as comparable to putting automobiles together on an assembly line. Let us say that we want to use a technique of tagging to determine when the tires are put on. First, we supply the assembly line with only black (untagged) tires and find that all automobiles come out with black tires. Small wonder. Next we supply red tires for one hour and then change back to black again. Seven hours later automobiles are produced wearing red tires! Therefore, we conclude that seven hours are required after the tires are put on before the automobile is finally ready. Using various automobile parts tagged in this way, we could determine when, for example, the carburetors are put on or the windshield wipers or any other part. In fact, we could even determine whether or not particular parts were used at all to manufacture automobiles. We could tag certain kinds of tires and not others, and we might learn that the tagged tires appeared only on trucks and not on cars. You can see from this analogy that without ever entering the automobile manufacturing plant, one can learn a great deal by selectively tagging the parts that enter it and the products that come out. Using radioactive substances, this same strategy is applied to cells, nuclei, and the chromosomes.

Several hours after the tagged thymidine molecules are incorporated, indicating that replication has occurred, the chromosomes condense and become shorter and thicker; when they do, each one is seen to be longitudinally double (Figure 4–4a, b). Of course, actually seeing that each chromosome is double gives us visual evidence of the event we detected earlier using the radioisotopes, namely the manufacture of new chromosomes. Presently a slight constriction called the centromere can be recognized somewhere along the length of each chromosome (Figure 4–4c, d, e). The exact position of the centromere is unimportant for our present discussions and may vary for

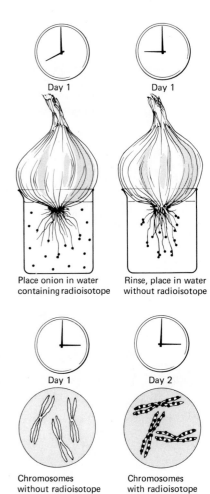

Place onion in water containing radioisotope

Rinse, place in water without radioisotope

Chromosomes without radioisotope

Chromosomes with radioisotope

FIGURE 4–3.

The time and mode of chromosome replication can be studied with an onion, as shown here. Place actively growing tissue (here it is the root tips) in a solution containing a radioactive material, which is necessary for chromosome replication. After one hour, rinse the onion with water, and periodically during the next several hours, examine the cells and chromosomes with a microscope to see which ones are radioactive. Radioactivity indicates that they were replicating at the time they were in the radioactive solution.

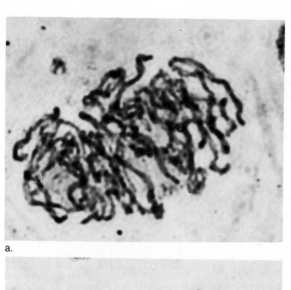

a.

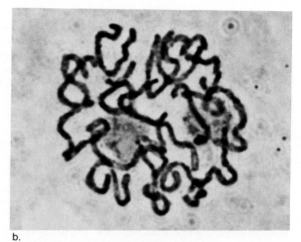

b.

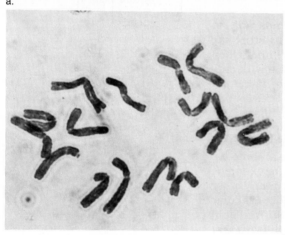

c.

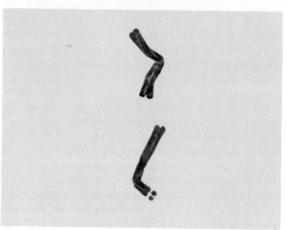

d.

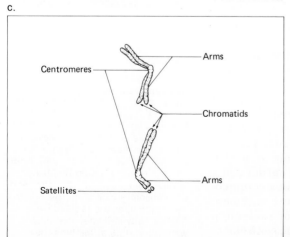

Centromeres

Arms

Chromatids

Satellites

Arms

e.

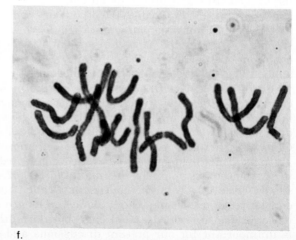

f.

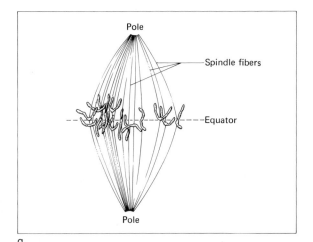

Pole

Spindle fibers

Equator

Pole

g.

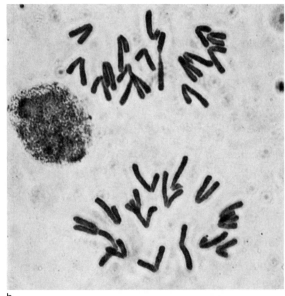

h.

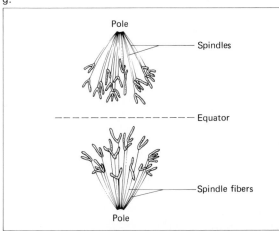

Pole

Spindles

Equator

Spindle fibers

Pole

i.

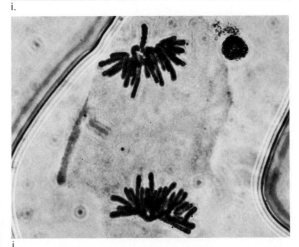

j.

FIGURE 4–4. MITOSIS

Mitosis in root tips of the cultivated onion (Allium cepa). Cytologists recognize five stages in mitosis (see text for details): (a,b) *prophase,* chromosomes long and thin, nuclear membrane intact; (c) *prometaphase,* nuclear membrane disintegrated, chromosomes compact, spindle fibers attach to centromeres; (f,g) *metaphase,* centromeres align along equator; (h,i) *anaphase,* sister chromatids separate and become daughter chromosomes which move toward opposite poles; and (j) *telophase,* anaphase movement completed, new membranes form around the daughter nuclei.

each chromosome. Its function, however, is critical, for it is the centromere which makes chromosome movement possible.

In the paragraphs immediately following, you will see that the chromosomes must move about in the cell if mitosis is to be completed, and we know that centromeres are required for chromosome movement from the following evidence. Chromosomes that have been broken by x rays or other agents yield two parts, one fragment with and one without a centromere. The part of a broken chromosome that contains the centromere moves normally during mitosis, whereas the part lacking a centromere does not move; it is simply left behind. The centromeres in most of the chromosomes in our figures are located approximately in the middle and thereby divide the chromosome into two *arms* of approximately equal length. Some chromosomes have a long and a short arm when the centromere is located nearer one end, and in some cases there is only one arm because the centromere is at the very end of its chromosome. Finally, the chromosome is longitudinally double, and each longitudinal half is called a *chromatid*. Thus, immediately before division a chromosome consists of two chromatids, and immediately after division each chromosome consists of one chromatid. The change from one chromatid per chromosome to two occurs at the time the DNA is replicated.

The early events of mitosis discussed so far take place within the nucleus. Indeed, mitosis is a nuclear process. At this point, however, just as the chromosomes reach their most compact state, the distinction between nucleus and cytoplasm becomes obscure because the nuclear envelope itself disintegrates. Unfortunately, you cannot see the presence or absence of this membrane in these photographs because the stain used stains only the chromosomes, not the membranes. In addition, the membranes of the cell are too thin to be seen with the light microscope used here, although membranes can easily be detected with an electron microscope. For those same reasons, in the photographs we cannot see another important structure of mitosis, the spindle, although this structure is indicated in the diagrams (Figure 4–4g, i). The spindle is a football-shaped organelle comprised of many protein fibers. With the electron microscope some of these can be seen attached to a centromere on chromosome fibers. Recent evidence suggests that the fibers stretching from the spindle to the centromere direct the chromosomes to the equator or midpoint of the spindle, which normally is in the center of the cell (Figure 4–4f, g). The chromosomes remain in this position a short while, and then, as you can see in Figure 4–4h and i, the two halves of every chromosome separate cleanly and with the help of the spindle move away in opposite directions so that we end up with two complete sets of chromosomes at opposite poles in the cell

(Figure 4–4j). Each set is then enclosed in a new nuclear envelope. Once the enclosing is accomplished, a membrane and wall form roughly down the middle of the cell and divides the cytoplasm. Now there are two cells where once there was one.

A point worth emphasis is that mitosis produces daughter nuclei, usually giving rise to two daughter cells identical to one another and to the parent cell from which they came. In some instances mitosis is not followed by cytoplasmic divisions, and several or many nuclei may be produced within a single cell. Another event concerns accidents or spontaneous changes in chromosome structure which occasionally occur; when they do, we call these changes mutations. Nevertheless, the primary role of mitosis is the exact duplication of the information centers, making possible cell multiplication in which all cells are identical in information content. With respect to asexual reproduction, this means that all members of one clone will be essentially identical over at least a few generations, and that circumstance is why highly productive kinds of some agriculturally important plants are maintained and propagated asexually. Deleterious mutations in such plants gradually change the picture through time, however.

Sex in the Biological Sense

Under special circumstances asexual reproduction may have certain advantages for a limited time; yet, in the longer term the advantages of sexual reproduction will certainly prevail. The two reasons for this are, first, that sexual reproduction allows harmful mutations to be sorted out and eliminated selectively and, second, variation is a normal consequence of sexual reproduction, and this aspect in turn allows plants to adapt to new and changing environments.

In the biological sense, sex means the union of two cells. This union is called fertilization and results in the formation of a special cell, the *zygote*. Since a single individual plant often produces both male and female sex cells, self-fertilization is possible in many species. In some of the algae, which you may recognize as pond scum floating on quiet water (Figure 4–5), the two cells which fuse in fertilization look just alike and can only be distinguished by special tests. In such cases we call the plants mating types, and two, three, or even more mating types can be found in a single species. In fact, sometimes three cells of different mating types can fuse in a single zygote all at once. In most species, however, the two sex cells are quite different: one large and stationary—the egg—and the other small and motile—the sperm. The actual fusion of these cells and the consequences with respect to chromosome number make up the next problem to be discussed.

Recall that in mitosis genetic information is exactly replicated, so that after a cell divides, each daughter cell includes a complete set of information in the form of DNA or genes. The correct distribution of this DNA to the daughter cells is assured and visibly verified because, as you can see in Figure 4–4 a–h, each chromosome in the nucleus replicates and divides during mitosis so that all the chromosomes originally present in the parent are also present in the two daughter cells. In fact, not only do all vegetative cells of any individual plant have the same number of chromosomes, but all individuals of any given species have the same chromosome number within their nucleus. The problem, then, is how do we reconcile or arrive at a constant chromosome number in species generation after generation in which sexual reproduction involves the fusion of cells each containing their own set of chromosomes. That is, if there are, let us say, n number of chromosomes in the egg and n more

a.

b.

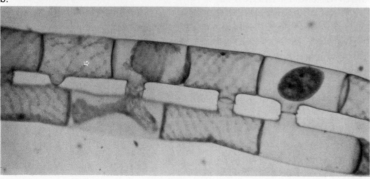

FIGURE 4–5.

a. Spirogyra is a common, filamentous pond scum which grows in still or very slow-moving water. b. In sexual reproduction two filaments lie side by side, and a tube forms between their cells so that the protoplast from one cell can flow into and fertilize a cell of the other filament.

Species	$2n$	Meiosis $\xrightarrow{}$ $\xleftarrow{}$ Fertilization	$1n$
Most pines	24		12
Corn	20		10
Peas	14		7
Cultivated onion	16		8
Tomatoes	24		12

TABLE 4–1.   CHROMOSOME NUMBERS

chromosomes in the sperm, then the zygote should contain $2n$ chromosomes.

If fertilization brings the chromosome number from $1n$ (in the sperm) and $1n$ (in the egg) to $2n$ (in the zygote, and thus the plant), how do the chromosomes go from $2n$ in the mature plant to $1n$ in the sperm and $1n$ in the egg? Clearly the number must be reduced sometime before the eggs and sperm are formed, or the chromosome number would double every generation. The answer is that a special kind of nuclear division occurs in special sex organs; it is called *meiosis*. Unlike mitosis, meiosis is *two divisions of the nucleus* but only *one division of the chromosomes*. The number of chromosomes in a cell which begins meiosis is always $2n$ or diploid, and after two nuclear divisions we find four daughter nuclei with the $1n$ or the haploid number of chromosomes because during the two divisions of the nucleus, there has been but one division of the chromosomes (Table 4–1).

In order to understand meiosis, it is very important first to understand the individual nature of chromosomes. Haploid cells contain some small number of chromosomes, usually 5–20 depending on the species, and each one of those chromosomes is unique. That is, the hereditary factors carried on one chromosome are not duplicated on any other chromosome in the same haploid cell. Other haploid cells produced by the same plant contain the same identical number of chromosomes, and likewise there will be one chromosome of each kind. Fertilization brings two such haploid cells together and therefore not only does just the *number* of chromosomes in the zygote become doubled to diploid, but in addition there are exactly *two of each kind of chromosome*. This is a very important point: *Haploid* cells contain *one of each kind* of chromosome, and *diploid* cells contain *two of each kind*. This point may be fairly easy to visualize in relation to fertilization, since one set of chromosomes added to a second set yields two sets. Beyond this basic level of understanding, you can now see that the role of meiosis is more than simply reducing the *number* of chromosomes from *diploid* to *haploid*; in addition, meiosis must guarantee that each *haploid* nucleus will contain *one of each kind* of chromosome. This very precise task is accomplished quantitatively and qualitatively in the following manner.

Meiosis

The First Step: Pairing

We know that the earliest step in meiosis is the replication of each chromosome at a time when it is invisible (as is also the case in mitosis). However, in meiosis some of the DNA (less than 1%) is left unreplicated, and it isn't until some time later that DNA replication is completed. Just as meiosis is to begin, the chromosomes become visible as long, very thin, single units. At this point an exceedingly important event occurs which makes possible all further critical stages which are unique to meiosis, including the subsequent production of fertile eggs and sperms. That event is the pairing of chromosomes. See in Figure 4–6a and b that each thread is actually double, and this doubleness is in fact because two chromosomes are lying side by side. You recall that diploid cells contain two of each kind of chromosome. It is those two chromosomes that now recognize each other as being alike; they come together so that their entire length is in close association. Each pair of chromosomes is now called a *bivalent*.

The Second Step: Chiasma Formation

During or immediately after the chromosomes pair, DNA replication is completed. This last little bit of DNA replication, less than 1% of the total DNA in the cell, is insignificant in quantity but critical to the subsequent success of meiosis. For example, if it is inhibited, as can be done with special experimental techniques, the chromosomes fragment into small pieces later on and the cells die. After the last DNA is replicated, each chromosome is seen to be visibly double, so that *each bivalent* now consists of two strands on each of the two chromosomes, or four strands altogether, as seen in Figure 4–6c–e. This doubling of the chromosomes is another very important event in meiosis because simultaneously two other changes occur which are probably related to DNA replication. One change is that the two paired chromosomes in each bivalent are no longer attracted to one another and they tend to drift apart, resulting in a looser, less intimate associations as seen in Figure 4–6e and f. The second change is the formation of *chiasmata* (Figure 4–6d, e). As shown in those illustrations, there are obvious points where two of the four chromatids in each bivalent cross over to the other chromosome in a reciprocal way, forming an X. That X is a chiasma, and chiasmata are very important for two reasons. First, chiasmata increase the genetic variation. Remember that genes are carried on the chromosomes, and so new combinations of genes occur as a result of these chiasmata which mutually exchange parts of chromosomes. Second, chiasmata hold together the two chromosomes in each bivalent and assure that the two behave as a coordinated unit during meiosis. If this coordination is lacking due to failure of chiasma formation, each chromosome behaves independently, meiosis is disrupted, and sterility results.

FIGURE 4–6. MEIOSIS

Meiosis in the anthers of cultivated onion (see text for details).

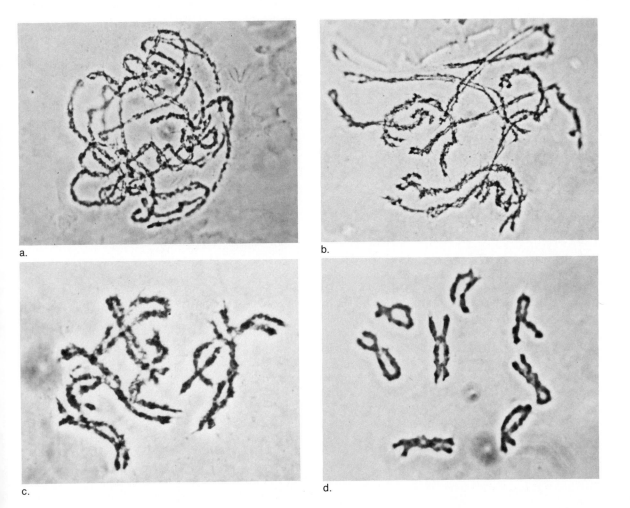

a.

b.

c.

d.

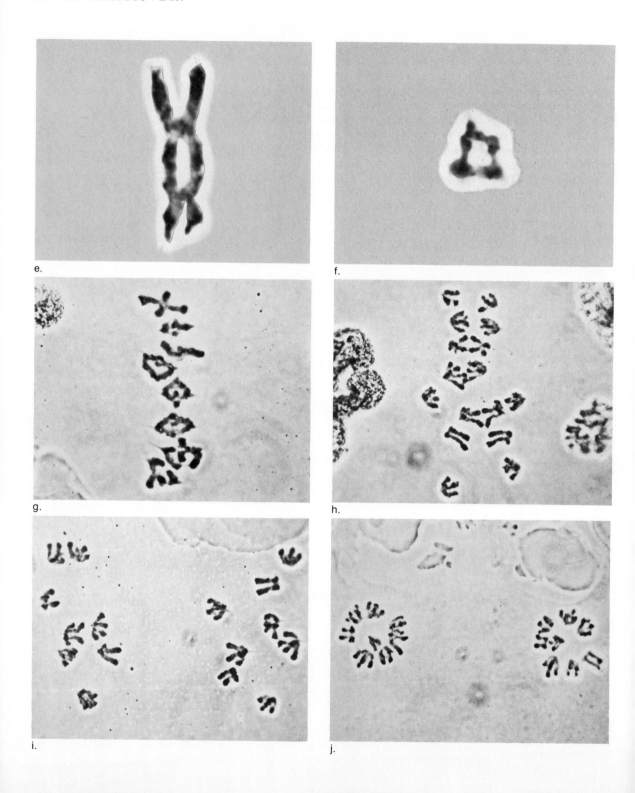

e.

f.

g.

h.

i.

j.

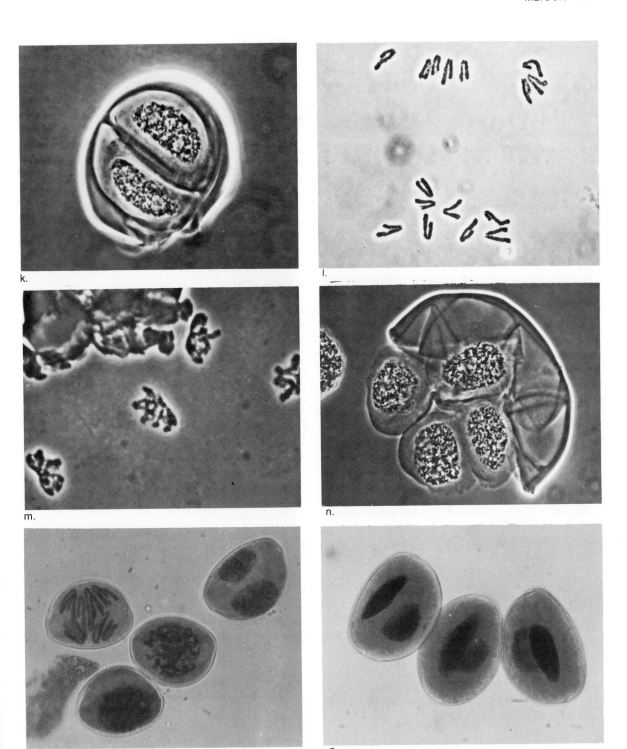

k.

l.

m.

n.

o.

p.

The Third Step: Sorting out the Products

You recall that meiosis consists of two nuclear divisions and one chromosomal division, and up to this point none of those divisions has occurred. They all do during the third and final stage of meiosis. In preparation for the divisions, the chromosomes continue to condense (Figure 4–6f) until they become very compact in comparison with their earliest appearance. Meanwhile, just outside the nuclear envelope, the spindle is forming, as it did for mitosis, by the aggregation of numerous protein fibers. Presently the nuclear envelope fragments into small pieces, at least some of which remain attached to the chromosomes for the remainder of meiosis. With the bursting of the nuclear envelope, the spindle fibers have a clear, unobstructed path to the centromeres of the chromosomes, and they now quickly attach. The manner in which the spindle fibers attach to the centromeres during meiosis differs in an important way from that of their attachment in mitosis.

You recall that during mitosis spindle fibers from *both poles* attach to *each centromere* so that each half centromere is attached to an opposite pole in order to move the sister chromatids to opposite poles as the cell divides. During the first division of meiosis, by contrast, *each centromere* is attached to spindle fibers of only *one pole*, and the centromere of its paired chromosome to the opposite pole. The immediate result of this is shown in Figure 4–6g, where the two paired chromosomes of each bivalent are oriented so as to separate and proceed to opposite poles at the first division. This sort of orientation assures that one chromosome from each bivalent will be sorted out to each daughter nucleus after this division. The only thing holding the two paired chromosomes together at this point is the chiasmata, and finally even they release, as shown in Figure 4–6h and i. Each chromosome, still made up of two chromatids, is drawn poleward (Figure 4–6i, j) and may go into a diffuse stage as the first nuclear division ends (Figure 4–6k).

The second nuclear division marks the end of meiosis and is rather similar to mitosis in some ways. The chromosomes, still comprising two chromatids, are aligned along the center of the spindle with spindle fibers from both poles attached to each centromere. Presently, the centromeres divide and pull the sister chromatids apart to opposite poles, as shown in Figure 4–6l. Because this happens in each of the two nuclei resulting from the first meiotic division, the end result is four haploid nuclei, each containing a full set of eight chromosomes (Figure 4–6m). Both of the meiotic divisions occur within the same cell wall, from which they now escape (Figure 4–6n). These haploid cells are called *spores*, and each spore contains a single haploid nucleus. Each spore usually goes through a mitotic division (Figure 4–6o), so that the resulting pollen grains contain two haploid nuclei (Figure 4–6p).

Synthesis of the Parts

Let us pause for a moment to review what has been said regarding reproduction in plants. First of all, we found out that asexual means of reproduction are easily found in wild and cultivated plants and that all means for asexual reproduction depend on mitosis for reliable, exact multiplication of cells. Second, we learned that sexual reproduction is the usual form of reproduction in most plants. It requires regular alternation between the diploid and haploid conditions, with the diploid condition being produced by the union of two haploid sex cells in fertilization, and with the haploid condition being produced by a special kind of nuclear division called meiosis. Figure 4–7 summarizes the roles of mitosis, meiosis, and fertilization in sexual life cycles. Furthermore, it shows that mitosis is the kind of nuclear division occurring throughout most growth and development and that meiosis is restricted to the one critical step in the life cycle when haploid cells are produced from diploid cells. Figure

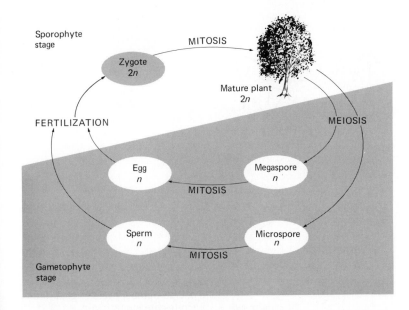

FIGURE 4–7.

The sexual life cycle of plants is summarized here, showing the alternation between haploid and diploid stages, and the roles of mitosis, meiosis, and fertilization. This will be covered in greater detail in Chapter 9.

4–8 summarizes the roles of mitosis, meiosis, and fertilization in the alternation of chromosome number between haploid and diploid.

Clearly, meiosis and fertilization are two critical events in the life histories of most plants, and so we should examine in more detail when and where they occur if we are really to understand plant reproduction. It may surprise you to learn that both these vital events usually occur within flowers. Of course, some plants, such as pines, ferns, molds, and algae (see Chapter 9), don't have flowers, but as we will see, even those plants have special structures where meiosis and fertilization take place. At this point, however, we'll consider only plants in which both fertilization and meiosis occur in flowers.

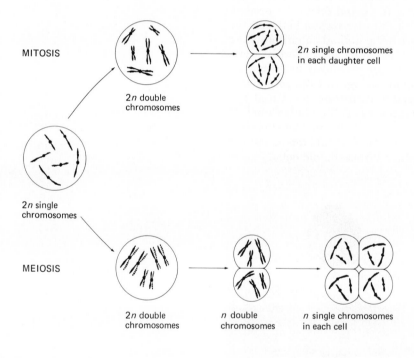

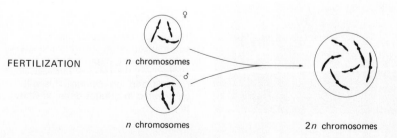

FIGURE 4–8.

The roles of mitosis, meiosis, and fertilization in relation to the chromosome-number cycle in plant-life histories are summarized.

The Flower's Role in Reproduction

A representative flower is the pea. Gregor Mendel used peas in his classic work on heredity over a century ago, and the pea family, Leguminosae, is worldwide in distribution and one of the largest in the plant kingdom. The flower parts are illustrated in Figure 4–9. Briefly,

> The calyx is the outside cup,
>   it holds the flower snugly up,
> Its sepals have been woven stout
>   to keep the cold and dampness out.
> The corolla is the colored part
>   that gladdens every childlike heart,
> Its petals wave open in the breeze
>   to summon butterflies and bees.
> The stamens next within the ring,
>   their anthers set on magic spring,
> These anthers store a generous mead of pollen,
>   needed to make the seed.
> The pistils in the center fare,
>   for they must have the greatest care;
> Their stigmas catch the pollen bead
>   which turn the ovules into seed.

The poem (source unknown) provides a nice introductory summary, but we will learn exactly how flowers function in sexual reproduction. Let us begin with the production of pollen. Meiosis in the anthers (see Figures 4–6a–n, 4–9) produces haploid cells called *microspores*. The nucleus in each microspore divides by mitosis, giving rise to two haploid nuclei called the *tube nucleus* and *generative nucleus*, and so pollen grains have two haploid nuclei (refer to Figure 4–6o, p). Later, when the pollen is deposited on the *stigma* and germinates, it grows a tube down the *style* with the tube nucleus leading the way. At this time the generative nucleus, following along, divides by mitosis, producing two haploid *sperms*. Thus, the pollen grain for a short time can be considered an independent little organism, structurally and functionally quite separate from the flower that produced it. A similar kind of thing happens on the female side, and because of that we have come to recognize two phases in the life cycles of nearly all plants. One is the haploid *gametophyte* stage, which produces *gametes* by mitosis (the word *gamete* means sex cell, either egg or sperm). In flowering plants the male gametophyte is a pollen grain, and the female gametophyte is called an embryo sac. The other phase of the life cycle is the diploid *sporophyte* stage which produces spores by meiosis. In seed plants, the gametophyte is much reduced in stature and

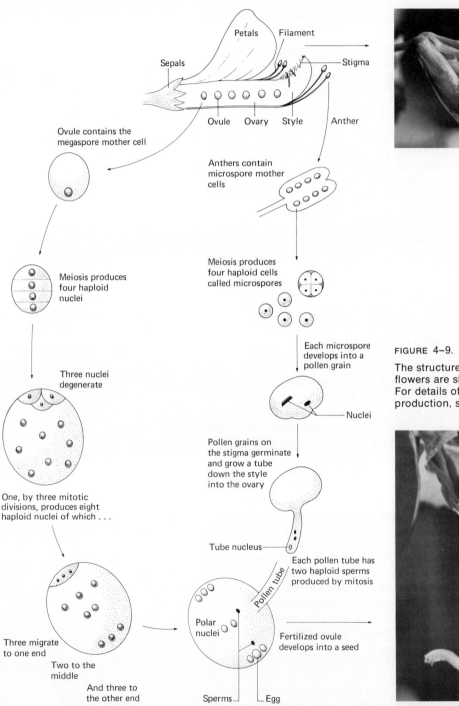

Petals   Filament

Sepals

Stigma

Ovule   Ovary   Style   Anther

Ovule contains the
megaspore mother cell

Anthers contain
microspore mother
cells

Meiosis produces
four haploid
nuclei

Meiosis produces
four haploid cells
called microspores

Three nuclei
degenerate

Each microspore
develops into a
pollen grain

FIGURE 4–9.

The structure and function of pea
flowers are shown in this life cycle.
For details of meiosis and pollen
production, see Figure 4–6.

Nuclei

One, by three mitotic
divisions, produces eight
haploid nuclei of which . . .

Pollen grains on
the stigma germinate
and grow a tube
down the style
into the ovary

Tube nucleus

Each pollen tube has
two haploid sperms
produced by mitosis

Pollen tube

Polar
nuclei

Three migrate
to one end

Fertilized ovule
develops into a seed

Two to the
middle

And three to
the other end

Sperms   Egg

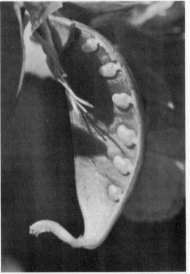

longevity as compared with the sporophyte. In certain other plants, mosses and some algae, for example, the roles are reversed, with the gametophyte being much larger and more dominant as compared with the sporophyte.

But let us return to reproduction in peas. The female side takes place inside small structures called *ovules* within the *ovary*. It is comparable to the male in that haploid (n) spores, called *megaspores*, are produced by meiosis, but it is somewhat more complex in that three mitotic divisions occur after meiosis and before fertilization. Consequently, we find eight haploid nuclei in the female gametophyte (see Figure 4–9), as compared with two in the pollen grain or three in the pollen tube. Of the eight nuclei, one is larger and called an *egg*, and this nucleus is fertilized by one sperm to form the diploid zygote, which is the first cell of the next sporophyte generation. The zygote then divides by mitosis and eventually develops into an embryo plant. Two other nuclei, called *polar nuclei*, are fertilized by the second sperm and form a $3n$ (triploid) nutritive tissue called *endosperm*, which is used by the developing embryo as a food source (see Chapter 6). The other five nuclei degenerate and develop no further. Most botanists believe that these five nuclei have no function at all, that they are present merely in a vestigial state in much the same way that our appendix may be. That is, the ancestors of flowering plants probably had a more elaborate female gametophyte, as found today in the gymnosperms and ferns, for example, and flowering plants have lost all except these few nuclei in the course of their evolutionary history. In any event, the remainder of the ovule forms the seed coat around each pea, and the wall of the ovary develops into the pea pod. Thus, in the botanical sense, not the culinary, peas are seeds, and the pea pod and the peas are the fruit.

Pollination and an Interlude

The only phase of the life histories of higher plants that remains to be discussed is the transfer of pollen from anther to stigma. At the outset you should keep in mind the distinction between *pollination*, which is transfer of pollen from anther to stigma, and *fertilization*, which is the union of gametes, discussed in the foregoing paragraphs. Insects, as well as many other animals, interact with flower structure to effect pollination, and many times this interaction is extraordinarily complex, even bizarre. Nevertheless, from the many cases that have been studied, a pattern seems to emerge. That is, flowers pollinated predominantly by one agent are structured so as to take full advantage of that particular mode of pollination. At this point we should review some of those patterns.

In certain flowers, including peas, pollen is usually deposited directly on the stigma of the same flower, and this is called *self-pollination*. In some cases self-pollination may occur inside a bud that never opens, and then there is no opportunity for the plant to be pollinated by a different plant. In other cases the flower opens and exposes its anthers and stigma to the environment; then there is opportunity for pollen from a different plant to be transported to the stigma. This method is called *cross-pollination* or *outcrossing*. But even in these instances selfing may occur with varying frequencies, depending on the flower structure and pollinating agents. In general, flowers that habitually self-pollinate are inconspicuous and small (Figure 4–10), and their anthers are held very near the stigma. Occasionally, a plant opens its flowers for outcrossing one day, and then the petals close at night in such a manner as to bring the anthers in contact with the stigma, thereby ensuring self-pollination in case its first option, outcrossing, failed.

Thus, there are two general modes of pollination in plants, self-pollination and outcrossing. Some plants are predominantly self-pollinated, but in most cases we find that outcrossing is encouraged or even obligatory. In sunflowers, for example, the pollen produced by one individual cannot fertilize the ovules of the same individual; these kinds of plants are called *self-incompatible*. A single sunflower can self-pollinate from now until doomsday and never produce a viable seed. However, if two individuals grow within a short distance of one another, they can pollinate each other, and full complements of viable seeds will be produced on both. This kind of self-incompatibility

FIGURE 4–10.

Self-pollinated flowers are generally small and inconspicuous, as is this *Cryptantha* from our southwestern desert.

is caused by special genetic mechanisms which prevent the pollen tube from growing down the style after self-pollination.

Outcrossing can be guaranteed in another way. In some plants all flowers on one individual plant are either male (pollen producing) or female (pistil only, no pollen). Such plants are *dioecious*, and examples are willows, pepper trees, and holly.

The above mechanisms for ensuring self-pollination or outcrossing are fairly common, but the more usual situation is that both male and female organs are produced on the same plant and the plants are self-compatible. In these cases we generally find that a special gimmick of some sort is exploited by the plants to encourage outcrossing at the expense of self-pollination. One such device is resorted to by unisexual flowers having both male and female flowers present but separate on the same plant. These are called *monoecious* plants, and an example is the castor bean (Figure 4–11). With anthers and pistils separated in that way, outcrossing is encouraged, although selfing would be possible between different flowers on the same plant.

Another device often found in plants having perfect flowers—that is, both stamens and pistil in the same flower—is a difference in timing between when the pollen is released and when the stigma is receptive. For example, we frequently find that during the first couple of days when a flower is open, its anthers are releasing pollen, but the stigma of that flower is not receptive until the third or fourth day. When the stigma matures and becomes receptive, there is no pollen left in the anthers of that

FIGURE 4–11. DANGER—POISONOUS PLANT!

The seeds of the *castor bean* are often eaten by children and can be lethal. The poison is a protein called ricin which is found throughout the plant but is most concentrated in the seeds. The plant is a good example of the monoecious type, having unisexual flowers with both male and female on the same plant. In the photograph you can see the female flowers (with all the spines) and the whitish male flowers below.

particular flower (Figure 4–12). Here again, selfing is possible between different flowers of the same plant, but outcrossing is encouraged.

Finally, outcrossing is often encouraged by the flower structure in relation to its pollinating agent. Some plants get along fine with only the wind (or water in the case of duckweed and other water plants) to assist in distributing their pollen (Figure 4–13). Normally, such wind-pollinated flowers are small and inconspicuous and have no odor or color to attract a pollinator. Lots of pollen is wasted, of course, and so a great abundance is produced by the anthers. The stigmas are constructed to catch pollen as it comes breezing by. Such stigmas may be featherlike (called plumose) as in grasses or sticky. Wind-pollinated flowers normally produce only a single seed, which seems a curious tactic to adopt in the competitive struggle for existence.

Animals frequently transport pollen from flower to flower and from plant to plant, and each kind of pollinator tends preferentially to visit certain kinds of flowers. For example, bees gather nectar and respond to a flower's fragrance; they do not hover in the manner of a hummingbird and are color-blind to red. Consequently, bee-pollinated flowers are white, blue, and yellow but generally not red, although they often turn red after being pollinated. Also, they grow in such a manner as to provide a landing platform for the bee (Figure 4–14); they produce nectar and are fragrant. This general description includes a lot of flowers, but then bees pollinate most of them.

FIGURE 4–12.

Outcrossing is encouraged if the stamens shed their pollen a few days before the stigma is receptive. a. *Fuchsia.* The younger flower on the right is just shedding pollen, but its stigma is not receptive. The older flower on the left has lost its pollen, but its style is elongated and its stigma is becoming receptive. b. Fireweed (*Epilobium*). The younger flower at the center of the photograph, just below the unopened buds, has pollen being released from its stamens. In the older flowers, just below and to the right, the anthers are empty but the 4-branched stigma is exposed and receptive.

a.

b.

If you happen sometimes to be watching common honeybees working on a group of flowers, keep your eyes on one particular bee as he moves from flower to flower. You'll find that each single bee tends to visit only one kind of flower, while other individuals are working different kinds. He will often by-pass what appear to be suitable flowers as he goes about his work. Thus, social bees tend to be species specific in their pollinating behavior.

Other animals also pollinate flowers. Moths often come out at night, and so moth-pollinated flowers are nocturnal and often white and fragrant. Consequently they are easily found in darkness. Moths, however, often hover while they suck nectar

FIGURE 4–13.

Wind-pollinated flowers, such as grasses, have inconspicuous flowers which produce abundant pollen and feathery stigmas to help catch the pollen as it is carried past by the wind.

FIGURE 4–14.

Bee-pollinated flowers must provide a landing platform and some nectar for the bee. a. Lawn clover. b. Turkey-Mullein *(Eremocarpus)*, a common roadside weed in California.

a.

b.

through their long proboscis (a long, narrow tube for sucking); correspondingly, flowers they frequent do not normally have a landing platform but produce abundant nectar at the bottom of a long tube (perhaps up to 30 cm in length). Flies pollinate flowers, and we all know what attracts flies. Find a flower that smells like dung or rotting flesh, and you probably have an example of pollination by flies or perhaps dung beetles (Figure 4–15).

In addition to insects other flying animals are effective pollinators. Among the birds, a hummingbird hovering in front of a conspicuously colored tubular, often pendant flower is a familiar sight to every nature lover. In addition to those characteristics, flowers that are pollinated by the hummingbird are normally medium to large, diurnal, and with nectar but no odor (Figure 4–16). Most hummingbird-pollinated flowers are red, and biologists have puzzled over that fact for a long time. Controlled experiments suggest that the birds are not particularly attracted to red in preference to other conspicuous colors such as yellow. Thus the question arises, Why are most bird-pollinated flowers red? One suggestion is that red does *not* attract

FIGURE 4–15.

The carrion flower *(Stapelia)* emits a characteristic stench which is particularly attractive to its primary pollinators—flies.

bees, and in the absence of bees more nectar is conserved for the birds.

Finally, bats pollinate a number of flowers. The characteristics of such flowers are the nocturnal habit, whitish drab to dull or dark color, abundant nectar, and often a musky odor. These flowers are large, or perhaps there are many small flowers in one large cluster. Here again, there is a strong correlation between the behavior of the pollinator and floral form.

There you have a short interlude involving the birds and the bees, floral form, and function. It should add a spark of interest to your next trip to the woods.

It is time now to shift gears in our discussion of reproduction and heredity, so it is good to pause for a moment to review. This chapter pointed out that asexual methods of reproduction are common in plants, and under certain circumstances they can be valuable to mankind. Nevertheless, the usual mode of reproduction for most plants is sexual. We discussed sexual reproduction and used peas as the illustrative example because peas were used by Gregor Mendel in his pioneering work on heredity. In Part Two of this chapter, we will look at this work and its applications.

FIGURE 4–16.

Tubular, pendant, generally red flowers with abundant nectar—as this *Fuchsia*—are often pollinated by hummingbirds.

PART TWO

Like Father, Like Son, or Two Peas in a Pod

The basic principles of heredity were discovered and published over a hundred years ago. However, it wasn't until 1900 that the scientific community came to appreciate the full significance of the original communication. The man who conceived the original work and carried it through with great perseverance was Gregor Mendel (1822–1884). By profession a priest, Mendel came from an educated family, which was something special in those days. Young Gregor showed promise in the early grades; at considerable personal sacrifice by various members of his family, he was able to move out of town in order to attend high school. Here again, his performance must have been outstanding because on completing high school he was selected to train for the priesthood in an Augustinian order of monks in Brno, in what is now Czechoslovakia. Mendel became an ordained priest in 1847 but continued his education in mathematics, physics, and natural sciences at the University of Vienna from 1851 to 1853 in order to be a teacher. Once back in Brno, Mendel, then 31 years of age, began his career as a teacher which lasted for 15 years; records indicate that he was extraordinarily successful.

At the same time, as with most other good teachers, Mendel took an active interest in his subject matter and initiated an experimental research program. He worked with a number of different materials in diverse fields of the natural sciences. For example, he recorded changes in positions and numbers of sunspots and maintained an active interest in meteorology to his final days. Apparently his interest in bees lasted all his life, and in later years he collected them from Europe, Egypt, and America in an attempt to hybridize different kinds. He maintained mice colonies for the purpose of studying hybrids between different strains. He bought his personal copies of all Charles Darwin's books.

For the purpose of testing the effect of environment on plants, he performed a number of transplant studies. Altogether his notes indicate experimental observations on at least 26 kinds of plants, many of them involving hybridization studies. One of these studies, that on hawkweed (*Hieracium*), was particularly perplexing because the hybrids produced seeds in the absence of fertilization and therefore bred true. This finding was contrary to all other studies on hybridization and baffled both Mendel and Professor Karl von Nägeli, the ranking authority of the time, with whom Mendel maintained correspondence for six years. The work for which he is best known, however, was conducted with pea plants.

Basically, Mendel noticed that some characteristics of peas tended to be the same from generation to generation (as one might expect with flowers that typically self-pollinate). He asked, What would happen to these characteristics if different kinds of pea plants were crossed? In order to find out, Mendel fertilized the flowers of one kind of pea plant with the pollen from a different kind. When the pea seeds matured, they were planted and a second generation was grown. He continued his experiments by pollinating these plants with their own pollen or with the pollen from one or the other parent. He carefully recorded the characteristics of all the plants that were produced. Let us review some of Mendel's experiments and attempt to rationalize the results. Although Mendel himself studied the inheritance of quite a number of traits and reported the results of seven, we'll use just two traits as examples: color and shape of the pea seeds.

Some pea seeds are green, while others are some shade of yellow. Some peas are smooth and round; others are angular and wrinkled (Figure 4–17). With two traits, then, four kinds of peas are possible. Peas might be green and round, green and wrinkled, yellow and round, or yellow and wrinkled. As it hap-

FIGURE 4–17.

The light-colored peas are yellow; the darker ones are green. Some are wrinkled; some are smooth and round. Part of Mendel's genius was in asking *how many* of each kind are produced in hybrid progenies. Other scientists of his time simply asked *what kinds* may be produced, and as a consequence they missed the whole idea of discrete hereditary particles being passed from parent to offspring.

pened, Mendel began his work with one variety or strain of peas which were round and yellow and with a second strain wrinkled and green. His question was, What is the mode of inheritance of color and shape of peas?

Peas will self-pollinate if left alone. This pollination occurs when the stamens mature and release their pollen directly onto the stigma of the same flower without the aid of insects or any other agent. Sometimes insects visiting the flowers for nectar may carry pollen around from plant to plant, and cross-pollination may occur this way. Seeds produced as a result of cross-pollination between different kinds of plants are called *hybrid* seeds, and plants grown from these seeds are hybrid plants. Therefore, hybrids carry the hereditary material, chromosomes and genes, from two different kinds of plants. Mendel's first question was, What would hybrids between the round-yellow strain and the wrinkled-green strain look like? Would they resemble one parent, or would hybrids be intermediate between the two parents? Mendel, of course, needed to produce hybrids between the round-yellow and wrinkled-green strains with no chance for error; he had to be sure. To do this he used forceps to remove carefully the immature stamens from a newly opened flower of the round-yellow strain, and then on the stigma he placed pollen from a flower of the wrinkled-green strain. The fruits were then left to develop in the usual way. This same general technique was used in making all his crosses and is used today in the experimental garden of the twentieth century. The crosses Mendel made and his results are shown in Figure 4–18.

It is important to analyze only one trait at a time, as Mendel did over a century ago. Let us look first of all at color, and for purposes of this discussion, we will introduce a few new terms. All the offspring of a particular parent plant are referred to as its *progeny*, and progeny produced by crossing between two different strains are called *hybrids*. We generally refer to hybrids as F_1 which stands for the first filial generation. These hybrids in turn give rise to progeny of their own, and they are second generation hybrids or F_2. In total, then, there are three generations; the parents, the first generation hybrids or F_1, and the second generation hybrids or F_2.

In the case of Mendel's work, one parent had yellow peas, the other green, and all F_1 hybrid peas were yellow (see Figure 4–18). We conclude as a beginning that color cannot be due to environment; it must be inherited genetically by these pea plants because two kinds, yellow and green, grew side by side in the same garden. Put another way, the traits must be determined by heredity and not environment because individuals grown in identical environments look different. Now let us carry this a bit further and ask, How might a hereditary factor pass from parent to offspring? What tissue or cell might perform this task?

The answer is that this *hereditary factor*, as Mendel called it and which we now call a *gene*, must be carried in the sperm and also the egg because these are the only two cells contributed by the parent generation immediately and directly to their offspring.

The next step in our analysis is to ask, How many genes for pea seed color might be carried by pollen and egg? *Answer:* one each. In the case described here, the pollen was produced by a green parent and carried a gene for green, and eggs were produced by a yellow parent and carried a gene for yellow-colored peas. These two cells fused to produce a zygote which developed into the F_1 hybrid. Thus, the F_1 hybrids themselves carried these two genes, one for yellow and one for green.

You may recall that in their inheritance genes behave similarly to the way chromosomes do. There are always two of each kind of chromosome in the diploid cells, and of those two chromosomes, one was contributed by the egg and the other by the sperm in fertilization. As the plant grows, new cells are produced by mitosis so that all cells in the whole plant contain similar sets of chromosomes. This changes with meiosis, however, because the chromosome number is reduced and all haploid cells contain only one of each particular chromosome. Because

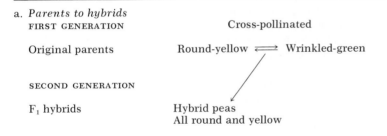

a. *Parents to hybrids*
FIRST GENERATION   Cross-pollinated

Original parents   Round-yellow ⇌ Wrinkled-green

SECOND GENERATION

F_1 hybrids   Hybrid peas
                            All round and yellow

b. *Hybrids to F_2 and backcrosses*
SECOND GENERATION

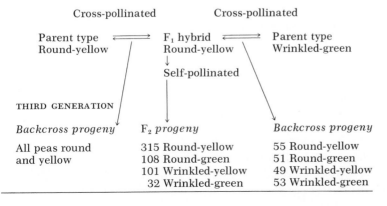

        Cross-pollinated   Cross-pollinated

    Parent type ⇌ F_1 hybrid ⇌ Parent type
    Round-yellow   Round-yellow   Wrinkled-green
                        ↓
                    Self-pollinated

THIRD GENERATION

Backcross progeny F_2 *progeny*   *Backcross progeny*

All peas round   315 Round-yellow   55 Round-yellow
and yellow   108 Round-green   51 Round-green
                    101 Wrinkled-yellow   49 Wrinkled-yellow
                     32 Wrinkled-green   53 Wrinkled-green

FIGURE 4-18.

The diagram summarizes Mendel's work with peas. a. The first year F_1 hybrids were produced from round-yellow and wrinkled-green strains of peas. b. The second year the F_2 progeny was produced from F_1 hybrids, and backcross progenies were produced between F_1 plants and parent types.

the genes are carried on the chromosomes, genes alternate between the $2n$ and $1n$ conditions in the same way that chromosomes do. Thus, two genes for each character are found in the zygote and in all other cells of a sporophyte plant. However, because of meiosis, only one gene for each character is found in the haploid nuclei of each pollen grain and each egg.

Now, if F_1 hybrids carry genes for both yellow and green, why are such peas not yellow-green in color, that is, why don't the traits blend? The answer is that the gene for yellow is expressed and the one for green isn't. This phenomenon is called *dominance*. That is, we define yellow as dominant and green as recessive because all the hybrids are yellow in color. At this point the distinction can be drawn between phenotype and genotype. *Phenotype* refers to the *visible appearance*, the *trait*. *Genotype* refers to the *inherited potential*, the *genes*. If the hybrids' phenotype were green, green would be the dominant trait and yellow the recessive, but that's not the way it is.

The concepts of dominance and recessiveness and of phenotype and genotype are among Mendel's major contributions to our understanding of heredity.

Now turn again to Figure 4–18. Note that in the second year these F_1 hybrid peas were planted and grew into mature plants themselves, and then they were used to produce three groups of progeny in the third generation. First, the F_1 hybrids were self-pollinated and thereby produced what we call the F_2 *progeny*. Second, the F_1's were crossed with the round-yellow kind of parent; this type of pollination produces individuals referred to as *backcross progeny*. Finally, a second backcross was made by crossing the F_1 plants to the wrinkled-green parent type. The results of all three crosses are explained according to what we have already learned about inheritance, namely: (1) Each individual pea and pea plant carried two genes for each trait. (2) Each pollen grain and each egg carried one gene for each trait. (3) Yellow is dominant over green. But let us look at each of the third generation progenies to see in detail how traits arise, and again for a moment we'll discuss only the inheritance of color.

First, the progeny from the backcross to the yellow parent type were all yellow. This can be easily understood. All pollen and egg cells from the yellow parent carried the gene for yellow. Therefore, it makes no difference whether the pollen or eggs from the F_1 carried a gene for yellow or green; these particular peas had to be yellow for the same reason that the F_1's were yellow, namely, yellow is dominant over green and all these peas carried a gene for yellow donated by the yellow parent.

The progeny from the backcross between the F_1 hybrid and the green parent type were half yellow and half green. The explanation for this 1:1 segregation is as follows. All the pollen and eggs from the green parent type will carry genes for green peas. Therefore, the peas produced from this cross will be green unless a gene for yellow is contributed by the F_1 hybrid. But exactly half the pollen produced by the F_1 hybrid will carry a gene for yellow, and half for green. The pollen population is half green carrier and half yellow carrier; the eggs are similar carriers. Thus, when the F_1 hybrid is crossed with the green parent type, half the progeny will carry two genes for green and will be green-colored peas. The other half will carry one gene for green and one for yellow, and those peas will be yellow because yellow is dominant. You see now that you have sound biological grounds for challenging someone who says, "As alike as two peas in a pod." Not so, sometimes.

Now let's see how the above discussion can help us to understand what happened in the F_2 progeny. Remember, the F_2 plants come from self-pollinating an F_1 plant or from crossing two different F_1's; the results are the same either way. Turn again to Figure 4–18 and you can see that when all the F_2 peas are taken together, pretty close to three-quarters were yellow and one-quarter was green. These ratios, incidentally, of three-quarters to one-quarter and half to half are the mathematical idealized model. Actual results only approximate the model because random chance causes some deviations. The explanation for the three-quarters to one-quarter F_2 segregation is summarized in Figure 4–19 and is based on the same facts that accounted for the backcross progenies. All F_1 plants carry two genes, one for yellow and one for green. Pollen cells, and eggs too, produced by these plants are haploid and therefore carry only one gene for color of peas. For convenience, a letter is generally used as a symbol for a gene, upper case for the dominant condition and lower case for the recessive condition. Half the pollen and half the egg cells carry a gene for yellow (A) peas, and half carry a gene for green (a) peas. Thus, when the flowers are pollinated, about one-quarter of the peas produced will carry two genes for yellow (AA), about half of them will carry one gene for yellow and one gene for green (Aa), and the remaining quarter will carry two genes for green (aa).

Notice that all green peas carried two genes for green (aa), whereas the yellow peas carried two genes for yellow (AA) or only one (Aa). Thus, there are two colors of peas but three combinations of genes. Here again we distinguish between *phenotype*, or appearance, and the *genotype*, or genes, carried by an individual. That is, they may look the same, but their potential for variation in future generations is different.

This concludes the discussion necessary to explain the inheritance of seed color, but what about seed shape? You can see by considering Figure 4–18 that essentially everything we said about color of peas applies equally well to shape. Which is dominant, angular or round? Note that all progeny from the backcross to the dominant parent type show the dominant phenotype,

a. *Parents to hybrids*
FIRST GENERATION

Original parents	Yellow $A//A$	Green $a//a$
Gametes	$A/$	$a/$

SECOND GENERATION

F_1 hybrids	$A//a$ F_1 hybrid peas All yellow

b. *Hybrids to backcrosses*
SECOND GENERATION

	Parent type Yellow $A//A$	F_1 hybrid Yellow $A//a$	Parent type Green $a//a$
Gametes	$A/$	Half $A/$ Half $a/$	$a/$

THIRD GENERATION

Backcross progeny	F_2 *progeny*	*Backcross progeny*
All peas yellow Half $A//A$ Half $A//a$	Three-fourths of peas yellow, one-fourth green (See part c below)	Half yellow, $A//a$ Half green, $a//a$

c. *Hybrids to* F_2
SECOND GENERATION

	F_1 hybrid Yellow $A//a$	F_1 hybrid Yellow $A//a$
Gametes	Half $A/$ Half $a/$	Half $A/$ Half $a/$

THIRD GENERATION

F_2 progeny: three-fourths of peas yellow (416), one-fourth green (140)	$A//A$ Yellow $A//a$ Yellow $a//A$ Yellow $a//a$ Green

FIGURE 4–19.

a. The diagram summarizes Mendel's first year's work. b. The second year he planted the F_1 hybrid seeds as well as other seeds of the original parent types, and made further controlled pollinations. c. The diagram explains the F_2 progeny.

Symbols used in this figure:
$A/$ = Gamete with a chromosome (/) carrying dominant gene for yellow.
$a/$ = Gamete with a chromosome (/) carrying recessive gene for green.
$A//A$ = Chromosomes in a pea plant with two genes for yellow.
$A//a$ = Chromosomes in a pea plant with one gene for yellow and one for green.
$a//a$ = Chromosomes in a pea plant with two genes for green.

and the progeny from the backcross to the recessive parent type segregated half dominant and half recessive phenotypes. Finally, three-quarters of the F_2 progeny showed the dominant phenotype and one-quarter the recessive. As you can see, the experimental design and results are comparable to those for inheritance of color, and therefore the explanation is the same.

Independent Assortment:
Separate but Equal

You probably noticed that more new combinations of pheno-types appeared in the third generation of the preceding example than did in the first and second generations. Whereas the orig-inal parents were of two types, *round and yellow* or *wrinkled and green*, some peas in the third generation were yellow and wrinkled, yellow and round, green and wrinkled, and green and round. Mendel noticed the same thing and called it *independent assortment*, referring to the phenomenon that the inheritance of color was independent from and not linked in any way to the inheritance of shape. The reason is that the two traits dis-cussed, color and shape of peas, are determined by genes which are carried on separate chromosomes. Think back to meiosis, now, in the F_1 hybrid. Two of the same kind of chromosome come together and pair; one of those two chromosomes carries the dominant A for yellow, while the second chromosome carries the recessive a for green. Thus, these two chromosomes are immediately and directly associated with the inheritance of color, but they have nothing to do with the inheritance of shape of peas. The genes for wrinkled and round are carried on a different pair of chromosomes. In peas $2n = 14$, and therefore there are seven pairs of chromosomes at meiosis. One of those pairs carries the genes for color, and on a different pair are the genes for shape—separate but equal. These two pairs are quite independent from one another as they proceed through meiosis to form haploid spores.

After making an effort to understanding the above, it should be apparent that the inheritance of traits via sexual reproduc-tion follows a given set of rules. First, each *individual sporo-phyte* plant carries *two genes* for each trait: one inherited from its maternal or egg parent and the other from its paternal or pollen parent. Second, each *sperm* and *egg* carries *one gene* for each trait. Third, when an individual carries genes that differ, such as one for green and one for yellow, or one for smooth and one for wrinkled, we frequently find the phenotype to be deter-mined by one gene and call this *dominance*. Fourth, the genes for one trait, such as color, are often *inherited randomly* with respect to the genes for a second trait, such as surface conformation of peas. This is due to the fifth rule, which is that genes are *carried on the chromosomes*. These rules acting together have important consequences. Let us discuss some of these consequences now.

Literally, life as we know it could not exist without sexual reproduction. First of all, sexual reproduction makes evolution possible because it gives rise to variation. In the example we discussed, the starting individuals for Mendel's experimental work were of two types, plants with round and yellow peas on the one hand, or wrinkled and green on the other. After two generations of sexual reproduction, we find round and yellow, round and green, wrinkled and yellow, and wrinkled and green—four types. Think for a minute what this means; consider that the same sort of recombination of characters is occurring with all other traits as well. For example, Mendel worked with seven traits. How many phenotypes might appear in the F_2 generation if seven traits are sorting themselves out independently? *Answer:* 128. With one trait, say pea color, there are two phenotypes, yellow and green. A second trait, such as surface conformation, gives two more phenotypes for each color. Seven traits, then, may combine for as many as 2^7 phenotypes, or 128. Clearly, evolution can proceed rapidly if selection has that level of variation to work with. Recently, as explained earlier in this book, Dr. Borlaug developed new strains of wheat which had short, sturdy growth habits and also resistance to disease. These two traits came from different wheat strains from different parts of the world. The strains were hybridized in the same manner as Mendel's peas, and from the subsequent progenies those plants that were blessed with both desirable traits were chosen to establish the new miracle wheats. Thus, evolution is possible because of sexual reproduction. Whether that evolution is natural or directed by man makes no difference to the genes and to the plants that carry them.

The second reason why sexual reproduction makes higher life forms possible is that deleterious genes are continually being weeded out so they do not accumulate to the point at which they eventually kill the organism. For example, let us again consider the navel orange which is continually reproduced by asexual means. Any deleterious (bad or harmful) mutation which arises will be perpetuated. Mutations happen all the time albeit at a low frequency, on the order of one or two in a million cells for each gene. But there are usually thousands of genes required to make a plant, and all genes follow the same rules. Therefore, over a period of time these deleterious mutations accumulate in those plants that are maintained only through asexual means. Each tree is beset with its own problems, its own set of deleterious mutations which have been accumulating little by little since that first bud variation arose in Brazil back around 1800. Today we find that navel orange orchards contain up to 75% substandard trees, and the problem worsens every year. If navels reproduced sexually, it would be possible to select from various combinations of characteristics those individuals which possessed the desirable and lacked the deleterious. Unfortunately, this is not the case, and so the original Washington strain is doomed to inevitable extinction as the deleterious mutations continue to build up and weaken the trees. The same is true for any organism that reproduces strictly by asexual means.

As a hypothetical example, let us take three traits associated with general plant vigor, such as rate of photosynthesis, rate of growth, and hardiness. For the sake of illustrating our point here, say that each of these traits is determined by genes that are independent from one another so that it is possible to have plants with all possible combinations of the traits listed below.

Genotype	Photosynthesis
AA	Greatest
Aa	Intermediate
aa	Least

Genotype	Growth
BB	Fastest
Bb	Intermediate
bb	Slowest

Genotype	Hardiness
CC	Hardiest
Cc	Intermediate
cc	Weakest

Further, we stipulate that initially natural selection has provided the best of all possible plants: genotype *AABBCC*. Now, let us compare the fates of these plants as they reproduce by asexual means on the one hand versus sexual means on the other. First, we know that our starting genes might mutate in some individuals, and again for the purpose of discussion, let us follow the fate of individuals which have mutated to genotype *AaBbCc*—that is, a plant which is not quite so vigorous and productive as its predecessors. Since mutations occur in all plants, there is no difference between sexual and asexual strains up to this point. However, here is where the story changes. With asexual reproduction these mutations will be perpetuated generation after generation, and new mutations will undoubtedly be added. A few of the deleterious genes might mutate back to the more advantageous form, but far more mutations will be from advantageous to deleterious simply because most of the genes at the start are for vigorous, healthy plants, and there is only one way to go from there. Little by little, generation after generation, asexually reproducing strains will decline in general vigor.

With sexual reproduction, by comparison, the genes are shuffled around, recombined as we say in the trade, and new genotypes are produced. The reason for this is based on what we learned from Mendel's work with peas. The results of this recombination are summarized in the following tabulation and compared with the results of asexual reproduction:

AA BB CC	mutations →	Aa Bb Cc	asexual reproduction →	Aa Bb Cc	
AA BB CC	mutations →	Aa Bb Cc	sexual reproduction →	AABBCC	best
				AABBCc AABbCC AaBBCC	good
				AABbCc AaBBCc AaBbCC	fair
				AaBbCc	so-so
				AaBbcc AabbCc aaBbCc	poor
				Aabbcc aaBbcc aabbCc	inferior
				aabbcc	worst

(Actually, 27 genotypes could result from sexual reproduction here; this lists only 15 for purposes of illustration.) With sexual reproduction, one can immediately see that all possible types are formed and are available for the plant breeder or nature to select.

In summary, sexual reproduction gives rise to variation, and as a direct consequence of that fact, two things are possible. First, new desirable types are produced, and second, deleterious mutations can be selected out.

Mendel Plus One Jolly-Green-Giant Step

The principles of inheritance first shown to us by the Austrian monk who taught mathematics to school children have been utilized by man in many ways and to his great benefit. We have developed new and better strains of agricultural crops, as we saw in Chapter 3 on the green revolution, and have strengthened existing strains against disease. The same principles are applied to animal breeding for the same purposes. Even to ourselves the application of Mendel's principles is proving very valuable indeed in understanding and preventing inherited diseases among children.

Unfortunately, some well-meaning individuals have come to expect more than is possible from genetics. These expectations range from spectacular increases in the yields of agriculture to producing or at least recognizing a superior race of humans. Clearly, such views are no more than groundless, wishful thinking at best; at their worst they have been the basis for the greatest carnage of human beings in the history of the world.

For these reasons it is important for us to understand more than just a smattering about heredity. Understanding is the first step toward control, and if we are to realize the true potential of heredity as well as its limitations, we must take a giant step beyond Mendel's basic principles. We must understand exactly what genes are and how chromosomes act as vehicles of heredity. We must see how genes work and how they produce a phenotype. We will do this in the next chapter. Using this knowledge and also our understanding of photosynthesis, we will be in a good position to assess adequately the green revolution and its potential for man.

Selected Readings

Allen, R. D. July 1959. Moment of fertilization. *Scientific American* 201(1):124–134.

Beadle, G. W. 1959. Genes and the chemical reactions in *Neurospora. Science* 129:1715. Lecture given by Nobel Prize Winner discussing the gene-enzyme concept.

Beermann, W., and U. Clever. April 1964. Chromosome puffs. *Scientific American* 210(4):50–58.

Dobzhansky, T. January 1950. The genetic basis of evolution. *Scientific American* 182(1):32–41. Easy-to-read article on an ever-popular subject.

Kamen, M. D. 1964. *A Tracer Experiment.* New York: Holt, Rinehart and Winston. An interesting and novel approach to the concepts of radioisotope use.

Kornberg, A. 1960. Biologic synthesis of deoxyribonucleic Acid. *Science* 131:1503. Nobel Prize lecture on DNA synthesis in vitro.

Kornberg, A. October 1968. The synthesis of DNA. *Scientific American* 219(4):64–78. Written by a leading authority on DNA.

Mazia, D. September 1961. How cells divide. *Scientific American* 205(3):100–120. (Offprint no. 93.) Nice review of mitosis.

Mendel, G. 1965. *Experiments in Plant Hybridization.* Cambridge, Mass.: Harvard University Press. A reprint in English of Mendel's original report. Interesting reading.

Michelmore, S. 1965. *Sexual Reproduction.* Garden City, N.Y.: Doubleday. Varied notes of sex in living things.

5   *The Trinity of Biology: DNA, RNA, and Protein*

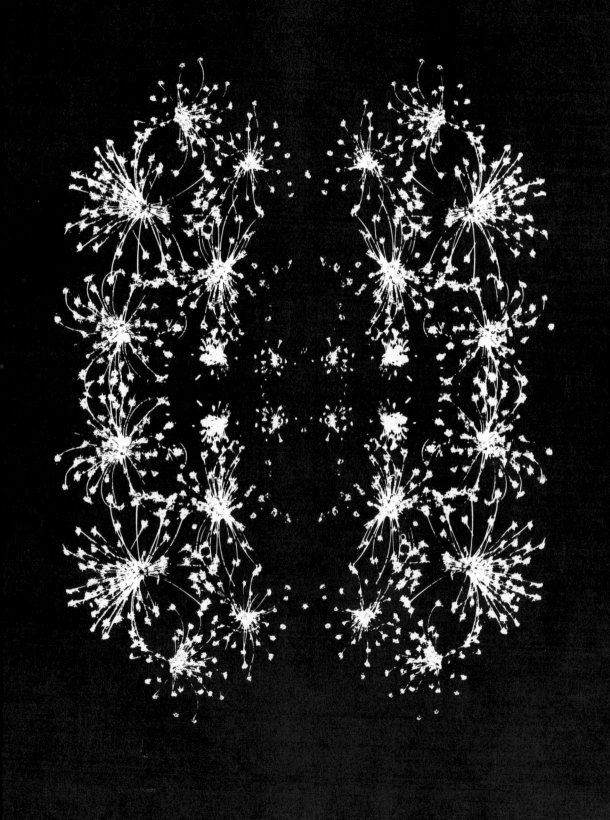

It is a large step to proceed from understanding Mendel's traits to understanding, in cellular terms, what traits are and how they are translated into recognizable features. The answers were a long time in coming and required the effort of many outstanding scientists. It is therefore rather surprising, now that many of the facts are in, to find that the basic principles are relatively simple to comprehend. Let's first consider briefly the basic requirements that must be met by the genetic or trait-carrying material. We will discuss each requirement in detail later in the chapter.

First and foremost, we know that this material must *carry information*, a great deal of information, because building and organizing even the simplest living organism must require thousands of separate steps. Second, it must be possible to *make copies* of the information contained in the genetic material, for the cell accomplishes this each time mitosis or meiosis takes place and the copying mechanism must be very accurate. Third, there must be some *mechanism for decoding* the genetic information and putting it to practical use in the cell. Last, this material must not be totally inflexible, that is, *changes* (mutations) *must be possible*, otherwise variation and evolution would be impossible.

To shorten a very long and involved story, it has now been established that the genetic material is deoxyribonucleic acid (DNA), and DNA, as you already know, is one of the materials that make up the chromosomes. Thus, chromosomes not only carry genes but are in fact made of genes. Genes are DNA. A gene or a piece of DNA, however, is not a trait itself. Rather, DNA determines which trait will appear in a particular individual. DNA is information, plans, potential; it is the light and the way, but it is not the trait itself. How is it that DNA can carry information regarding the structure of a plant or animal but cannot actually build the organism itself? The DNA (located in the nucleus) codes for another kind of molecule, messenger RNA (mRNA). mRNA can then migrate out of the nucleus into the cytoplasm where it directs the assembly of all cellular proteins (Figure 5–1). Proteins directly or indirectly give rise to traits and so, to understand how genes determine characteristics, we must understand the trinity of biology: DNA, RNA, and protein. We can best approach some of the details involved by first considering the nature of the DNA molecule itself.

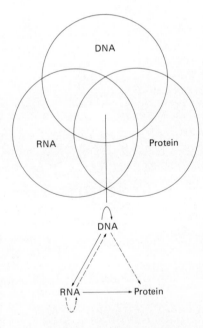

FIGURE 5–1.

The basis of present-day biology can be illustrated by the interrelationship between DNA, RNA, and protein. The solid arrows show general transfers of information; the dashed arrows show special transfers or suspected transfers.

DNA: Long and Slender but with a Right-Hand Twist

DNA is very large as molecules go. A schematic and somewhat stylized diagram is shown in Figure 5-2. The molecule is long, slender, double-stranded, and helical. More precisely we can say that DNA is composed of two "railings" made of sugar-phosphate units, and these railings are cross-linked in staircase fashion. The cross-linking units are nitrogenous bases, and there are only four kinds found in DNA: adenine (A), cytosine (C), guanine (G), and thymine (T) (Figure 5-3). A simple but useful analogy to DNA structure is a twisted ladder or spiral staircase—but of course on a molecular scale.

Imagine that you have molecular models of the various components that make up DNA and that you begin building a model of a short DNA segment. First, you construct the sugar-phosphate railings, which should prove to be relatively easy; then you begin adding at random the various nitrogenous bases to one side. Again this should not provide any difficulty, and you would find that there is no problem in arranging the bases (A, T, C, and G) in any order. The fit is equally good, for example, if you use AGCTAATCCAA or CTCCCCGCGGTA. In other words, you have a tremendous amount of flexibility in the "words" you can spell using this four-letter alphabet. Literally thousands of variations are possible. Now suppose you begin assembling nitrogenous bases on the adjacent railing and attempt to make them fit with the previously assembled half to create a complete model similar to that in Figures 5-2 and 5-3. You would find that this is not easy. Many pieces won't fit without severe straining or bending; your job might appear hopeless. But, by carefully studying Figures 5-2 and 5-3, you see there is a pattern to DNA assembly. The arrangement of the bases on the second railing cannot occur randomly. Rather, in building the second half of the molecule, attention must be given to the order of bases on the previously constructed half. For example, if in your original half you started with A, then the adjacent chain must start with T. If the next base is G on the original, then following T you must insert a C. Pairing is critical: A always pairs with T, and G with C; likewise T with A, and C with G. Then, and only then, do the pieces match up making model construction feasible. Thus, we have rediscovered the most important feature of DNA: Base pairing. This phenomenon was discovered in exactly the same way we have done—through model construction—by James D. Watson and Francis Crick in 1953. Subsequently they were awarded the Nobel Prize for this work. Let's pursue this matter of base pairing, and in doing so you will come to understand why Watson and Crick's original discovery of this phenomenon was worthy of the Nobel Prize.

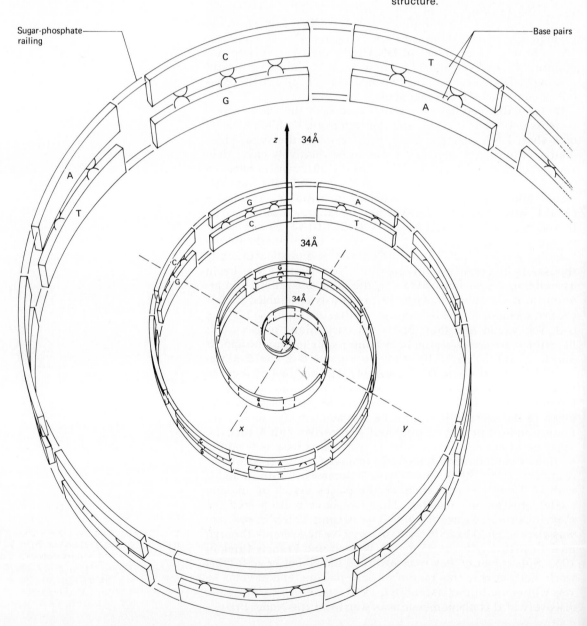

FIGURE 5–2.

If one could look down into a DNA molecule, it would appear as a spiral structure.

Information Storing and Replication of the Messages

First and foremost, DNA must be an information carrier, a blueprint, and by constructing a model (mentally at least) we have seen that any sequence of the base pairs AT, TA, GC, or CG can exist in a given segment of DNA. Now let's assume that these base pairs are equivalent to a cellular code that can be translated by the cell. With this assumption in mind, we can visualize how a number of different messages can be spelled by simply altering the sequence of base pairs. How many messages are possible? Billions! For example, let's assume that the average gene contains about 1500 base pairs. Theoretically, with four possibilities at each of 1500 sites, the number of different kinds of genes possible would be 4^{1500}. So the information potential in DNA is very large even for one gene, and organisms are composed of many genes. In the relatively simple bacterium *Escherichia coli*, about 64 genes have been identified and studied in some detail. A rough estimate of its total number of genes is about 1000. Higher plants have about 500–5000 times as much

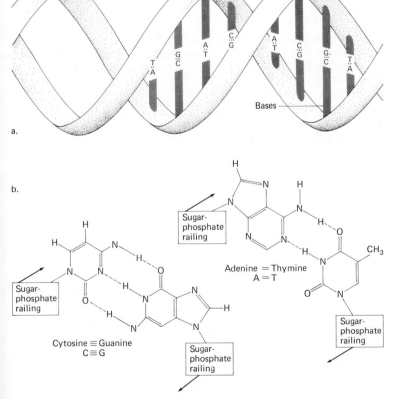

FIGURE 5–3.

a. As viewed from the side, a DNA molecule appears as shown here. b. Specific bonding occurs between the base pairs C and G, and A and T. The dashed lines between the base pairs indicate the weak hydrogen bonding that is responsible for specific pairing.

DNA in their cells as *E. coli* does, and so the number of genes in higher plants is very large indeed. Obviously DNA meets the first criterion we listed for the genetic material—it has the information potential. But what does this information code for? DNA contains the information of how to assemble all the protein molecules in the cell. We will see exactly how this is possible shortly, but let's first take care of the second requirement for the genetic material—the replication of information.

The model for DNA with its specific base pairing immediately suggests a possible mechanism for replication, and this mechanism was first recognized by Watson and Crick. At the proper time the two axes of DNA can separate zipper fashion, and then each single axis makes a complementary partner (see Figure 5–4 for further details). In this way, two molecules are produced where once there was one, and most important, both molecules are exactly alike and have the same information content.

Transcription—Getting the Message out

The third requirement states that there must be some mechanism for decoding the genetic material so that the information is useful to the cell. The explanation of the way in which this happens is a little more complicated than are the explanations of the first two requirements. Our first problem is that DNA is in the nucleus, and protein synthesis occurs in the cytoplasm. Obviously some mechanism is needed to get the instructions on how to assemble a particular protein out of the nucleus and into the cytoplasm. This particular process is called *transcription*.

As transcription begins, the two complementary strands of DNA separate slightly, exposing the nitrogenous bases. A new complementary strand is then made by base pairing onto a por-

FIGURE 5–4.

Replication of genetic information occurs when the DNA molecule unwinds and, via base pairing, a new copy of each half is made. The complete process is obviously carefully controlled and occurs in steps regulated by enzymatic proteins. The net result is two copies where once there was one.

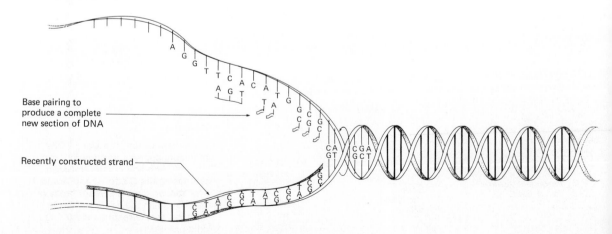

Base pairing to produce a complete new section of DNA

Recently constructed strand

tion of one of the pre-existing DNA strands. This new chain of nitrogenous bases is messenger RNA (mRNA) (Figure 5–5). mRNA is complementary to that portion of the DNA strand that codes for its production, with one exception. Uridine (U) is used in RNA in place of DNA's thymine (T). mRNA is unlike DNA, however, in that it remains single-stranded. Furthermore, it is known that a single mRNA molecule does not represent the total information in the entire DNA molecule. Rather, there are start-and-stop codes along the DNA directing the production of mRNA. Because of these punctuation marks, many species of mRNA are formed, each containing the information in a particular portion of the DNA. It is important to realize that these start-and-stop signals on the DNA strand are not just scattered at random but are spaced in a very specific way. Each species of mRNA contains the information of how to build a single, or perhaps several, protein molecules. Of course, the DNA molecule itself contains in a linear order the information on how to build hundreds of different protein molecules. Thus, mRNA production would start on a signal at the beginning of the instruction for building, say, protein A and would end on a signal after all the information for constructing protein A was transcribed in an mRNA chain.

Another mRNA molecule would be made either concurrently or at some other time, carrying the information on how to build proteins B, C, D, and so on. In other words, in mRNA synthesis discrete information bits are parceled out. We know that the size of these information bits is a few hundred to, say, several thousand nitrogenous bases arranged in a linear sequence. The exact length would depend on the complexity of the message. Thus, while the entire message in an mRNA molecule can be long and complex, the individual "words" within the message are rather short, only three bases (letters) each. The idea is that

FIGURE 5–5.

Transcription of the message. DNA can separate slightly to allow for base coding of the mRNA molecule. Note that in RNA, U substitutes for T.

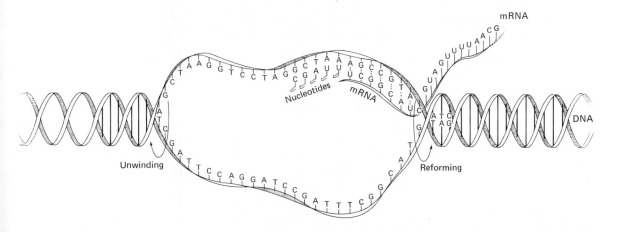

each three-letter word in mRNA is the code for a particular amino acid. For example, UCU codes for the amino acid serine, GGU for glycine, GCU for alanine, and so on for all twenty of the amino acids used in protein production (see Appendix Two). Since a given sequence of mRNA may have several hundred or so of these three-letter codes, it is easy to see how such a molecule can contain the information for assembling amino acids in a given order. This point is critical because the order in which the amino acids are linked together in a protein determines the function of that particular protein.

The subtleness of the arrangement of amino acids and how their order can alter protein function is easily illustrated by considering the basis for the human blood disorder called sickle-cell anemia. This blood disease alters the blood's oxygen-carrying capacity. The disease results because the protein molecule hemoglobin, which carries oxygen, has a single amino acid (glutamic acid) replaced by another (valine). It is also known that this disease can be traced back to the DNA and to the mRNA which codes for the production of hemoglobin. That is, the mRNA for hemoglobin in the diseased individual is exactly like the mRNA for hemoglobin in a normal individual except that somewhere on that molecule there is a three-letter sequence that says UGU instead of AGU—a single nitrogen base is different out of perhaps 1000 total bases in the mRNA species that codes for hemoglobin. This seemingly small mistake can be deadly because it alters the placement of one amino acid, and the entire structure of the hemoglobin molecule is changed in such a way that it no longer functions properly in carrying oxygen.

Once an mRNA molecule is made, it does not remain long within the nucleus but moves through the pores in the nuclear membrane into the cytoplasm. Once in the cytoplasm the mRNA arranges itself in an orderly fashion on special cellular assembly lines, called ribosomes, where the actual business of assembling amino acids into proteins occurs. In order to understand exactly how this works, we have to understand the function of yet another kind of RNA called transfer RNA (tRNA).

Translation—Getting the Message Read

Transfer RNA (tRNA) is a smaller molecule than mRNA, typically consisting of only 70–80 nitrogenous bases. Many different kinds of tRNA molecules exist within a cell, and among this population are types that can attach specifically to each of the twenty amino acids found in plants. Thus, each amino acid attaches to a specific tRNA, and in such a form the amino acids are delivered to the site of protein synthesis (Figure 5–6).

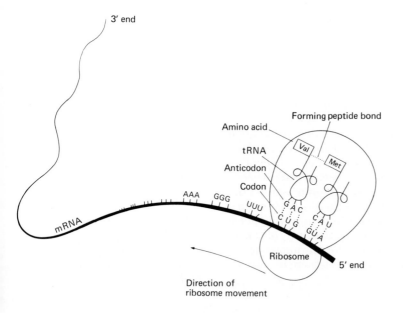

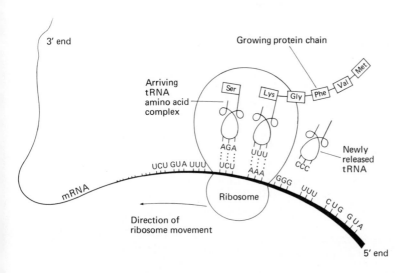

FIGURE 5–6.

The sequence of amino acids in a given protein is determined by the sequence of three-letter words *(codons)* in an mRNA molecule. Initially, each amino acid is complexed with a specific tRNA molecule which has a three-letter code *(anticodon)* complementary to the codons on the mRNA molecule. Thus, amino acids can be arranged in a specific order, and peptide bonds can form between adjacent amino acids, forming a protein. When the tRNA molecule is released—minus its amino acid—the process is repeated.

Remember, we already have mRNA attached to the ribosome.
At the point where the mRNA touches the ribosome, the tRNA
with its attached amino acid settles down on the mRNA chain.
Again, this is a directed process, not a random one, and it is
directed by base pairing. As the ribosome rolls along down the
mRNA molecule, the tRNA eventually detaches itself from the
mRNA and leaves the amino acid it was carrying behind. As this
is occurring, a second tRNA settles into position with its own
specific amino acid, and the two amino acids become attached to
one another via a covalent (peptide) bond. When this union is
completed, the second tRNA minus amino acid is removed, and
the process continues with a third, a fourth, and so on. The free
tRNA molecules migrate back into the cytoplasm, are "re-
charged" with another amino acid, and are then able to function
again in protein synthesis. (See Figure 5–7 for a summary of the
entire process.) The important thing to remember, however, is
that each three-letter word in the mRNA (called a codon) codes
for 1 of the 20 kinds of tRNAs. As these tRNAs pair against the
mRNA, they bring their associated amino acids into position in a
specific way. In the course of time a long chain of amino acids
can be produced with the position of each one carefully planned.

Thus the original plan is in the DNA on the chromosomes in
the nucleus. This plan is carried to the cytoplasm by mRNA

FIGURE 5–7.

Summary of protein synthesis.
mRNA is produced in the nucleus,
tRNA is coded for in the nucleus, and
proteins are synthesized in the
cytoplasm.

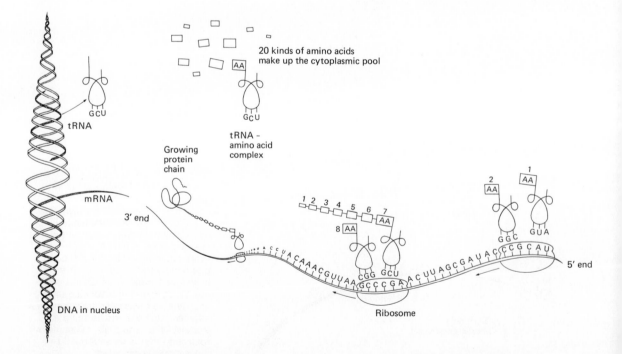

where tRNA and mRNA together guide the process of protein synthesis, with a supporting cast of ribosomes, enzymes, and other cofactors. The final result is a very large protein molecule which serves to facilitate some cellular process.

Protein to Trait: The Final Step

How can the production of a particular protein lead to a specific trait? As an example, normal corn grows to 5 feet or more at least partially in response to the growth hormone gibberellin (see Chapter 6). Dwarf corn grows to only 2–3 feet because it lacks this particular growth-stimulating hormone. The reason for the presence or absence of gibberellin is ultimately due to the message or code contained within the DNA molecules inside the nucleus of the cells in the corn plant. The cells of a normal, tall corn plant contain many protein molecules which make possible, via their catalytic activities, all the cellular reactions necessary for normal life. Among these enzymes are the seven or eight responsible for the synthesis of gibberellin (Figure 5–8). These latter enzymatic proteins, as well as all the others, are coded for by DNA as previously discussed. Dwarf corn, on the other hand, makes all the enzymes present in normal corn except for one. It lacks one of the enzymes responsible for gibberellin production. With no gibberellin, normal growth isn't possible, and we have the physical phenomenon dwarfism. Why does dwarf corn lack an enzyme in the pathway leading to

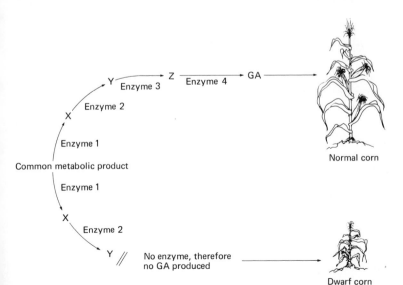

Normal corn

No enzyme, therefore no GA produced

Dwarf corn

FIGURE 5–8.

The dwarf condition in corn is due to a single gene mutation. This mutation results in the absence of an enzyme (here it is enzyme 3) in the pathway leading to gibberellin (GA) synthesis.

gibberellin production? The answer is that dwarf corn has suffered a mutation, and a mutation, as you know, is simply a slip of the tongue, so to speak, in DNA. One of the three-letter codons has gotten twisted around so that it perhaps says CAT instead of CAG or TAC instead of TAG. As a consequence of this "nonsense," a nonfunctional protein is made instead of the functional enzyme, no gibberellin is produced, and the trait dwarfism is apparent. Or put another way, the presence of all the correct enzymes gives rise to the trait of tallness. So it is, more or less, with all traits.

Mutations, Flexibility, and Evolution

The fourth and last requirement of genetic material is that it must have some mechanism for changing. That is, it must be flexible enough to allow for variation and, therefore, evolution. Examples of variation are dwarf corn and sickle-cell anemia, described above. Variation or, more specifically, new traits arise via mutations which are changes in the DNA code, and thus the fourth requirement is satisfied in principle. However, it does perhaps give one the false impression that variation and the appearance of new traits are a simple matter. To correct this situation and at the same time give you a better idea of the work behind the Green Revolution, let us consider again some of the experimentation that is involved in breeding more highly productive plant strains.

The Green Revolution—Again

Productivity is a function of time and space, both of which are limited. The goal of a plant-breeding program is really to produce more food in less time and space. What heritable characteristics of plants might help us toward that goal? Consider just the sizes of plants. Other things being equal, a smaller plant is better because more plants can grow in a given space. Also, smaller plants normally reach maturity earlier than larger plants do, and so more crops can be harvested per year. Shorter plants are less likely to fall over and ruin a crop, a most important consideration in wheat (see Chapter 3). But what are the other things that must remain equal? There are a fair number of factors, such as productivity of each individual plant and resistance to disease. These factors are often correlated with plant size. Our problem is to find new traits that increase productivity and at the same time to retain or improve on other desirable traits.

Where do the new traits come from? In general there are two sources. One is from other related strains or species. Let's say that we have a strain of wheat established in an experimental garden; it is not particularly productive but is resistant to some disease that is becoming a serious problem in agriculture. The plant breeder makes hybrids from plants of the resistant strain and those of the highly productive but susceptible strain. All possible combinations of traits will appear among the F_2 progeny from these hybrids. From the F_2 progeny it is possible to select those individuals which are both highly productive and also resistant to the disease and to use those plants to produce seed for distribution to the farmers in the areas where the disease is a problem. In actual practice, of course, the breeding program is more complex, involving F_3, F_4, backcross progenies, and so on. The principle in all cases, however, is to combine desirable traits derived from several different strains into a single true-breeding strain through a program of hybridization and selection.

Mutation is the second source for new traits. All genes of course had to come from somewhere, and that somewhere is mutation. The resistant strain in the experimental garden mentioned in the preceding paragraph is an example. Mutations normally occur at a low rate all the time, and most can be tolerated with scarcely any noticeable change in the plant's functioning. Others, however, are more critical, depending on the particular function that they affect. Should a desirable mutation arise, the plant breeder makes every effort to maintain that new gene in a population of individuals until it can be utilized in agriculture. However, when plant breeders are seeking a particular kind of mutation that does not yet exist, they can artificially increase the rate at which new mutations arise by using x rays and other so-called mutagenic (i.e., mutation-causing) agents. And thus, with lots of work, perseverance, and a fair dose of luck, the desired mutation may fall into the right hands and eventually find its way to your plate.

The above may not sound all that difficult, but as it turns out there are a number of difficulties. To begin with, most traits are inherited in a quantitative way. The examples from Mendel's work are really exceptions; he was incredibly lucky in his choice of traits. You generally find that pea plants and wheat plants (and also people) cannot be grouped into only two categories such as tall and short. Some are quite tall, others are very short; but there are also all degrees in between. All heritable traits that occur as a gradual series of intermediates rather than just as the two discrete extremes are examples of *quantitative inheritance*. Thus, growth rate is not simply fast or slow but is all intermediate rates as well; time of maturity and flowering is not simply early or late but also every day in between.

Resistance to diseases is also usually inherited in a quantitative way and so are most other traits that concern us here. The mechanism for quantitative inheritance is in principle the same as for Mendelian traits, differing only in that *several genes* affect a single trait. One gene contributes a portion, two genes a little more, and three more yet. What this means to the plant breeder is that he cannot simply take an afternoon and expect to pick out of a segregating F_2 progeny those individuals that show the desirable combination of traits, because those specific traits may only appear once in several thousand individuals, if they appear at all. If we are looking for a physiological rather than a morphological trait, such as resistance to disease or a higher rate of photosynthesis, then special assay techniques may be required to detect it. Thus, selecting desirable genotypes based on quantitatively inherited traits is a tedious and expensive proposition.

Second, mutation, one of the key tools of the plant breeder, is not easy to direct at a particular trait because mutations occur at random. That is, a breeder *cannot* bring about mutations in one particular gene without finding that others have mutated too. As a consequence, not only the single trait being sought but all others as well must be carefully selected and screened in order that undesirable side effects are not established, having come along for the ride as it were.

Third, most mutations are deleterious. They do not improve the strain; rather they cause it to degenerate. We are talking now on the order of 99% or so of all mutations being deleterious, only the rare one being of any value. The plant breeder does not simply induce a dozen or so mutations. discard those that are undesirable, and use the rest to improve his strain. Rather he induces hundreds or thousands of mutations in hope of finding one or two that may be of some value.

Finally, genes tend to be inherited in groups all tied together in packages which are only occasionally re-sorted into new groups or packages. Small wonder, for genes are part of chromosomes, and each chromosome tends to be inherited as a single unit. Thus, all genes carried on a single chromosome, probably on the order of 500 or more for most standard-sized chromosomes, tend to reappear generation after generation as a recognizable group. This phenomenon is called linkage, and the number of linkage groups corresponds to the haploid number of chromosomes.

Reviewing these points we can say:
1. Most traits are determined by many genes acting together in an additive way. That is, most traits are inherited quantitatively.
2. Mutation is rare and random.
3. Most mutations are deleterious.
4. Genes number in the thousands. Groups of genes can be linked and thus inherited as a unit.

These four points together mean that undesirable genes are often linked to desirable ones, and the task of selecting beneficial genotypes of quantitatively inherited traits is monumental indeed.

The Essence of Modern Biology

As we have seen, a plant (or animal for that matter) normally arises from a fertilized egg by cell divisions. First, there is one cell but later there may be millions. It has been emphasized that during this process all the cells of the organism receive the same genetic material and therefore contain the same information. But stop and think about this for a moment. If each cell in a particular plant contains the same blueprint, wouldn't one expect that all the cells in the plant would be constructed in an identical manner? It's logical, but at the same time it obviously doesn't fit well with what we can see with our own eyes. Visual facts tell us that somehow, using the same plans, a plant can produce a great variety of cells and organs. A plant may have leaves but also flowers, roots, and stems. And to complicate matters further, in addition to just producing different "looking" cells, we also know a plant produces cells that have different biochemical capacities. This paradox involving *identical information* but *nonidentical cells* is a central problem in biology today. We have some reasonable ideas on how all this works, but as you will see, we are still rather far from total comprehension.

Let's begin to probe this paradox with an example which can perhaps bring this problem down to its most basic conceptual level. Suppose you have an especially nice plant growing in your garden, and you wish to propagate it. You want ten plants instead of just one, but you don't want to be bothered with collecting seeds. You could, with a little luck, simply remove 10 or 15 young shoots, place them in moist sand, and wait. Before long, roots appear. What does this mean? It means you now have additional plants, not just cuttings, and that stem cells, under proper conditions, can change their function and form. They don't normally develop roots, *but* they do have the potential. Therefore, it can be said that the information on how to build roots was present in at least some of the stem cells. This information is normally masked or turned off, and some switch is necessary to turn it on. This switch can apparently be thrown by simply cutting the stem and keeping it in moist soil, although the rooting process can generally be accelerated by chemical treatment (Figure 5–9). There are many other examples of regeneration phenomena in the plant world.

But does this necessarily mean that every cell has a full information complement? You might argue, for example, that it

FIGURE 5–9.

The rooting of camellia cuttings can be promoted by treatment with *auxins*. In the photograph, the plant on the left has not been treated; the plant on the right has been treated. Auxins are plant-growth hormones; their activities are discussed in detail in Chapter 6.
From *Control Mechanisms in Plant Development*, by A. W. Galston and P. J. Davies, Fig. 3.9, p. 65. © 1970 by Prentice-Hall, Inc., Englewood Cliffs, N.J. Courtesy of the Boyce Thompson Institute for Plant Research, Yonkers, New York.

still is impossible to say for sure that a single cell in the stem has the ability to build a root. It does seem likely, but after all, you started with a mature plant organ (a stem) which itself is a population of many different types of cells. It could be that several of the cells of the stem contributed bits of the overall plan for building a root. In other words, we haven't strictly proved as yet that *one* cell contains the necessary set of genetic instructions. What we really need, in order to prove our point and make our argument regarding the information content of single cells really strong, are data showing that a *single isolated plant cell* contains a complete set of plans (DNA) for building an entire plant. Such data are available, but some background information is useful to a discussion of the experimental details.

Tissue Culture and Callus Formation

In the 1940s and 1950s many research groups became interested in the culture of isolated plant parts. The strategy is simple enough. One simply surgically removes, under sterile conditions, a bit of stem tissue, root tissue, or perhaps a young plant embryo and tries to keep it alive by providing the nutrients believed to be essential. When all the necessary nutrients are supplied, one hopes that the isolated tissue will begin to grow. As with much of science, the theory is relatively simple, but the technical details are difficult. When this work was initiated, scientists were unaware of all the factors necessary to make a truly complete nutrient solution. It was possible to provide sugars, various mineral salts, and vitamins, and the isolated plant part would remain alive for a time. The cells would not, however, divide mitotically, and thus there was no continued growth. This inability to initiate cell division in isolated parts stimulated researchers to begin a search for a cell-division factor, a chemical that would "wake up" the isolated cells and cause them to divide and grow.

A few clues indicated the presence of a cell-division factor in plants, but basically the search was a trial-and-error process. The hope, of course, was that someone somewhere would have the desired ingredient and eventually discover its usefulness. Often such searches have an unhappy ending, but in this case they were successful. Carlos Miller found the cell-division factor in a bottle containing an aged sample of herring sperm DNA. A little of this mixed with mineral salts and nutrients caused isolated pieces of plant stems to divide and grow. The final result was usually a mass of thin-walled, unspecialized, and practically identical cells. Such a lump of tissue is called a *callus*.

It was found eventually that the sample of herring sperm DNA contained, among other things, a compound subsequently named *kinetin* (Figure 5–10) and that kinetin was the *cell-division factor*. Kinetin itself has never been found as a natural constituent of plant cells. However, a structurally similar class of compounds capable of inducing cell division was rapidly isolated from the plant tissue. These compounds are called

1. A small bit of plant tissue is removed under sterile conditions.

2. The tissue is placed on a nutrient medium containing sugars, mineral salts, vitamins, auxin and kinetin. Agar is usually added to the medium to form a solid support.

3. The cells take up the nutrients from the medium, and an unorganized cell mass develops. From the resulting tissue mass, plant regeneration is sometimes possible.

The cell-division factor

Kinetin

The final result

FIGURE 5–10.

When establishing a callus culture, it is necessary to maintain sterile conditions as the callus is growing. This procedure means that the medium must be previously sterilized and that the plant parts used must also be sterile. Otherwise, the medium will quickly become covered by either bacteria or fungi, and these microorganisms will deplete the nutrients in the medium and thus eventually cause the plant tissue to die.
Photograph by G. Corduan, Ruhr University, Bochum, Germany.

cytokinins; they are discussed more fully in Chapter 6. The major point to remember here is that, after the discovery of kinetin and cytokinin, it was possible to grow isolated plant cells under culture conditions, and thus the relatively new field of biology known as plant tissue culture developed.

It is possible to produce plant tissue cultures from almost all plant species and all actively growing plant parts, although most commonly the fleshy parts of the plant are used. Many laboratories are currently utilizing tissue culture techniques. The techniques seem to have good potential not only for the study of cellular processes (Figure 5–11), but also as a commercially important tool. It is interesting to note that callus growth is not just a laboratory curiosity and that callus like growths or tumors sometimes develop on otherwise normal plants (Figure 5–12). Such growths usually arise as a result of fungal, bacterial, or viral infections which, in some as yet unknown way, trigger uncontrolled and unregulated cell divisions. Often these plant tumors are malignant and spread, resembling in certain aspects cancer in animals. If we could fully understand plant callus growth, it might eventually tell us something more about the nature of cancer in animals.

The Culture of Single Plant Cells

Our objective in the preceding discussion was to introduce you to plant tissue culture and to provide sufficient background so that we could finally attack directly the problem of the information content of a single cell. Does a single cell have all the information necessary to construct an entire plant? The answer is yes. It is possible to isolate single plant cells, and such cells

FIGURE 5–11.

Tissue culture techniques are now practiced on a grand scale. a.,b. Special nipple-shaped flasks have proved very useful for the liquid culture of single cells and/or cell clumps. c. Other containers of various shapes are useful for the culture of a particular plant species or for a special experimental technique. Photographs by M. H. Zenk, Ruhr University, Bochum, Germany.

a.

b.

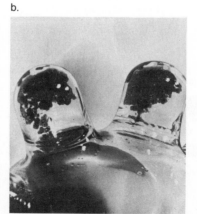

c.

under controlled conditions can produce an entire plant. However, a very high degree of technical skill and knowledge is required. Part of the problem is that plant cells tend to stick together and form clumps, even when a callus is grown in liquid media. Nevertheless, with agitation it's occasionally possible to jiggle a single cell loose. Under carefully arranged conditions, such a single cell will begin to divide and eventually will form a small clump of cells that resembles a normal plant embryo as it might look inside a seed. The embryo continues to develop, and finally a complete plant is formed (Figure 5–13). Therefore, we can now say for certain that at least some cells contain the information necessary to construct an entire plant, and it seems most likely that all cells contain a full complement of information. We hedge a little bit here because not all cell types are equally suitable for these experiments. Some cells within a plant are extremely specialized for a specific function (see Chapter 7), and it would appear that they are incapable of cell division and growth. This doesn't necessarily mean they do not contain the proper information; rather it would seem more likely that they are simply unable to express that information due to their previous history of development.

FIGURE 5–12. THE DISEASE CYCLE OF CROWN GALL

Crown gall is more or less typical of plant tumors; it is thought to be caused by a specific bacterium, *Agrobacterium tumefaciens*. Crown gall is a worldwide problem, and more than 60 plant families are susceptible to it.
From *Plant Pathology*, by George N. Agrios, 1969, Academic Press, Inc., Fig. 61, p. 346.

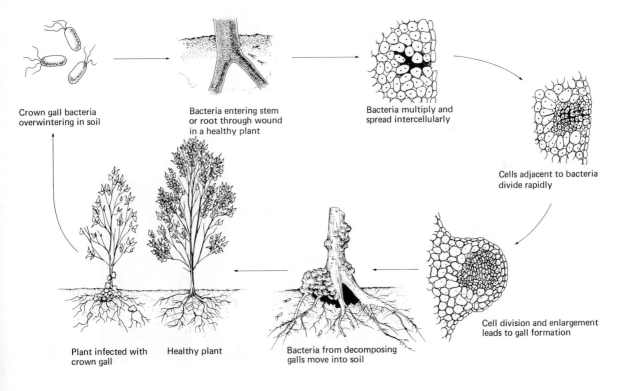

Crown gall bacteria overwintering in soil

Bacteria entering stem or root through wound in a healthy plant

Bacteria multiply and spread intercellularly

Cells adjacent to bacteria divide rapidly

Cell division and enlargement leads to gall formation

Plant infected with crown gall

Healthy plant

Bacteria from decomposing galls move into soil

At this point you might ask why single cells, or even small clumps of 2–10 cells, sometimes develop and behave like embryos in culture while a callus culture does not. The answer is not really completely clear, but for some reason it would appear that when cells are growing in large groups, they interact with their neighbors and the environment in a manner different from that of single, isolated cells and that this interaction somehow prevents them from expressing their intrinsic potential. This point can be made clear when you realize that callus cultures do not *always* behave as just clumps of cells but can be manipulated. For example, it is possible simply to change the ratio of two specific components of the medium, and then the cultures behave quite differently (Figure 5–14). We could say that although all cells have the same information content, they are not always able to express or use this information. When cells are in groups, as they are in a callus or in an organism, their environment (i.e., oxygen tension, hormone level, and so on) is determined by their geometric position in the mass. The microenvironment somehow prevents some instruction from being translated, even though it facilitates the translation of other instructions.

Let's go through this again, using our original example of roots developing on shoot cuttings. The cells present in a plant stem have the information potential to produce roots. They don't usually make use of this information, since only rarely do you see roots growing from the upper part of an intact seedling shoot. When a cutting is made, the environment of the shoot is certainly changed. In some way, this operation causes an environmental change which physically alters the internal workings

FIGURE 5–13.

a. The culture of single cells, showing cell division. Note the large nucleus and various cytoplasmic strands. First there is one nucleus, later there are two. b. Various embryo-like structures begin to form; first there is one nucleus, then there are many. c. Eventually these embryos develop into normal plants.
(a) From "Growth and division of single cells of higher plants in vitro," by L. Bergmann, in *Journal of General Physiology* 43:843, Fig. 12 (March 1960). By permission of The Rockefeller University Press. (b) Photographs by Walter Halperin, University of Washington, Seattle.

a. First one nucleus, then two (cell from bean)

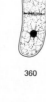

0   30   120   360

Time in minutes

b. Embryo development in cultured cell (wild carrot)

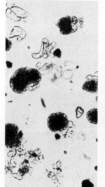

c. And finally . . . . . .

of some of the shoot cells, and they begin dividing and producing roots. Exactly what changes are involved we don't know. Furthermore, we do not fully understand how all this works on the level of DNA, RNA, and protein, but it is undoubtedly related in part to the phenomenon of masked genes and regulation of DNA translation.

Masked Genes and Integrated Information Flow

We know that physical traits are produced via the activities of various protein molecules and that the presence of a given protein is related to the information content of the DNA. Previously we were concerned with phenotypes that are expressed on an individual organism (short or tall, red or white, and so on), but it should be obvious that traits are actually expressed at the cellular level. That is, a cell looks as it does and functions in a specific manner because it is receiving instructions to do so from the DNA, and this information is put into use via the synthesis of specific proteins.

To this knowledge we can now add another idea or two. Within any cell there is a vast amount of information in the DNA; however, only part of this information gets translated into mRNA. The rest of the DNA (genes) is (are) somehow masked so that transcription or translation isn't possible. For example, in leaf cells the genes for the production of the enzymes for the Calvin-Benson cycle are obviously *not* masked because the leaves photosynthesize. Now consider the situation in root cells. The genes are present but masked, and thus we can't find the enzymes for the Calvin-Benson cycle there. This makes sense because roots can't carry on photosynthesis (no light), and thus it would be a waste of energy to make such enzymes. But, fortunately or unfortunately, plants can't think so they don't have to go through this kind of reasoning; nevertheless, the fact remains that roots don't make certain enzymes. Why?

FIGURE 5–14. A DEVELOPMENTAL SPECTRUM

One can alter the amount of two plant hormones, *auxin* and *kinetin,* in a tissue culture medium and, with luck, obtain the following results: (a) no growth, (b) callus with roots, (c) callus only, and (d) callus with shoots. These data indicate that the starting material has a great range of developmental possibilities, but that the expression of this intrinsic potential can be regulated by the external environment.

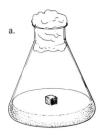

a.

No auxin
No kinetin

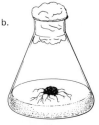

b.

2mg/L auxin
0.02mg/L kinetin

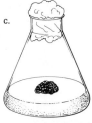

c.

3mg/L auxin
0.2mg/L kinetin

d.

0.03mg/L auxin
1mg/L kinetin

Presumably because the microenvironment of a leaf cell is different from that of a root cell, and this difference triggers some biochemical process which unmasks the genes for enzymes in the Calvin-Benson cycle in the leaf and/or causes them to become masked in the root. Unfortunately, we don't understand exactly how this masking and unmasking process works in higher organisms; perhaps portions of the DNA are covered or uncovered by a protein coat. The mechanism involved isn't too important for us here; rather we want simply to emphasize we are certain that the phenomenon of masking occurs. By using such information, we can now understand in at least a general way how such a process can lead to the production of many cell types and organs. Leaves are leaves because the information on leaf production is not masked, although the instructions for flower formation, for example, are masked. Similar reasoning applies to other parts of the plant. There are microgradients or microenvironmental differences which occur between the cells as an organism develops, and these differences directly or indirectly control the masking or unmasking of various genes in an ordered pattern. The result is a functional organism. At the biochemical level the precise details are still unknown, but the pursuit of this knowledge is the essence of modern biology.

And Lastly, Some Examples— Tissue Culture for Fun and Perhaps Profit

Besides being simply a laboratory curiosity that proved an important point (i.e., the information content of a single cell), is there any future in tissue culture? The answer appears to be yes. Tissue cultures are being eyed by businessmen as well as by some scientists for their commercial and scientific value. The reasons are obvious. First, we have already seen that the uncontrolled growth habit of plant tumors or callus cultures resembles cancerous growths in animals. Therefore, people are interested, for practical as well as theoretical reasons, in knowing about what regulates callus growth. Second, the mass producing of certain seedlings or plant products via the culture of single cells could be extremely rewarding in the commercial sense. Third, one can isolate mutants from tissue stocks more easily than from a population of field-grown plants, and some of these mutants might prove to be a useful tool in the production of rare and expensive compounds.

Let's look in a little more detail at the second and third points because they provide good examples of how pure science eventually contributes to applied science.

One Rare Orchid but the Potential for Thousands

Fortunes have been made in orchids, and lives have been lost searching for new and exciting varieties. Societies of local, national, and international scope exist for the study of the propagation of orchids. In the early days of orchid exploration, competing firms employed cunning, intrigue, espionage and even murder in their quest to obtain the first, only, or best specimens of a particular kind of orchid. Exploration still goes on today, but laws prevent orchids from being exported from many countries on the scale that was once practiced, and so much of the zeal for finding new types has dissipated. Nevertheless, orchidologists, ranging from the local neighborhood hobbyist to the scientist and commercial grower, are still among the most ardent devotees.

There are good reasons to be enthusiastic about orchids. They are as interesting as they are beautiful. They comprise the largest family of flowering plants, with over 20,000 species which amounts to almost one-tenth of all known kinds of flowering plants in all 300 families.[1] Orchids have been the subject of scientific writings beginning with the Greek father of botany, Theophrastus (fourth and third centuries B.C.) and the Roman physician and botanist, Dioscorides (first century A.D.). Charles Darwin's book *On the Various Contrivances by Which Orchids Are Pollinated by Insects* is available in many libraries.

Indeed, the pollination mechanisms of orchids are truly extraordinary, and it is just these improbable pollinating gimmicks that are partly responsible for the story told here. One example will illustrate how devious and contrived orchid pollination can be. The example, a favorite among biologists, concerns pseudocopulation which is known to occur in several species of orchids in different parts of the world. The orchid flowers involved in pseudocopulation bear striking resemblance to the *females* of the wasp species which pollinate them. Coloration, size and shape, little fuzzy hairs, and odor are all almost an exact copy of the female wasp. That ought to explain sufficiently, but the life history of the wasp contributes to the whole picture. It seems that the orchid-pollinating wasps come into sexual maturity in two phases: first the males and then a month or so later the females—hence the problem. All these male wasps are flying around with no receptive females in sight, but those attractive orchid flowers are available. Thus the pollen-covered male wasp pollinates the female orchid flowers (Figure 5–15).

FIGURE 5–15.

One example of pseudocopulation is seen here as the orchid *(Caledenia lobata)* is visited by its wasp pollinator. In pseudocopulation, each orchid species attracts only males of a specific wasp.
Photograph by Warren Stoutamire, University of Akron, Ohio.

[1]The numbers of species of orchids and other flowering plants and the number of plant families vary, depending on the authority. The figures given here are rough estimates.

Pseudocopulation illustrates an important point. Most species of orchids are pollinated by a single species of insect, and its flowers are highly specialized for accomplishing that end. It is, in fact, that extreme specialization of flower structure which has given the many variations seen in orchids. This has far-reaching consequences. To begin with, it means that hybrids between species of orchids almost never occur in nature. In cases in which hybrids do form, it has generally been found that such specimens are not so well adapted to their environments as the parent species. For these reasons hybrids do not generally occur in the wild. However, orchidologists can by-pass these problems by artificially cross-pollinating two kinds of orchids. As it turns out, hybrids are easily formed in this way, and what's more, many of them grow vigorously and flower profusely under the sheltered conditions found in greenhouses.

The most highly prized of all orchids are generally hybrids and hybrid derivatives, that is F_2 and backcross progenies. As a consequence of recombination, these lovely orchids which appear in hybrid progenies are all different (see Chapter 4). An additional complicating factor is that it takes 7 years for most orchids to flower when they are grown from seed. Thus, at least 14 years are required from the initial hybrid cross until the first F_2 flowers appear. The seeds, incidentally, are extremely small, numbering from several thousand to more than a million *per flower*, and lack any stored nutrients of their own.

In short, special techniques and many years of work are required to produce the most desired orchids (which are few in number), and normal vegetative propagation is slow. It is clear why prize orchid plants can sell for several hundred dollars or more. Furthermore, the prize orchids are themselves hybrids and will not breed true. One can imagine why tissue culture could be of some use. Once a new hybrid is established, more of this hybrid orchid can be propagated rapidly via tissue culture and subsequent embryo culture.

Tissue culture in orchids varies in some details, depending on the particular species being propagated. However, the general procedure is the same in all cases. First, a small piece of shoot meristem (see Chapter 6) is excised from a plant and sterilized. The piece of meristem is then placed in a culture medium containing all the minerals, nutrients, and hormones necessary for growth. Under these conditions the tissue grows rapidly and is divided and subcultured as it grows. In cymbidiums, for example, the tissue is cultured in a liquid medium which is continually agitated. Under these conditions the plants grow as undifferentiated tissue clumps which can be divided into smaller pieces and placed back into culture. When this division is repeated several times at one- or two-month intervals, and each time 10–20 new cultures are started for each original one, you can see that very quickly all the laboratory and greenhouse space will be occupied.

The next step is to allow the various cultured tissues to develop normal roots and shoots. In cymbidiums this is accomplished by simply removing them from the liquid culture medium and placing them on a firm stationary substance. Within a few weeks roots, stems, and leaves appear. The young plants can eventually be transplanted to individual containers, and one is assured of a plant virtually identical to the original parent from which the apex culture was begun. Other plant species have also been grown with good results from tissue culture.

Oil Palms from Tissue Culture

Unilever, a British-based company, has made use of tissue culture techniques to produce high-yielding oil palms. The program started, at least partially, after it was noted that every once in a while one particular tree (perhaps one in ten thousand) on Unilever's palm plantation was an extremely fine specimen. It was a good oil producer with the correct balance of saturated and unsaturated molecules, it grew rapidly, and it was resistant to disease. Obviously business would really boom if one had an entire plantation devoted to such trees, but there are a few problems involved. First, it takes a long time for an oil palm to reach a reproductive stage (perhaps 15–20 years), and thus a breeding program for such trees is difficult. Second, and of even more importance, is the fact that oil palms are self-sterile. That is, it is not possible to pollinate the superior tree with its own pollen. To produce viable seedlings, the superior tree must be pollinated with pollen from another tree, and that other tree is by definition inferior. The result of such a cross is offspring with only a few of the good traits that the breeder was after. Somehow, the reasoning goes, one must skip the sexual process and generate offspring with the genotype of only the superior tree. In other words, asexual reproduction is desirable. Unfortunately, the usual methods of asexual reproduction don't work with oil palms. In this case, however, tissue culture and the subsequent production of young plantlets were successful. Unilever scientists can make tissue cultures from superior trees and then arrange the conditions so that embryos and, eventually, normal plant seedlings are formed. All of these seedlings (excluding mutations) are exactly like the parent. Similar work involving other crop plants is also taking place in universities and research stations around the world (Figure 5–16).

a. Cotton flower on the morning of anthesis. Ovary wall, bract, and petal are removed to show ovules.

b. Fiber initials (epidermal cells) on the morning of anthesis. Their elongation phase is just beginning. Magnification 200×.

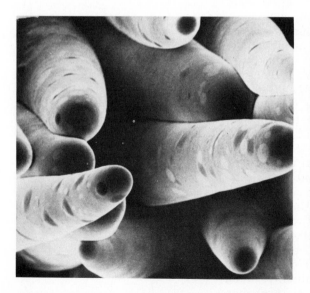

c. Elongating fiber cells on the second day post-anthesis, showing images of organelles through the thin primary wall. Magnification 1600×.

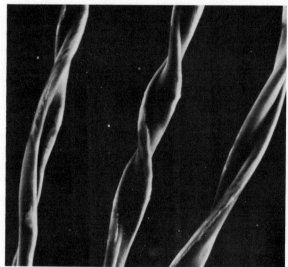

d. The nonliving, mature, marketable cotton fiber, 60 days postanthesis. Magnification 575×.

e. Fertilized cotton ovules placed into culture on the second day postanthesis and grown in vitro for two weeks. Although not necessary for fiber elongation, gibberellic acid at 5 μM (micro molar) was included to promote growth.

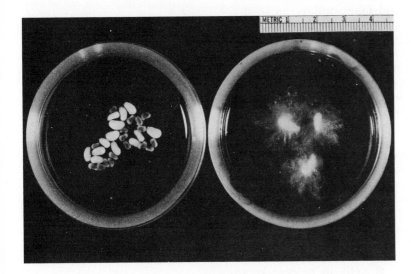

f. Unfertilized ovules placed into basal medium (left) and basal medium plus 5 μM indoleacetic acid and 0.5 μM gibberellic acid (right) on the morning of anthesis. Photograph was taken after 14 days of culture in vitro.

FIGURE 5–16. TEST-TUBE COTTON

Dr. C. A. Beasley and co-workers have taken a large step toward the development of a system for growing test-tube cotton. a.–d. Cotton fibers are hair-like filaments, each consisting of a single cell, that elongate from the outermost tissue layer (the epidermis) of a cotton seed after the ovary is pollinated. e. Dr. Beasley has been able to isolate fertilized cotton ovules and place them in culture. Under such conditions, normal cotton fiber production is possible. f. It is also possible to initiate fiber development in the absence of fertilization when the correct balance of hormones is established in the medium. (Compare the left and right sides of part f in the figure.) Using such "in vitro" techniques, it may be possible to produce more fibers per seed and improve the overall quality of the cotton fibers.

Photographs by C. A. Beasley, University of California, Riverside.

Natural Products via Tissue Culture

Many commercially valuable compounds, such as perfume oils and drugs for the treatment of human disorders, can be obtained only from plants. The high cost of some of these products reflects the fact that the plants which produce them are exotic and/or that the yield of the desired product is very low. You can perhaps see the potential for growing such plants in tissue culture and for the recovery of their commercially important products. This discussion of perfume oils and drugs may seem rather straightforward after the two previous success stories with orchids and oil palms; however, it isn't and a rather different strategy is involved.

Perfume oils are normally obtained by laboriously harvesting certain flowers and plant parts by hand. The total yield is low, and the man-hours invested are high. To complicate matters many good oil-producing species are especially fussy about climate. For example, perfume manufacturers in North America may have no choice but to import, again at an added cost, the already extracted oil. It would obviously be an advantage if a company could get plants producing perfume oil into callus culture *and* if such cultures could then be induced to produce the costly oils. The flaw in the scheme is that callus cultures don't produce the desired oil. In the case of perfume oils, only special parts of the plant or special plant organs make the desired product. These particular structures cannot be cultured, although the non-oil-rich tissues of the plant can be. It's frustrating but perhaps not hopeless. We know that all cells have the same information content; therefore the cells that can be cultured *do* have the potential to produce the oils, but unfortunately they don't. The strategy then is to select an environment, in this case a particular chemical medium, which initiates the unmasking of this information. With respect to perfume oils, people are still searching.

The problem encountered with drug production via tissue cultures is similar, although some modest progress has already been made. For example, there are several species of *Datura* (jimson weed) which produce various useful drugs such as scopolamine. (Scopolamine is used as a central nervous system depressant, an anti-Parkinson's disease agent, and as a motion-sickness preventive.) *Datura* can be cultured, and the cultures will produce scopolamine. But still the technique may not be practical. Raising large quantities of cells in culture is an expensive proposition, and at present the tissue culture yields of this drug do not offset the costs. It must be remembered, however, that the technique is relatively new at the industrial level, and therefore costs will probably come down. Second, one must also keep in mind that production cost is only relative to the cost of the final product. Scopolamine is relatively inexpensive, but other drugs cost much more. Thus, it still may be possible that in the future drug synthesis via tissue culture chemistry will be an unqualified success.

Selected Readings

Kendrew, J. C. 1966. *The Thread of Life*. Cambridge, Mass.: Harvard University Press. This brief account of molecular biology was the outgrowth of TV lectures.

Steward, F. C. 1970. Totipotency, variation, and clonal development of cultured cells. *Endeavour* 29:117–124. Views of a pioneer in the area of plant tissue culture.

Taylor, J. H. 1965. *Selected Papers on Molecular Genetics*. New York: Academic Press. Collection of original research papers reporting unusually important findings.

Watson, J. D. 1969. *The Double Helix*. New York: New American Library. A highly recommended behind-the-scenes account of the work of Watson and Crick. Reads like a novel.

Watson, J. D. 1970. *Molecular Biology of the Gene*. 2nd ed. New York: W. A. Benjamin. Strongly recommended for more information on genes, chromosomes, and protein synthesis.

6   *Plant Growth—Seed to Seedling*

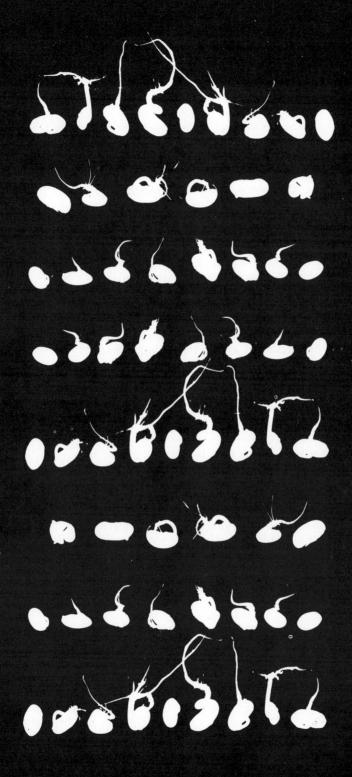

Suspended animation is a term usually associated with science fiction books or perhaps talk of the future. However, plants have made use of a similar principle for millions of years. Viewing a mature bean seed, for example, one may naively think it is simply a rather uninteresting, dry, shriveled-up piece of plant material. Perhaps you have even stored seeds on the garage shelf for years and never given them a second thought or noticed any change in their outward appearance. Yet, there must indeed be something special about seeds since they can readily give rise to a living organism when placed in a suitable environment, even after years of storage. Seeds are, in fact, in a state which is not unlike suspended animation, and the transition from the resting state to a dynamic and growing seedling is an appropriate point for us to begin our study of plant growth.

Seeds and Germination

Despite its outward appearance, a seed is a living unit composed of three basic parts: an embryo, a source of stored food, and an external covering called the seed coat. In Figure 6-1 two main seed types which differ in their method of food storage are illustrated. In the bean and certain other seeds, food reserves are contained within the expanded embryonic leaves (cotyledons) of the embryo itself. In wheat, however, the embryo is embedded in a separate food storage tissue called the *endo-*

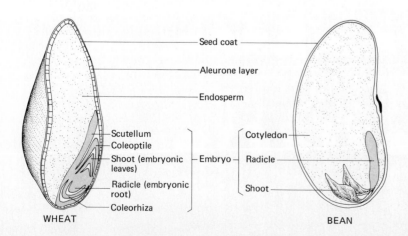

WHEAT

BEAN

FIGURE 6-1. ANATOMY OF THE SEED

The primary difference between the wheat seed and the bean seed is the manner in which the reserve food materials for future embryo growth are stored. In wheat the endosperm is the chief storage area, while in bean the reserve material is packaged largely in the cotyledons or embryonic leaves of the plant embryo itself. Adapted from *Plant Anatomy,* by A. Fahn, 1967, Pergamon Press Limited, p. 478. By permission of Pergamon Press Limited, Oxford, England.

Sugar maple	Less than one week
Willow	Less than one week
Wild oats	About one year
Alfalfa	6 years
Yellow Foxtail	10 years
Chrysanthemum	30 years
Clover	90 years
Loco weed	100–150 years
Mimosa (sensitive plant)	220 years
Indian lotus	1040 years

TABLE 6–1. THE LIFE SPAN OF SOME SEEDS
Note the extreme variability in how long seeds remain viable. There is no particular pattern (such as seed size, amount of stored food, etc.) which one can use to predict from outward appearances the longevity of any given species.

Source: Modified from F. B. Salisbury and C. Ross. 1969. *Plant Physiology,* Belmont, Calif.: Wadsworth Publishing Company, Table 25–1. By permission of Wadsworth Publishing Company, Inc. and Pergamon Press Ltd.

sperm. Thus seeds can vary in the exact way in which food is stored; however, in truly fundamental respects all seeds are remarkably similar.

As long as a seed remains in a relatively dehydrated state, the living embryo remains in a state similar to suspended animation. It is living but all the life processes are greatly slowed down (Figure 6–2). The length of time that the embryo can remain in such a state varies, depending on the type of seed. Table 6–1 lists the life span of just a few seeds in order to illustrate just how great this variation can be.

With most seed types, to activate the plant embryo is a simple task. We can simply saturate the seed with water, place it in a suitable environment, and wait a few days for a seedling to appear.

While the simple relationship of

Seed + water + suitable environment = seedling

holds for many seed types, there are exceptions. For example, some seeds such as the stone fruits (peach, cherry, etc.) require a long pretreatment (i.e., 4 weeks or more) at cold temperatures (0–10°C) before they will germinate. The reasons for this requirement are not entirely clear. However, in at least some cases reduced temperatures may be required for the final stages in embryo maturation. Other types of seeds are difficult to germinate because they are enclosed in an extremely hard, tough seed coat which may actually prevent growth or restrict the uptake of water. Such seeds will readily germinate, however, once the seed coat is broken or worn away. In natural situations this can occur in various ways; such as, simple mechanical abrasion, microbial action, or even partial burning of the seed coat.

The germination of some seeds can also be prevented by the internal or external presence of certain agents called germination *inhibitors*. To visualize one such situation, consider the seeds within a ripe tomato. Here we have seeds, water, and a favorable temperature, yet how many times have you seen seeds germinating within a salad tomato? The seeds are certainly mature and even seem ready to germinate; if they are removed from the fruit, the germination process begins straight-away. The currently favored explanation for the internal inhibition of germination is that a kind of chemical warfare takes place between the tomato fruit and the seeds. It is postulated that the fruit contains certain chemical compounds that actively prevent seed germination. Presumably, once the seeds are removed from the fruit and rinsed, the level of the inhibitors decreases to a point at which they are no longer effective.

It is interesting to speculate on the reasons why some of the mechanisms regulating germination have developed. Perhaps you can suggest some ideas. For example, can you think of some reasons why it would be advantageous for stone fruits to require a cold treatment before they are capable of germinating? What would be the advantage of seed coats so tough that mechanical abrasion is necessary before germination can occur?

a.

b.

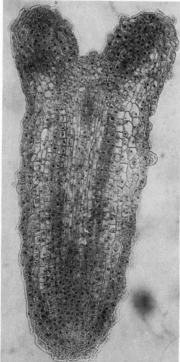

FIGURE 6–2.

Embryos of wild carrot. a. Seed embryo. b. Embryo derived from cultured leaf tissue. c. Cultured embryos at an early (globular) stage of development. The three embryos are connected by a common mass of tissue or "suspensor."
Photographs by Walter Halperin, University of Washington, Seattle.

c.

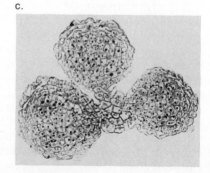

Early Events in Germination and the Hormone Concept

Once conditions necessary for seed germination are met, what are the events that lead up to the emergence of the young seedling? Clearly an early requirement for seedling growth would be an adequate source of food for the embryo. Remember that a germinating seed is normally under the soil so that there is no light available for photosynthesis, and therefore the plant embryo initially must rely on its own stored food reserves (see Figures 6–1, 6–2). Oftentimes these reserves are in the form of stored starch and as such are not directly useful to the plant. To become useful, the starch must be broken down into the small sugar units which form the starch molecules. Plants have adapted some rather sophisticated means of "communication" between the food reserves and the embryo proper which appear to accomplish this end. That is, systems have evolved which allow for some coordination between the embryo's needs and the breakdown of the storage material. In the barley seed, for example, the starch reserves are located in the endosperm. When water is added, the embryo is activated and begins producing messenger substances known as *gibberellins*. The messenger molecules diffuse from the embryo to a specialized tissue known as the aleurone layer (Figure 6–3). In response to the arriving gibberellin molecules, the aleurone layer secretes a number of digestive enzymes. One of these is an enzyme that can break down starch (α-amylase). This breakdown of the

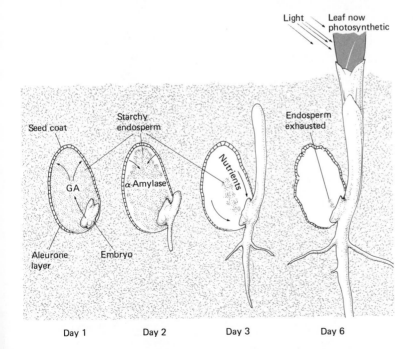

Day 1   Day 2   Day 3   Day 6

FIGURE 6–3. THE GERMINATION OF A BARLEY SEED AND THE ROLE OF GIBBERELLINS

On day 1, the seed is planted and takes up water thus activating the embryo. The embryo begins producing gibberellins (GA) which diffuse to the aleurone layer. On day 2, the aleurone layer begins secreting α-amylase which, in turn, starts to digest the starchy endosperm. By day 3, the embryo has grown considerably, utilizing the nutrients obtained from the digestion of the endosperm. By day 6, the endosperm has been completely digested; however, the plant is now no longer dependent on endosperm-derived nutrients because the first leaf has emerged from beneath the soil and is now photosynthetic.

Adapted from *The Living Plant,* 2d ed., by Peter Martin Ray. Copyright © 1963, 1972 by Holt, Rinehart and Winston, Inc. Reprinted by permission of Holt, Rinehart and Winston, Inc., Fig. 10.3, p. 181.

starch produces sugars which diffuse into the embryo and are essential for growth. Thus the embryo can send messenger molecules (gibberellins) to another area in the seed (aleurone layer) and induce the initial breakdown of food reserves (starch) which the embryo can then utilize for growth.

Gibberellins, as we will see later in this chapter, are not the only messenger molecules that a plant cell uses to control its own development. In fact, there is a general term for such molecules, *hormones*. A hormone is a substance, produced by cells in one area or tissue that can, at very low concentrations, cause an effect in another area or tissue. The tissue can be, and usually is, some distance from the original site of synthesis. Thus, in the barley system, gibberellin, a hormone, is produced in the embryo, and the target tissue is the aleurone layer.

Growth is More than Just Cell Divisions

Let us assume that we have met the necessary requirements for the germination of a particular seed and that somehow we have mobilized sufficient reserves to nourish the embryo plant. The

FIGURE 6–4.

a. The position of the apical meristems in the seed. b. The seedling. c. The general cellular organization of the meristems. The meristems are active areas of cell division and thus continually provide new cells which will eventually elongate and specialize. Adapted from *Botany, An Ecological Approach,* by William A. Jensen and Frank B. Salisbury. © 1972 by Wadsworth Publishing Company, Inc., Belmont, California 94002. Reprinted by permission of the publisher, Fig. 24–2, p. 417.

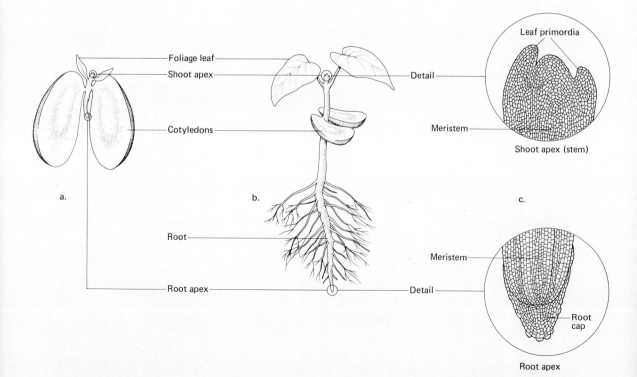

first event normally observed in a germinating seed is the emergence of the root. This is shortly followed by the expansion and appearance of the seedling shoot. Growth of both the root and the shoot then proceeds rapidly, with both organs making use of the same general scheme for achieving an increase in size. In the top portion of the shoot and in the tip area of the root, there are specialized areas called *meristems* (Figure 6–4). Within the meristem new cells are continually produced by cell divisions. One might think that this production of more cells could lead to growth of the shoot or root; but this isn't so. Dividing a cell into two daughter cells doesn't necessarily increase the total size of the plant. Growth normally occurs only when cell enlargement or elongation is coupled with the production of new cells. Therefore we can say that in most rapidly growing areas of the seedling, virtually the entire increase in length is achieved via cell elongation of cells originally produced via cell divisions in the meristems. It is most important to realize that the above isn't just a hodgepodge of divisions and cell enlargement; it is well organized. In plants the area of cell division, the meristem, is *physically* different from, although adjacent to, the area of maximal cell enlargement. Cell elongation occurs in a special *zone of elongation* located directly behind the meristem; it is here that a cell may increase many, many times in length. To put it another way, all the cells in a young seedling are not dividing and elongating—rather, only those cells located in specific regions of the plant axis, the meristems and zones of elongation, respectively.

This point can be illustrated in several ways; perhaps the most graphic is with the aid of a simple experiment (Figure 6–5). This experiment requires a young seedling, some India ink, a small paintbrush, and a ruler. Beginning at the top of the seedling (a bean plant will do nicely), marks are made with the ink at 2 mm intervals until the soil level is reached. The seedling is set aside for two or three days, and then the distance between each of the marks is measured. After this time interval, one should find that the uppermost intervals are still about 2 mm; however, the next few marks should have spread apart considerably. These results would mean, first, that most of the seedling's growth occurred in a localized region slightly below the tip or meristem. Since there are few cell divisions in this area, the increase must be due almost entirely to the elongation of pre-existing cells, Second, the cells in the lowermost portion of the shoot must have already reached their final length since there is no change in the marked intervals. In this latter region we have relatively mature cells which are becoming specialized to perform specific tasks. Furthermore, we should point out that while we used the plant shoot as an example in our marking experiment, we could do exactly the same experiment with the emerging plant root and obtain similar results.

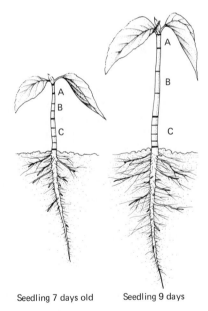

Seedling 7 days old   Seedling 9 days

FIGURE 6–5.

Experiment to illustrate zone A, cell division; zone B, cell elongation; and zone C, cell maturation or specialization. Cells produced in zone A provide a continuing supply for subsequent elongation and hence growth in zone B. Fully elongated cells in zone C specialize to perform specific functions.

In the root, however, there is an extra feature, the *root cap*. The root cap is a cup-shaped mass of cells that surrounds and protects the meristematic area of the root (see Figure 6–4). The functional significance of this area can readily be appreciated by imagining that the tip of the root is being forced downward through the soil by elongating cells behind the meristem. Without some protection from the soil, abrasion would quickly wear away the tender meristem. Instead, however, the cells of the root cap are being worn away or sloughed off. The cap remains functional because new cells are continually being added as a result of divisions in the meristem. Thus, the root meristem has a dual function: to produce new cells for subsequent root growth and to produce new cells for the maintenance of the root cap (Figure 6–6).

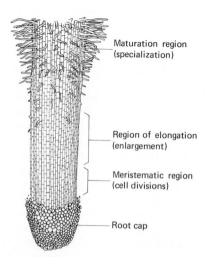

Maturation region
(specialization)

Region of elongation
(enlargement)

Meristematic region
(cell divisions)

Root cap

FIGURE 6–6.

Growth and development of the root. Note position of root cap, and compare with growth of the shoot in Figure 6–5.
Adapted from *Plant Form and Function: An Introduction to Plant Science*, by G. J. Tortora, D. R. Cicero, and H. I. Parish, Fig. 7.10, p. 187. © Copyright, the Macmillan Company, 1970.

Contrasting the Growth of Plants with Growth in Animals

The concept of plant growth proceeding in localized areas is one of the unique features of plants. Animals tend to have no such specialized zones of cell division and cell elongation; instead growth occurs more or less throughout the entire body, although in a regulated way. At this time it also seems appropriate to point out another rather fundamental difference in the growth patterns of plants and animals. Some plants have what we loosely call unlimited or indeterminate growth potential, while the growth of virtually all animals is strictly determinate or limited.

The determinate nature of growth in animals is a natural consequence of the cessation of growth and the processes of aging and death. Many plants, however, do not seem to have a developmental cycle which normally terminates in the cessation of all growth followed by aging and death. This, of course, implies that plants have the potential to live forever, and indeed some do seem to have this potential. However, remember it is just a potential and can probably never be realized since factors other than the functioning of the meristem will ultimately become limiting. For example, a plant can only achieve some finite height because there are physical limitations on the height to which water can be drawn upward from the soil (see Chapter 7). Yet despite the obvious theoretical limitations on plant height, certain trees can achieve great age and at least give the impression of immortality. Two examples immediately come to mind: the bristlecone pines and the giant sequoias. Many of these trees are several thousand years old. In fact, the longevity record is held by a bristlecone which was discovered in Nevada in 1965. Its estimated age is 4,900 years, and it is still going strong (Figure 6–7).

FIGURE 6–7.

The bristlecone pine can be found at high elevations in the White Mountains on the California-Nevada border.
Photograph by Robert L. Hays, San Diego State University, California.

Light and Plant Growth: More than Just Photosynthesis

There is at least one other significant difference in the growth and development of plants as compared to those of animals. This has to do with the importance of light in regulating the development of plants. We can best illustrate this point by using the bean sprout as an example, although the same general phenomenon holds for most species of plants. The bean sprouts that one purchases in the supermarket are young seedlings that have been grown in the *dark* or at least under very dim lighting. As a result their appearance is quite different from plants grown in the light. The first thing you might notice about dark-grown plants is their color—generally white or slightly yellow but certainly not green. The reason for this is that visible light is necessary for the final assembly of the chlorophyll molecule. A further comparison of light- and dark-grown bean plants would indicate that the shoot is much longer in the dark-grown plant than in a light-grown plant of the same age. Also, the leaves are smaller, and the apex or top portion of the dark-grown plant is hooked (Figure 6–8). Let's pursue the question of light and

a.

b.

FIGURE 6–8.

Bean seeds were germinated and allowed to grow for (a) 8 days in the light or (b) 8 days in the dark. Compare height, leaf size, and apex.

plant growth by considering a standard question biologists are fond of asking.

What wavelengths of light are responsible for "converting" dark-grown plants into light-grown plants, and what molecule absorbs the effective wavelengths? Perhaps the way to approach this problem is to use the same sort of reasoning and experimentation that served to implicate chlorophyll in the photosynthetic process. This procedure involves doing an action spectrum on dark-grown plants in order to determine the wavelengths of light that are effective in changing the appearance of the dark-grown plants (see Chapter 2). Next one would attempt to isolate and characterize a molecule from dark-grown plants that had an absorption spectrum matching the previously obtained action spectrum.

When these experiments are actually done, it is found that red light is primarily responsible for the morphological features (appearance) of light-grown plants. That is, dark-grown plants maintain their particular set of characteristics because they have not been exposed to red light. Furthermore, it has been shown that there is a specific pigment molecule, *phytochrome*, which absorbs the red light. The situation is considerably more complex than one might first suspect because it is also known that the effects of red light can be canceled by a subsequent exposure to far-red light (light of a wavelength slightly longer than red). This phenomenon initiated experiments which eventually led to the concept that phytochrome exists in two different forms, a red-light-absorbing form (P_R) and a far-red-absorbing form (P_{FR}). Furthermore, we now know the two forms are interconvertible. In dark-grown plants the red-absorbing form of phytochrome (P_R) is in abundance. However, when the plant is irradiated with red light, P_R is converted to P_{FR}. P_{FR} then presumably initiates a set of reactions which change the appearance of the plant. If a plant in the dark is given a flash of red immediately followed by far-red, the P_{FR} formed by red light is immediately converted back to P_R (our original form in dark-grown plants), and there is no possibility of initiating the change of events that convert dark-grown plants into light-grown plants (Figure 6–9). Unfortunately, at the present time very little is known about the reactions initiated by P_{FR}, but this is the subject of considerable ongoing research.

A Summary of Events and More

Before continuing, let us briefly summarize what we have learned about the growth of plants from the seed to the seedling stage. The seed contains a source of stored food and also a living embryo which exists in a state not unlike suspended animation while in the dehydrated condition. In most seeds (although there are exceptions) the simple addition of water activates the embryo and growth begins. These early events in seedling growth normally take place underground, and the young seedling must make use of the preformed food material within the seed itself. Some seeds have evolved a hormonal system to regulate the flow of nutrients into the growing embryo. The actual growth of the young seedling results from localized cell divisions in areas called meristems and from the subsequent elongation of the newly formed cells in the adjacent zone of elongation. We have also seen that there are certain special morphological features which a young seedling has as a result of growing in darkness under the soil. In the dark-grown seedling the shoot is hooked (presumably to protect the meristem from the abrasive action of the soil), and the cells are elongating more rapidly than they would under lighted conditions. When the shoot emerges from the soil, light initiates chlorophyll synthesis. Furthermore, sunlight (which contains red wavelengths) causes the P_R form of phytochrome to be converted to the P_{FR} form. This pigment transition somehow initiates a chain of events which cause the hook to straighten out and the leaves to expand, and reduces the rate at which the cells elongate.

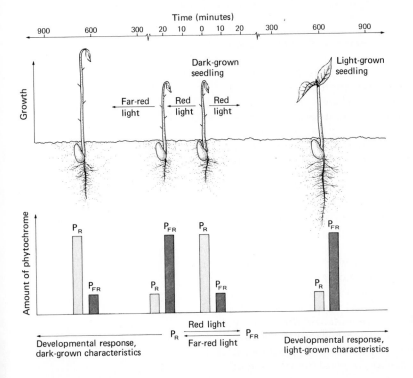

FIGURE 6-9.

Phytochrome interconversions and the developmental characteristics of a pea seedling. See text for details of phytochrome physiology.

Clearly the early (underground) events in the growth of a seedling are under both internal (hormonal) regulation and external (presence or absence of light) control.

As a seedling develops further, there are still other control mechanisms which a plant utilizes to regulate its growth and its orientation with respect to the environment. Perhaps one of the most obvious responses is the tendency of plants to orient themselves with respect to a light source. Certainly you have noticed this if you have a house plant placed near a window. Such plants will curve toward the window (light source) and must be rotated every few days to maintain reasonably straight stem growth. This phenomenon of bending in response to light is called *phototropism*. Phototropism can also be observed in the field, and consideration of the behavior of a sunflower plant adequately illustrates this point. In the morning the apex of a young sunflower stem and some leaves can be seen bent toward the east. However, as the sun moves across the sky, the apex and leaves appear to follow, so that in the late afternoon these structures are bending to the west (Figure 6–10). One can rather easily rationalize the development of such a mechanism since it would tend to present the maximum photosynthetic surface (absorbing area) to the rays of the sun. But what are the cellular mechanisms that account for phototropic behavior?

Yet Another Hormone: Auxin

The contortions referred to previously as phototropic behavior as well as certain other growth responses in plants are controlled by a hormone known as *auxin*. Auxin is produced in meristematic areas and in the young leaves of a plant and is actively transported to the elongating regions in a plant where it acts to regulate the rate of cell extension. Furthermore, shoot cells respond to auxin in direct proportion to the amount of auxin present; that is, within limits more auxin means greater elongation. It is important to realize that auxin regulates the rate and extent of cell elongation not only in seedlings which have already reached the soil surface but also in the underground growth of a seedling. (We mention auxin at this point rather than at an earlier stage simply for convenience.)

The discovery of auxin resulted from the work of many scientists who were interested in the phototropic behavior of seedlings. The earliest experiments were performed many years ago by Charles Darwin. Darwin discovered that the curvature response produced by light could be prevented if the tip of a young seedling was covered with a small cap. From these data as well as from other observations, it was correctly postulated that only the plant tip could perceive an unequal distribution of light (as from unequal illumination) and that a stimulus was somehow transferred from the tip to the growing zone, causing curvature to take place. A considerable amount of time elapsed before the nature of Darwin's "stimulus" was established. In

a.

b.

c.

FIGURE 6–10.

The sunflower's stem and leaves are oriented in different positions through-out the day, and this orientation is related to the position of the sun.
(a) Morning, (b) noon (sun is directly overhead), and (c) afternoon.
Photographs by Hiroh Shibaoka, University of Tokyo, Japan.

about 1927 Fritz Went discovered that one could place seedling tips on small blocks of agar, a gelatinlike substance, and recover a substance in the blocks that could cause bending when placed asymmetrically on a decapitated seedling (Figure 6–11). The substance that Went recovered in his agar blocks was called auxin and later was chemically identified as indole-3-acetic acid (IAA; structure shown in Figure 6–11). When different amounts of IAA are incorporated into agar blocks and the blocks are then tested for curvature-producing ability, it can be shown that extremely minute amounts are effective and that the amount of curvature is dependent on the concentration of auxin in the block.

With this information it becomes obvious that an unequal distribution of auxin in an intact plant shoot could lead to curvature, and thus in order to explain the phototropic response, we have only to reconcile the role played by light. Unfortunately the details are not yet known, but it seems likely that there are light receptor molecules (pigments) located in the shoot tip. As a

FIGURE 6–11.

Went's original experiments that eventually led to the discovery of auxin were performed in essentially the manner illustrated here, although he used a different seedling type (the oat) and a slightly more sophisticated experimental technique. The amount of bending or angle of curvature depends on the concentration of auxin (the substance) in the agar block and thus, within limits, is proportional to auxin concentration.

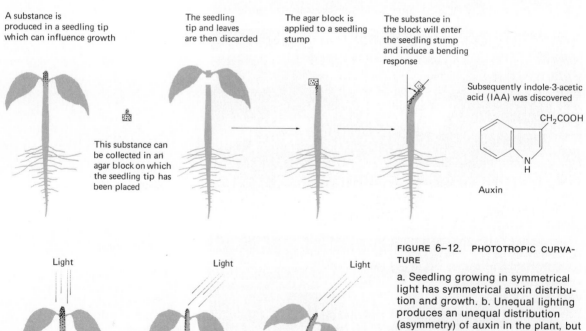

A substance is produced in a seedling tip which can influence growth

This substance can be collected in an agar block on which the seedling tip has been placed

The seedling tip and leaves are then discarded

The agar block is applied to a seedling stump

The substance in the block will enter the seedling stump and induce a bending response

Subsequently indole-3-acetic acid (IAA) was discovered

CH_2COOH

Auxin

FIGURE 6–12.   PHOTOTROPIC CURVATURE

a. Seedling growing in symmetrical light has symmetrical auxin distribution and growth. b. Unequal lighting produces an unequal distribution (asymmetry) of auxin in the plant, but the exact mechanism is unknown. c. Asymmetry in auxin distribution leads to asymmetry in growth, which results in curvature.
Adapted from Arthur W. Galston, *The Green Plant,* © 1968. By permission of Prentice-Hall, Inc., Englewood Cliffs, N.J., Fig. 4.13, p. 66.

Light

Light

Light

a.

b.

c.

result of unequal lighting, the pigment molecules on the side receiving the most light initiate a chain of events within the tip that eventually leads to a preferential transport of auxin to the darker side. Auxin is then actively pumped downward into the region of elongation. The net result is that the elongating cells on the darkened side receive a larger dose of auxin and thus elongate more than the cells on the lighted side. This then leads to curvature (Figure 6–12). Unfortunately we are not absolutely certain which kind of pigment molecule absorbs the light responsible for phototropic behavior, but blue light is most effective. Chlorophyll and phytochrome are definitely not involved. Similarly, we know virtually nothing about the reactions of the excited pigment which leads to the redistribution of auxin within the tip.

Seedling Growth and Gravity

The growth of seedlings is also strongly influenced by gravity. If a growing seedling is uprooted and placed in a horizontal position, within a few hours the root begins growing downward again and the shoot starts to bend upward. This type of bending response, called *geotropism*, is also due to unequal cell elongation. Let us first consider the upward bending of the shoot. When a seedling is placed horizontally, it has been found that auxin accumulates on the lower surface. This results in accelerated growth on the lower side of the shoot, and therefore the shoot will bend upward. Once the shoot is in the vertical position, auxin is evenly distributed around the stem, the amount of growth on all sides will be equal, and the shoot will grow straight up. If we follow the same reasoning for roots, what would be the result? Obviously there must be some difference in the way root and shoot cells respond to auxin. There is. It has been found that root cells are extremely sensitive to even small concentrations of auxin and that they are normally operating at concentrations close to or beyond an optimal level. This means that when a root is placed in a horizontal position, those cells on the lower surface are subjected to a greater than optimal level of auxin, and this subjection results in a growth inhibition. That is, in roots an increase in the auxin level actually *reduces* (inhibits) the growth rate of the cells on the lower side, and as a result the root grows downward (Figure 6–13).

We know very little about how a plant actually senses gravity. However, we do know that gravity does not directly "pull" the auxin molecules to the lower surface. Rather it is thought that some large particles within the cell, possibly starch grains, fall to the lower surface of the cell in response to gravity. When

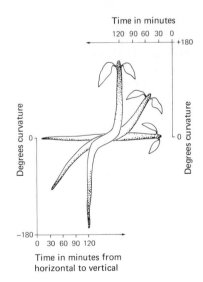

FIGURE 6–13. GEOTROPIC CURVATURE

Geotropic curvature is due to a gravity-induced unequal distribution of auxin in the plant. Note that auxin affects roots and shoots in the opposite manner.

these particles hit the lower membrane, some unknown mechanism is then activated which serves to redistribute the auxin so
that the lower surface is enriched.

A tremendous amount of time and effort has been expended
in trying to elucidate the precise mechanism of the action of
auxin in the cell elongation process. Unfortunately we still don't
have all the details; however, current evidence favors the
following explanation. Plant cells are bound by a rigid structure,
the cell wall (see Chapter 1). In the region of cell elongation,
the properties of the wall are such that it will slowly yield in
response to the internal pressure (turgor pressure) of the protoplast. That is, the protoplast is continually pushing outward
against the cell wall; the wall is gradually yielding to this force,
and the cell enlarges. When such cells are bathed in auxin, there
appears to be a change in the physical properties of their walls
such that they yield more readily under force. This means that
if the internal pressure against the wall is maintained (and it is),
cells bathed in auxin will elongate more rapidly since the resist-
ing structure, the wall, is more pliable. It is known that auxin
itself does not directly affect the cell wall. Rather it appears that
auxin must first bind to the cell membrane. This phenomenon
initiates events which eventually lead to the activation and/or
synthesis of some factors, presumably enzymes, which break
cell wall bonds, thus making the wall more pliable.

It is important to realize that auxin not only affects cell
elongation but also directs many other aspects of growth and
development. However, since many of these effects are of prime
importance in mature plants, they will not be dealt with here
but at appropriate places in subsequent chapters.

FIGURE 6–14.

Summary of events which implicate a
fungal infection and the subsequent
secretion of gibberellins by the fungus as a cause-and-effect phenomenon in this seedling disease. The
active substances secreted by the
fungus are gibberellins.
Adapted from Arthur W. Galston, *The
Green Plant*, © 1968. By permission of
Prentice-Hall, Inc., Englewood Cliffs, N.J.,
Fig. 4–17, p. 70.

From abnormally tall
rice seedlings a
fungus was obtained

The fungus can be
grown on a synthetic
nutrient medium

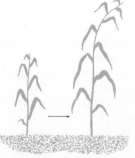

The fungus can be
applied to healthy
plants and they
develop the disease

When an extract of
the synthetic medium
is applied to healthy
plants, they develop
the disease.

Gibberellins and the Foolish Seedling Affair

We have already seen that a class of hormones known as the gibberellins can participate in the mobilization of food reserves in certain seeds. While this particular aspect of gibberellin action is important in seedling growth, there are still other physiological effects of the gibberellins that must be discussed.

At about the same time that Went was conducting his experiments which ultimately led to the discovery of auxin, scientists in Japan were studying a strange disease of the rice plant. The symptoms of this disease were excessive, spindly shoot elongation, failure to mature, and the inability to flower. The disease, termed "foolish seedling disease," was ultimately traced to a fungal infection within the seedling. The infecting fungus was isolated and found to be capable of producing and secreting a particular substance that could give rise to disease symptoms in a previously healthy plant (Figure 6–14). The active ingredient was called gibberellin after the fungus that produces it, *Gibberella fujikuroi,* but years passed before the chemical structure of the gibberellin molecule was elucidated.

Scientists in the Western world eventually became interested in gibberellins also. It was discovered that the production of gibberellin-like compounds was not restricted to fungal pathogens and that such compounds were naturally occurring regulators in higher plants. Oddly enough, there is not just one gibberellin in the fungus or in plants, but rather there is a family of over 40 naturally occurring compounds, all closely related in structure, which possess biological activity (Figure 6–15).

As it is with auxin, the level of gibberellins within a seedling is one of the factors that normally regulates growth rate. In fact, there is some evidence that rapid cell elongation requires the presence of both auxin and gibberellins and that neither regulator is fully active in the absence of the other. The growth-promoting action of gibberellins is perhaps most dramatic when it is applied to certain dwarfed plants. Dwarf corn, as you know from Chapter 5, contains a mutation that deprives it of the metabolic machinery necessary to produce its own active gibberellins; therefore its growth is abnormal. However, if dwarf plants are sprayed with gibberellin, they begin to grow rapidly and eventually become undistinguishable in height from normal corn.

While such a response may be interesting in itself, its real value perhaps is that it provides us with a very sensitive *bioassay* for gibberellins. Gibberellins, like auxin, occur in such minute quantities that it is difficult, if not impossible, to detect them by ordinary chemical means. Suppose, for example, that you noted some form of abnormally rapid growth in your prize

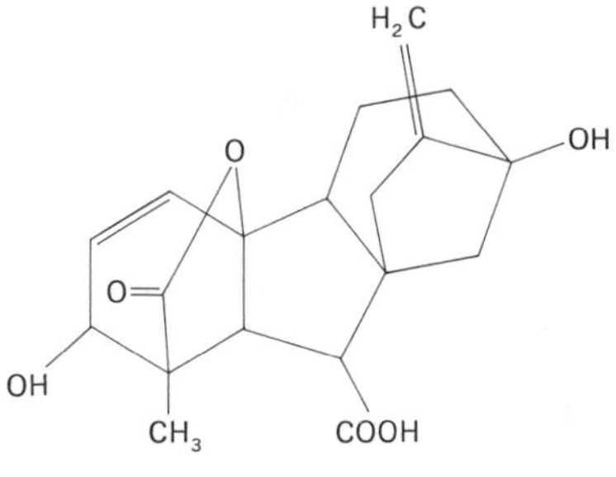

FIGURE 6–15.

The chemical structure of gibberellin A_3. All gibberellins have similar structures.

tomato plants and suspected that this "disease" was due to abnormally high levels of gibberellin. You were able to obtain an extract from the diseased plants which produced the same symptoms in normal tomatoes, but unfortunately the amount was too small for chemical analysis. How do you proceed? One of the options open to you would be the dwarf corn bioassay. The procedure is as follows. Several concentrations of an active gibberellin from a known source are prepared and applied to dwarf corn seedlings. To another set of dwarf corn seedlings, you apply an unknown extract. After waiting several days, check the results. If the extract did indeed induce rapid growth, you have good reason to conclude that it contained a gibberellin because no other regulator will cause such a response in dwarf corn. Furthermore, by comparing the elongation with that produced by known concentrations of gibberellin, you can at least estimate the amount your sample contained and determine if this amount was unusually high (Figure 6–16).

You might wonder at this point if gibberellin applications would not be useful in increasing the size and vigor of crop plants. Unfortunately this idea has not proved to be as exciting as one might at first suspect. It turns out that much of the added growth induced by gibberellins is due to increased water content without a concomitant increase in food value. Furthermore, treated plants usually have other undesirable characteristics including poor storage capacity and top heaviness. On the other hand the overall picture is not totally bleak. There are other physiological effects of gibberellins that *have* proved to be useful in agriculture. One example is the use of gibberellins in the grape industry. If the hormone is applied at the right time and in the correct dosage, the results are striking. Grape bunches elongate dramatically and hence are less tightly packed, thereby increasing quality and decreasing the likelihood of spoilage due to fungal infections. In addition, the

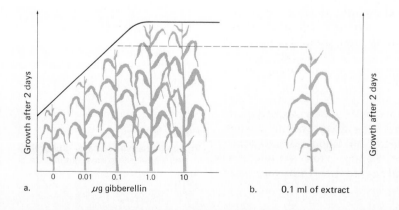

a.   μg gibberellin

b.   0.1 ml of extract

FIGURE 6–16.

The dwarf corn system can be used as a bioassay for gibberellins. Auxin does not influence growth in this system. a. Various amounts of a known gibberellin are applied to dwarf corn seedlings of the same size. The growth of each seedling is measured 2 days later. b. A known amount of extract is applied to a dwarf corn seedling. Growth data indicate that the extract contains 0.1 μg gibberellin per 0.1 ml of extract.

grapes are also larger and more marketable. Gibberellins can also induce certain types of plants to flower at an earlier stage than they would under normal conditions. This phenomenon is of use to the plant breeder as well as the flower grower. Gibberellins can induce the formation of an enzyme (amylase) in the barley aleurone layer. This effect has proved useful to the beer-brewing industry where naturally occurring gibberellins are routinely used in malting (Figure 6–17).

To summarize, we can see that gibberellins, like auxin, have many effects on plant development. They can stimulate elongation, induce enzyme formation in certain seeds, stimulate flowering, and affect the development of fruit. How does one group of related compounds go about influencing so many seemingly unrelated events? We simply don't know, and furthermore we don't even have many good guesses. Perhaps the best lead we have is some suggestive evidence that gibberellins may alter cells' genetic information so that previously masked portions of the DNA are available for transcription, thus giving rise to enzymes that were not present in the absence of optimal gibberellin. Another idea receiving considerable attention presently is that gibberellins may alter the pattern of membrane synthesis, and it may be that this structural modification of the cell could have far-reaching physiological ramifications.

The Cytokinins

We have already seen in Chapter 5 that cytokinins are extremely potent promoters of cell division in tissue culture systems. But what about intact seedlings? The evidence was rather slow in coming, but we now know there are naturally occurring cytokinins in plants, and it seems fairly certain that they do influence cell division within intact seedlings and mature plants, possibly in concert with auxin.

Perhaps we should stop at this point and mull over the ramifications of this finding to see if perhaps it could be related to the extreme longevity of some plants. It is certainly possible that there is a connection between the long life span of some plants and their ability to maintain a functional meristem and continually to produce new cells. Perhaps longevity is related to cytokinin activity. In a way, the plant is really a mixed population of cells, each at different ages, often with different functions. As some cells age, others are maturing and will eventually replace aging nonfunctional ones. Is it possible that cytokinins have the ability to keep certain cells young, that cytokinins are an antiaging hormone or a juvenile hormone? It would be hard to build an airtight case for any of these notions; however, we

1. Barley seeds are allowed to imbibe enough water to activate gibberellin production in the embryo.
2. The germination process is then allowed to proceed for a sufficient time to allow the gibberellin produced by the embryo to reach the aleurone layer and initiate α-amylase synthesis.
3. The entire seed is then baked at 70° C for a few hours and allowed to dry. This procedure results in the death of the embryo but does not inactivate the enzyme α-amylase. This process is called *malting*.
4. The dried barley seeds are then ground and added to a vat of water along with yeast cells.
5. In the vat α-amylase begins breaking starch (originally the barley endosperm) into small sugar units which are then utilized by the yeast cells. It is important to note that yeast cells themselves do not produce an extracellular α-amylase, and therefore, without the malting process (steps 1 to 3 above), beer production utilizing starch would not be possible.
6. Yeast cells take up the sugar units from the medium, and via the process of fermentation, alcohol is produced.

$$C_6H_{12}O_6 \xrightarrow[\text{tension}]{\text{low O}_2} 2C_2H_5OH + 2CO_2 + \text{energy}$$

7. The accumulation of alcohol eventually inhibits the activity of the yeast cells. Various strains of yeast have different tolerances, thus beers can have varying alcohol concentrations.

FIGURE 6–17.

Beer brewing and the aleurone system. Malting is an important early step in the production of beer. Photograph courtesy of Anheuser-Busch, Inc.

can offer some rather intriguing and as yet unexplained observations.

Cytokinins have the ability to slow down the aging process in leaves. That is, when cytokinins are applied to a leaf, or to a portion of a leaf, the breakdown of chlorophyll and protein, which normally occurs as the leaf ages, is greatly slowed down. In fact, sometimes the leaf even becomes greener! This effect does not depend on cell divisions but rather appears to be a separate and distinct physiological response to cytokinins. Correlated with this rejuvenation response is the "mobilization" of various nutrients into the cytokinin-treated areas. Amino acids and other compounds seem to flow from the untreated areas into the treated areas (Figure 6–18). All the responses to cytokinins are as yet unexplainable on a molecular basis, but they are attracting increased attention for rather obvious reasons.

Ethylene—The Different Hormone

If you somehow feel that the above discussion of growth regulators is a bit remote and unassociated with your everyday experiences, you might find ethylene more to your liking. Ethylene can and perhaps already does have a significant impact on your eating habits. For example, let us assume you belong to that strange breed who are fond of persimmons, and proceed with the following imaginary demonstration. First, you secure four nice, hard, inedible persimmons, easily found in most supermarkets. Two are placed in a relatively airtight container along with a

FIGURE 6–18.

Cytokinins: production, structure, and effect on the retardation of leaf aging.

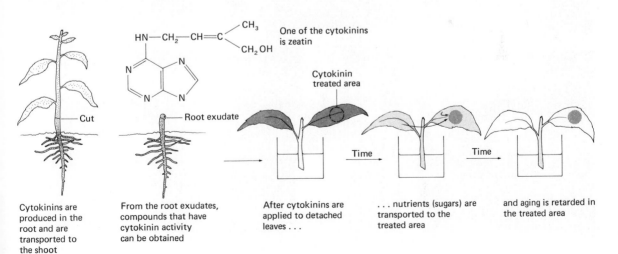

One of the cytokinins is zeatin

Cytokinin treated area

Cytokinins are produced in the root and are transported to the shoot

From the root exudates, compounds that have cytokinin activity can be obtained

After cytokinins are applied to detached leaves . . .

. . . nutrients (sugars) are transported to the treated area

and aging is retarded in the treated area

slice of apple; the other two go on the windowsill. In several days you will observe that the persimmons in the closed container are ripe and soft, while the controls on the windowsill have ripened very little. The factor responsible for the observed difference is ethylene ($H_2C{=}CH_2$). The cut apple placed in the container produces this gas in relatively large amounts, and to a lesser extent so do the persimmons. Since the gas cannot escape, its concentration within the container builds up. Because one of the effects of ethylene is to hasten fruit ripening, it's no surprise that the enclosed fruit ripens faster than the fruit left on the windowsill where the gas is free to diffuse.

Ethylene, of course, is now known to be produced by many ripening fruits and is a causal agent in the ripening process. Ethylene hastens the ripening of apples, bananas, oranges, lemons, and melons, to name just a few. This particular phenomenon, as you might imagine, is very important commercially and has been increasingly exploited. Bananas, for example, at one time presented tremendous problems in transport. When bananas were sealed in the holds of ships, ethylene produced by the ripening fruit would accumulate, and often-

FIGURE 6–19.

High concentrations of ethylene can cause abnormal seedling development as shown here. Note particularly the lateral swelling of the stem and inhibited growth. (a) Control, (b) treatment, and (c) result.

a.

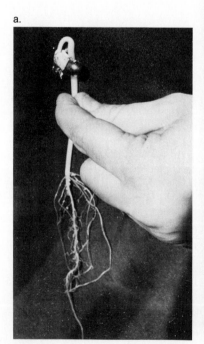

b.

c.

times the cargo would be overripe before it ever neared port. We are now aware of this problem, and today bananas can be kept for months, provided they are kept cool and constantly bathed in fresh, ethylene-free, humid air. When the ship arrives in port, the cargo can be ripened at will by artificial treatment with ethylene. Ethylene treatment is also used commercially to ripen citrus fruits, although the quality may suffer slightly as a result.

While the fruit-ripening role of ethylene is perhaps the best understood of its physiological effects, its role in the control of seedling growth should not be forgotten. Ethylene, like auxin and gibberellins, is produced in the meristems of young seedlings in minute amounts (perhaps 1 part in 10 million parts of air), but nevertheless, such concentrations are sufficient to initiate physiological responses. For example, ethylene may participate in the development and maintenance of the apical hook (see Figure 6–8). Ethylene also antagonizes certain auxin responses such as cell elongation and may act as a brake in the regulation of cell elongation. In fact, at high levels ethylene causes cells to expand in all directions rather than simply in a longitudinal direction. This characteristic can lead to swelling and odd developmental patterns as shown in Figure 6–19. The interaction between auxin and ethylene is subtle and complex. For example, it is known that auxin itself can stimulate ethylene production; and as a result of this knowledge, some responses previously attributed solely to auxin have had to be re-evaluated.

It is a bit embarrassing to say once again that we do not understand how this particular regulator functions, but it is true. While much is known about the physiological responses to ethylene and how to turn these to our advantage, the mechanism of ethylene action is obscure. But lack of knowledge doesn't mean there is any lack of interest in the mechanism of hormone action. Quite the contrary; it is an active and stimulating field. Nevertheless, there are still many things to be discovered, and this is perhaps somewhat refreshing in our modern world.

The Seedling as a Population of Cells: Adding More Detail

Up to this point we have considered a seedling to be rather like a simple stack of cells with areas of cell production (meristems) at both ends of the stack. We have introduced the concept of hormones, zones of cell division, elongation, and maturation, but nevertheless the discussion here has taken a simplistic view of the growing seedling. In actual fact a young plant is much more than just a mass of cells. It is a complex population of cell types and tissues, each of which has specific function, physiology, and appearance. Therefore, before we conclude this chapter, we should at least briefly examine some details of seedling anatomy, although many of the functional problems will not be discussed until the following chapter.

The Root

Let us begin by cutting a plant root down the middle and carefully examining the exposed cells, beginning at the bottom and working upward (Figure 6–20). First you will recognize the root cap. Cell divisions in the meristem, you will recall, continually produce new cap cells to replace those worn away by the abrasive action of the soil. The meristem itself is a rather localized tissue composed of small uniform cells, which extend only a millimeter or so from the root tip. Moving from the meristem toward more mature tissue, the cells become more elongated, and there are no apparent cell divisions. This area is the zone of elongation. All the cells that compose this zone were derived from the meristem itself, but now the meristem has grown down and away from them. That is, with time the cells making up the zone of elongation have been displaced to their present developmental position, and as more time passes, they will become even farther removed from the root tip. It is interesting to note also that *only* those cells at the very tip of the meristem remain

FIGURE 6–20. ANATOMY OF THE ROOT

a. Cross section of the root. b. The root in longitudinal detail. For simplification, the number of cells shown is somewhat less than that in an actual root. Differentiation of only the first-formed sieve tube and vessel is shown.

(a) Adapted from *Botany: A Functional Approach,* by Walter H. Muller, 1963, Macmillan Publishing Company, Fig. 8–2, p. 84. © Walter H. Muller, 1963. (b) Adapted from *The Living Plant,* 2d ed., by Peter Martin Ray. Copyright © 1963, 1972 by Holt, Rinehart and Winston, Inc. Reprinted by permission of Holt, Rinehart and Winston, Inc., Fig. 8.2, p. 125.

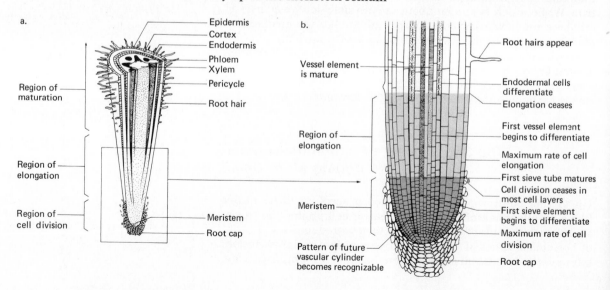

meristematic indefinitely. All the daughter cells produced from these few permanently meristematic cells may themselves also divide many times, but eventually *every* such cell and its off-spring will be displaced as the root tip grows downward.

Progressing through the zone of elongation, we notice that the root becomes more complex, anatomically that is. We begin to encounter cells unlike anything we have seen before. These atypical-looking cells were not produced by new cell divisions in this area but rather are cells that were initially produced in the meristem, that passed through a cell elongation stage, and that then matured and changed in various ways, depending on their spatial position within the root. More will be said in a moment about the specific cell types found in plant roots and the function of each, but for now it should be emphasized that cells can and do change in their outward appearance at specific locations within the plant. This fundamental aspect of cell development is called *differentiation*. Certain cells differentiate at various places above the meristem, and this maturation process becomes more and more obvious as our examination proceeds away from the root tip. Thus, beginning at the tip, we can see that we have a developmental progression which includes cell division, cell elongation, and cell differentiation.

The first cells to become obviously different from their neighbors are called *sieve tube elements* (Figure 6–21). Prospective sieve tube elements actually differentiate, unlike other cell types, very near the meristem. The first sieve tube elements to be found are not isolated units but rather are connected in stackwise fashion to older sieve tube tissue known collectively as the *phloem* (see Figure 6–20). Sieve tube elements function in the long-range transport of food materials, primarily sugars, from the photosynthetic parts of the plant into areas where such nutrients are needed. Since the root meristem is quite active metabolically, but photosynthetically inactive, it is rather easy to rationalize why such specialized cells would appear at so early a developmental stage in the plant root.

The next specialized cell type to be encountered is usually first found in the zone of elongation. These cells belong to a tissue called *xylem*. The cells that make up the xylem are primarily *tracheids*, *vessel elements*, or both (Figure 6–22). Some plants have mostly tracheids (more "primitive" plants), some mostly vessel elements ("advanced" plants), and others have a mixture of both. Tracheids are long, narrow cells with oblique end walls covered by a membrane. Vessel elements are shorter in length but greater in diameter and at maturity have no end walls. Both types of xylem cells are dead when they reach *functional maturity*. Such dead cells comprise the long-distance water-conducting system within the plant body.

The first xylem elements encountered when we move from the

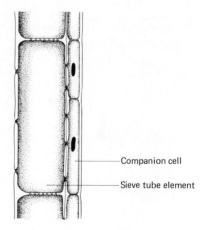

Companion cell

Sieve tube element

FIGURE 6–21.

The sieve element and associated companion cell are part of the food-conducting system in plants known collectively as the phloem. The structural details of these cells are fully discussed in Chapter 7.

root apex toward more mature parts of the plant have bands of secondary wall thickenings rather than smooth and uniform secondary walls. It is after these bands are formed interior to the primary wall that the cell contents autodestruct. We are left with a more or less tubular pipe of primary cell wall rein-

a.

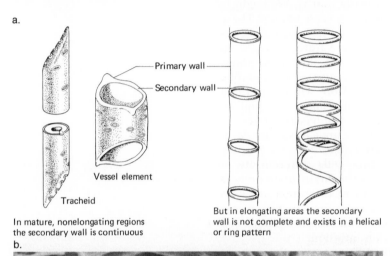

Primary wall

Secondary wall

Vessel element

Tracheid

In mature, nonelongating regions the secondary wall is continuous

But in elongating areas the secondary wall is not complete and exists in a helical or ring pattern

b.

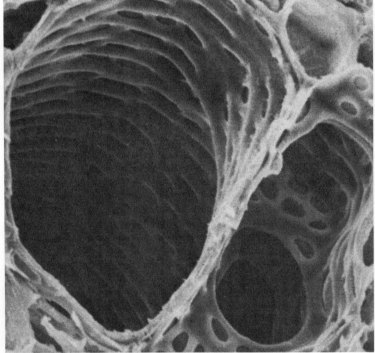

FIGURE 6–22.

a. Cell types found in xylem. b. The scanning electron microscope provides this view of a vessel in a cucumber root. Note pitted appearance of the secondary wall.

(a) Adapted from *The Living Plant*, 2d ed., by Peter Martin Ray. Copyright © 1963, 1972 by Holt, Rinehart and Winston, Inc. Reprinted by permission of Holt, Rinehart and Winston, Inc., Fig. 8.5, p. 129. (b) Photomicrograph from *Probing Plant Structure*, by J. H. Troughton and L. A. Donaldson, 1972, McGraw-Hill Book Company, Plate 60. Courtesy of the Department of Scientific and Industrial Research, Lower Hutt, New Zealand.

forced or strengthened by the secondary wall bands. Of course, if strength were the only objective, it might be asked why the plant does not lay down a complete secondary wall. The answer seems straightforward enough if you simply consider the location of these water-conducting pipes. They first appear in populations of other cells that are still elongating. Thus they must be able to stretch, and stretching is possible only if the cell has helical or ringlike secondary wall thickenings. Interestingly, those cells that differentiate into xylem *after* they and their surrounding cells have become fully elongated do not have stretchable secondary thickenings but rather develop a continuous secondary wall and resemble the "typical" vessel elements or tracheids shown in Figure 6–22.

As we progress out of the zone of elongation, you will notice more and more differentiated cells. However, the pattern is far from random; we don't just find a sieve tube element here and a vessel element there, but rather a characteristic pattern develops which is much the same for the roots of all higher plants. This pattern can be seen most easily in a cross section of the root (Figure 6–23). The xylem occupies the centermost position in the root cylinder with the sieve tube elements positioned outside this central core. We have already seen that cells differentiate to give rise to this pattern through time and space. First, some cells differentiate into phloem. Next, in slightly more

FIGURE 6–23.

Developmental progression of tissues in the root. The diagrams on the right are cross sections taken at the indicated positions on the root.
Adapted from *The Living Plant,* 2d ed., by Peter Martin Ray. Copyright © 1963, 1972 by Holt, Rinehart and Winston, Inc. Reprinted by permission of Holt, Rinehart and Winston, Inc., Fig. 8.4, p. 128.

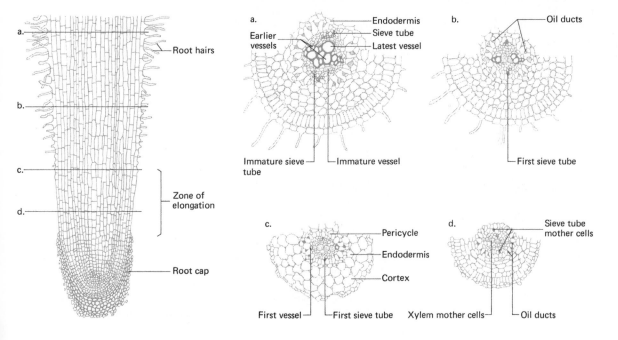

mature tissue, interior to the phloem, elements of the xylem appear. As we reach more and more mature tissue, cells interior to the original phloem elements and interior to the original xylem elements differentiate to form sieve tube elements and tracheids or vessel elements, respectively.

Within this zone of differentiation or maturation, still other cell types become obvious. The cell layer immediately adjacent to the vascular cylinder (xylem and phloem) is the *pericycle*. This cell layer retains a meristematic potential and is capable of dividing, if given the proper signal, to form lateral or branch roots. Immediately outside the pericycle is the *endodermis*. This particular cell layer regulates the flow of materials into the vascular cylinder. Outside the endodermis is a tissue known as the *cortex*. These cells are as close to our original generalized cell as you are likely to find in a plant (see Chapter 1); they function primarily as food storage centers. Finally, the root is covered by a cell layer known as the *epidermis*. In certain areas the long, thin cells of the epidermis produce extensions called root hairs. Root hairs greatly increase the surface area of the root and the ability of the root to absorb water and nutrients (Figure 6-24). Root hairs are only found over a limited portion of any root—just above the zone of elongation. Perhaps you can think of a reason why it would be impractical for these fragile structures to be formed in the elongating region. The fact that we only find root hairs in specific regions must mean they are

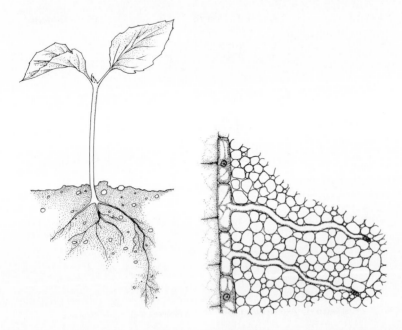

FIGURE 6-24.

Detail of root hairs, viewed with increasing magnification.

constantly destroyed and replaced by newly developing epidermal cells..This is indeed true; the average life expectancy of any given root hair is only a few days.

The Stem

The anatomy of the stem is, in general terms, comparable to that of the root. Yet it's more complex, primarily because of the added complication of developing leaves and buds. Remember that we are presently concerned with seedlings and therefore will omit for the present any consideration of expansion in girth and of the functional physiology of any of the tissues. Our purpose here is only to describe the visible structures so we will have a common vocabulary and descriptive knowledge. The next chapter deals more extensively with form as it relates to function.

For our discussion of the stem, let us take a slightly different approach in our anatomical study and begin with an examination of the tissues in the area well below the zone of elongation (Figure 6–25). We will begin with the outermost tissue, the epidermis, and progress inwardly. This tissue is similar to that in roots; however, in the aerial portions of the plant, the outer sides of the epidermal cells are covered by a waxlike substance (the cuticle) which prevents water loss. Inside the epidermis we again encounter the cortex—as we did in the root. Still farther inward, we note the absence of a pericycle and the endodermis. Furthermore, we see that the vascular tissue is arranged in a slightly different manner. In a young seedling there are distinct vascular bundles with phloem to the outside and xylem to the inside. The relative spatial positioning, therefore, is similar to the root, but the centermost position in the stem is occupied by living cells similar to those found in the cortex. This innermost tissue is called *pith*.

Knowing the ultimate arrangement of the cell types within a young stem, we could now trace the various tissue types upward into the younger portions of our seedling. This would be a simple enough task if it were not for the fact that stems, even young ones, have leaves or leaf primordia and sometimes have small branches, each containing a vascular system which is—or will be—continuous with that of the stem.

The origin and formation of the embryonic leaves are of special interest in the development of a seedling. As you know, the tip of the shoot is a region of active cell division which contributes new cells which subsequently enlarge. On the sides of the shoot meristem are rather special meristematic areas which

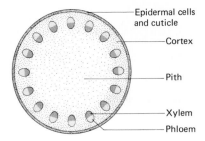

FIGURE 6–25.

The primary tissues of the stem are shown here somewhat simplified. Compare with the root tissues in Figure 6–23. In grasses and their relatives, the arrangement of the vascular tissue (xylem and phloem) is not as precise as indicated here. Rather, the bundles are scattered throughout the cortex.

at certain times rapidly divide and give rise to the embryonic leaves. Furthermore, these zones of leaf formation are not static but rather with time move around the meristem. Thus leaves occur not just in stacks, one over the other, but oftentimes spiral down the stem (Figure 6–26). As a young leaf develops in size, cells near the base differentiate to form sieve tube elements. Shortly thereafter elements of the xylem appear. As time passes, cells above and below these new vascular elements also differentiate, so that as the vascular system develops, it appears to spread toward the leaf tip and also downward where it will eventually become continuous with the vascular system of the stem.

There is another event of critical importance that occurs near the apex: the development of *lateral buds*. These buds (Figure 6–27) appear in the angle between the embryonic leaf and the stem. Lateral buds are capable of becoming meristematic under certain conditions; if they do, lateral branches develop. Most such buds, however, remain dormant until the leaf is mature. In fact, most of these buds never become active, as a careful examination of almost any young seedling will clearly show.

The overall pattern of seedling development presented here is adequate for many, many plants; however, there is one ex-

FIGURE 6–26.

a. Leaf development begins with cell divisions on the apical flanks of the meristem. b. With time the zones of division move around the stem apex in a regular way, as indicated by the numbers.

(a) and (b, center) From *Plant Anatomy*, 2d ed., by Katherine Esau, 1965, John Wiley and Sons. (b, left) From *Apical Meristems*, by F. Clowes, 1961, Blackwell Scientific Publications, Ltd. By permission of Blackwell Scientific Publications, Ltd., Oxford, England.

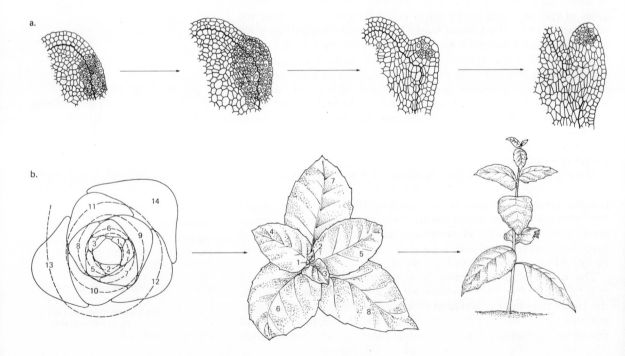

ception so glaring that the pattern will be commented on briefly. This aberrant pattern is best characterized by grass seedlings, and with a little thought you should at once see the problem. If the growing zone in a typical lawn grass were at the tip only, what would happen the first time you mowed the lawn? Obviously, it would also be the last time. However, such grass seedlings contain a *basal* or *intercalary* meristem. Thus new cells can be produced at the base of the leaves and stem even when the top is cut off. This type of growth pattern is important in selecting food sources for grazing animals; it is probably an adaptation to survive the continual onslaught of herbivorous beasts.

Retrospection

The growth of seedlings is indeed a complicated process with many things going on simultaneously. Cell division is required,

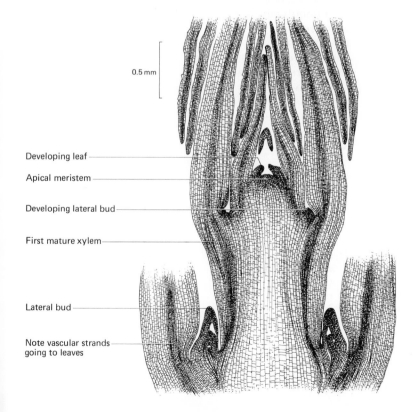

0.5 mm

Developing leaf

Apical meristem

Developing lateral bud

First mature xylem

Lateral bud

Note vascular strands going to leaves

FIGURE 6–27.

Longitudinal section of a shoot tip. Note the location of developing lateral buds.
From *The Living Plant,* 2d ed., by Peter Martin Ray. Copyright © 1963, 1972 by Holt, Rinehart and Winston, Inc. Reprinted by permission of Holt, Rinehart and Winston, Inc., Fig. 8.6, p. 131.

but so is cell elongation. Naturally, neither of these processes could occur without the differentiation of specialized tissues which aid in support and serve to furnish water and also food material which is derived from the mature and functional areas. We have seen that cells do not differentiate in a haphazard fashion. Rather their development is progressive and controlled, giving rise to specific tissue patterns which are remarkably constant from plant to plant. We haven't mentioned at all in this chapter how a given cell "knows" to differentiate into a sieve element rather than into a vessel or epidermal cell, but it is a fascinating, although poorly understood, phenomenon. In fact, much of modern botany and biology is concerned just with this problem: How is differentiation controlled and regulated? One might suspect from this chapter, as well as from the information in Chapter 5, that hormones are somehow involved. Indeed, at least four such regulators can greatly influence seedling development: auxin, gibberellins, cytokinins, and ethylene. In addition, we have seen that the external environment, light for example, also plays a significant role in influencing seedling development. We have only begun to understand how each regulator operates alone, but undoubtedly the situation is infinitely complex within an intact seedling where subtle combinations and interactions are at work. Perhaps, if nothing else, the next time you have occasion to see a plant grow, you will observe it more carefully and will realize that seedling growth is indeed a remarkable and complex process.

Selected Readings

Burg, S. P., and E. A. Burg. 1965. Ethylene action and the ripening of fruits. *Science* 148:1190–1195. Interesting but somewhat dated article by the two scientists who pioneered much of the research on ethylene.

Esau, K. 1965. *Plant Anatomy.* 2d ed. New York: John Wiley. Classic in the field.

Galston, A. W., and P. J. Davies. 1970. *Control Mechanisms in Plant Development.* Englewood Cliffs, N.J.: Prentice Hall. Advanced text but could be understandable to the motivated student.

Phinney, B. O. 1957. Growth response of single-gene dwarf mutants in maize to gibberellic acid. *Proceedings of the National Academy of Sciences* 43:398–404. Original report of the genetic nature of dwarf corn.

Ray, P. M. 1971. *The Living Plant.* New York: Holt, Rinehart and Winston. Highly readable small book on plant growth and its regulation.

van Overbeek, J. July 1968. The control of plant growth. *Scientific American* 219(1):75–81. (Offprint no. 1111.) A somewhat dated but still useful review of plant-growth hormones.

Wittwer, S. H., and M. J. Bakovac. 1958. The effects of gibberellin on economic crops. *Economic Botany* 12:213–255. Early thoughts on the potential value of a growth hormone on crop plants. Interesting to check on predictions.

7 Plant Growth: Seedling to Maturity and More

PART ONE

In this chapter we will continue our study of the growth of plants by considering how plants expand in girth or diameter. This will include a reasonably detailed look at wood and a slight detour to examine something about papermaking. We will then depart from a strictly anatomical approach and place more emphasis on how plants carry on the various physiological processes necessary for existence on the land. This will necessitate understanding gas exchange in leaves, the long-distance transport of water and food molecules, the factors controlling flowering, and lastly, a consideration of aging.

Secondary Growth and the Production of Wood and Bark

Up to this point our discussion of growth in plants was restricted to a consideration of the *primary tissues*. That is, the development of cells and tissues that arise directly from the apical meristems. This allowed us to consider the longitudinal growth of a seedling rather thoroughly but avoided lateral expansion. This latter type of growth in a plant shoot or root is called *secondary growth*. The addition of secondary tissues is a characteristic common to many vascular plants (exceptions being certain herbs and monocots, such as grasses) and is of enormous economic significance. We can easily distinguish and define the difference between primary and secondary tissues since primary tissues arise from divisions of the apical meristems while secondary tissues arise from cell divisions in two special nonapical meristems called the *vascular cambium* and the *cork cambium*. Let us now consider how and where the cambium layers arise and function in the shoot, keeping in mind that the same general pattern of development also occurs in most roots.

As discussed in the previous chapter, the differentiation of primary xylem (i.e., xylem arising from cells derived from the apical meristem) proceeds from the interior of the stem outward, eventually meeting the developing primary phloem. However, we omitted one very important fact, and that is that between the primary xylem and primary phloem there also develops a permanent layer of rather special cells which retain the ability to carry on cell divisions even though they are well separated from the apical meristem. This tissue is called the *vascular cambium*. The vascular cambium is composed of two cell types: long, thin cells called *fusiform initials* and shorter, more boxlike cells called *ray initials* (Figure 7–1).

The vascular cambium can first be detected as a functional entity in the lowermost part of the zone of elongation or in the upper part of the zone of maturation. It is in this area that the cambial cells begin dividing and thereby increase the total number of cells squeezed between the previously formed primary xylem and primary phloem (see Figure 7–1). The new cells which are produced by the fusiform initials (we'll consider ray cells later) do not all remain as fusiform initials, but some differentiate according to their spatial position in the plant axis. Those new cells, which are produced interior to the vascular cambium, develop into *secondary xylem*, while those cells produced exterior to the cambium become *secondary phloem*. This means that the secondary xylem forms in contact with the exteriormost cells of the primary xylem, and the secondary

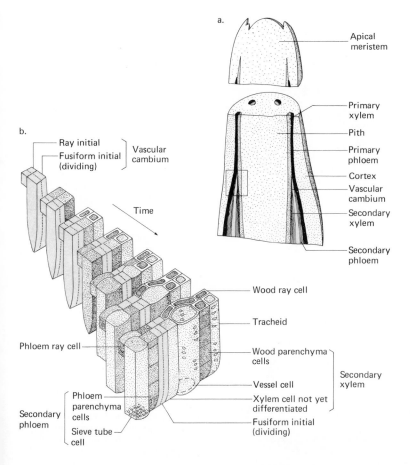

a.

Apical meristem

Primary xylem

Pith

Primary phloem

Cortex

Vascular cambium

Secondary xylem

Secondary phloem

b.

Ray initial

Fusiform initial (dividing)

Vascular cambium

Time

Wood ray cell

Tracheid

Phloem ray cell

Wood parenchyma cells

Secondary xylem

Vessel cell

Phloem parenchyma cells

Xylem cell not yet differentiated

Secondary phloem

Sieve tube cell

Fusiform initial (dividing)

FIGURE 7–1.

a. The area of development of the vascular cambium. b. Details of how the cambium, with time, gives rise to secondary xylem and phloem. The entire illustration is somewhat stylized.

(b) Adapted from *The Living Plant*, 2d ed., by Peter Martin Ray. Copyright © 1963, 1972 by Holt, Rinehart and Winston, Inc. Reprinted by permission of Holt, Rinehart and Winston, Inc., Fig. 8.9, p. 137.

phloem develops interior to but continuous with the primary phloem.

Unfortunately, we do not know exactly what governs the differentiation of the cells produced by the fusiform initials. It is interesting, nevertheless, to consider the subtle nature of the factors that must be involved. After all, the same cell type divides to give rise to cells that will develop into xylem *or* phloem within fractions of a millimeter of one another, and mistakes are almost never made. What factors could possibly be different exterior to the cambium where cells develop into phloem as compared to those interior to the cambium where xylem is formed? We simply don't know, but suggestions ranging from hormone gradients to oxygen gradients have been proposed.

You should not get the idea irreversibly fixed in your mind that the vascular cambium is limited for all time to just the region between the primary xylem and phloem. Rather, as you can see in Figure 7–2, cells differentiate in the cortex *between*

a. Primary tissues

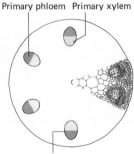

Primary phloem   Primary xylem

Cambium begins forming

b. Vascular cambium develops within and between bundles; new vascular tissue is produced

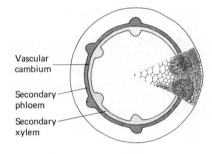

Vascular cambium

Secondary phloem

Secondary xylem

c. Production of new vascular tissue forces external primary tissue outward; internal primary tissue is crushed

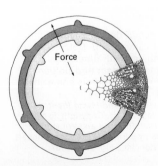

Force

d. Eventually all primary tissues are crushed or sloughed off

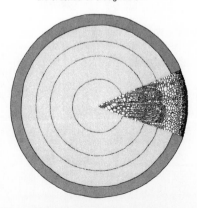

FIGURE 7–2.

The vascular cambium forms initially in the cells between the primary xylem and primary phloem. Somewhat later the cambium develops between the vascular bundles, and a complete ring of secondary vascular tissue is possible (parts a and b). With time the continued addition of secondary tissue crushes the previously existing primary tissue or forces it outward where it is sloughed off (parts c and d). Note that eventually only secondary tissues are present in a trunk cross section.

the vascular bundles and form a cambium which encircles the entire stem. Since these new cambial cells function exactly like the ones previously described, it should be obvious that eventually the stem will no longer contain isolated vascular bundles. Rather, a continuous ring of vascular tissue will eventually form, resulting in an increase in the stem's girth.

Perhaps you're wondering at this point what happened to the ray initials which also are part of the vascular cambium. They also divide (see Figure 7-1) but produce cells which elongate laterally into the secondary xylem and phloem. Such cells remain alive but relatively undifferentiated (again we don't know why). It is thought that these cells function in the transport of food materials from the phloem into some of the living cells within the xylem and pith and/or act as food storage areas (Figure 7-3).

Let's consider what happens as a result of continued cambial activity for several years. The diameter of the stem increases as we previously said, but there are many associated problems which must be considered as well. For example, as the addition of secondary xylem occurs, the vascular cambium itself must increase in diameter, or it would soon be cracked and discontinuous. This problem is overcome rather easily by continued radial divisions which increase the girth of the cambium as necessary. Similarly, the increasing accumulation of secondary xylem within the stem will push the primary and secondary phloem outward (return to Figure 7-2 and then look at Figure 7-3). This pressure can and does often result in crushing some of the outermost tissues of the stem and if unchecked would result in large cracks in the phloem. Large cracks don't develop due to certain adjustments described below and to the activity of yet another tissue called the *cork cambium*.

The cork cambium forms in the cortex at a point very near the outside of the stem at about the same time that the vascular cambium develops. The function of the cork cambium is to produce new cells called *cork* to the outside of the stem (Figure 7-4). Cork cells form an extremely compact tissue and by virtue of the waxy material in their cell walls, seal the stem rather effectively against water loss. Furthermore, they have the important property of being reasonably elastic and can stretch to a certain extent as the diameter of the stem increases. However, simple elastic expansion isn't enough to compensate entirely for yearly increases in girth, and eventually the cork cells and the original cork cambium do crack and are sloughed off. As this occurs, a new cork cambium is produced from unspecialized cells within the secondary phloem.

Eventually, continued expansion causes the tissues exterior to the *new* cork cambium also to peel or flake away. This tissue loss would include older cork cells and the primary phloem. With time yet another cork cambium will be formed interior to

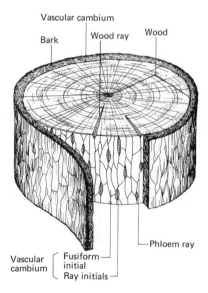

FIGURE 7-3.

Cross section of a tree trunk. Note that proliferation of the vascular rays occurs in both the secondary xylem and secondary phloem. This phenomenon prevents cracks from forming as the trunk expands due to cambial activity. Cracking is further prevented by the activities of the cork cambium. Adapted from *Botany: An Introduction to Plant Biology*, 3rd ed., by W. W. Robbins, T. E. Weier, and C. R. Stocking, 1964, John Wiley and Sons, Fig. 7.31A, p. 100.

the last, and so it goes. Remember that the old cork cambium and phloem will *continually* be lost as the stem expands. We usually call this cracking dead tissue the *bark*. The process can continue indefinitely because the cork cambium is continually regenerated within the secondary phloem which in turn is continually formed from the vascular cambium. Furthermore, cracking deep within the secondary phloem is further controlled by the periodic proliferation of ray cells which divide and expand to keep pace with the middle-age spread of the entire stem (see Figure 7–3).

If we continue this line of thought, it should be apparent that a cross section of the lower trunk of a six-year-old tree (Figure 7–5) will appear quite different with respect to internal anatomy from the stem cross section described in the previous chapter. It is interesting to note the ringlike appearance of the secondary xylem. Such rings are apparent because the xylem cells produced by the vascular cambium in the spring are rather large; those produced in the summer are much smaller. This is presumably the result of the relative abundance of water during the spring, the lack of water in the summer, and the subsequent effect on xylem differentiation. These rings are called *annual rings* and have been useful in determining the age of a tree (Figure 7–6).

What does a botanist mean by the term *wood*? Wood is nothing more than secondary xylem. The most interior portion of a woody trunk consists of the original pith, primary xylem, and older secondary xylem. These cells are usually crushed and eventually die (if they are not dead already) and are referred to as *heartwood*. The younger and functional secondary xylem

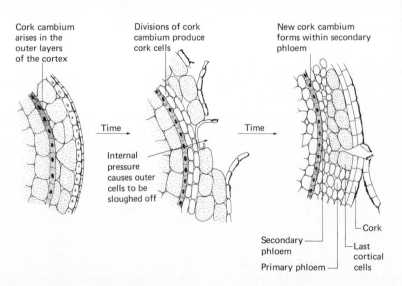

FIGURE 7–4.

The cork cambium forms first in the cortical cells and then is sloughed off due to the continued activity of the vascular cambium. As this occurs, another cork cambium forms deeper within the trunk, in the secondary phloem. Two species that have especially active cork cambial activity are Canary Island pine (top) and cork oak (bottom). The cork oak (*Quercus suber*) is a native of Europe and is the species from which we normally obtain cork.

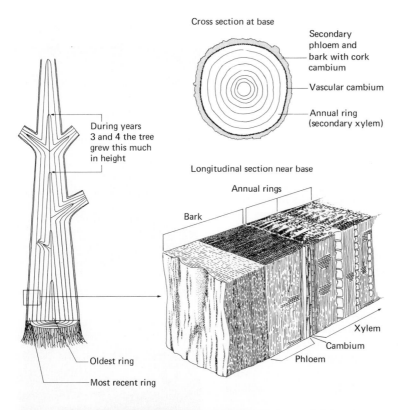

During years 3 and 4 the tree grew this much in height

Oldest ring

Most recent ring

Cross section at base

Secondary phloem and bark with cork cambium

Vascular cambium

Annual ring (secondary xylem)

Longitudinal section near base

Annual rings

Bark

Xylem

Cambium

Phloem

FIGURE 7–5.

Annual rings result from a size difference in the xylem cells that develop during the growing season—spring cells are larger than those made in late summer. In the diagram of the upright tree, note that the number of rings varies with the distance above the ground.

(left) From *Botany: An Introduction to Plant Biology,* 3rd ed., by W. W. Robbins, T. E. Weier, and C. R. Stocking, 1964, John Wiley and Sons, Fig. 7.47, p. 110. (right) Adapted from Biological Sciences Curriculum Study. *Biological Science: Molecules to Man,* Houghton Mifflin Company, Boston, 1968, Fig. 14.8, p. 347.

FIGURE 7–6.

The annual growth rings in this cross section of a trunk of the California big tree, *Sequoiadendron giganteum,* indicate that the tree was nearly 1700 years old when felled. The tree grew faster when young, and so the growth rings near the center are larger than those near the perimeter. When the seed giving rise to this tree germinated, the Roman Empire was not yet at its peak.

(located to the exterior of the heartwood) is called *sapwood*. Man's economic utilization of wood and the forestry industry are discussed in Chapter 10, but let us now examine briefly how wood is used in making paper.

Paper Is Secondary Xylem

Over 500 lb, or about 1.4 lb per day, is a close approximation of the paper used annually by each person in America. Figured on a per capita basis, we use paper products at nearly twice the rate of the next highest consumer, Canada, and at better than 25 times the rate of the Communist bloc nations taken as a whole.

Immediately, of course, we think of paper as a means of communication; historically, that was nearly its exclusive use for over a thousand years. Paper appears to have been invented in China around A.D. 105. Paper remained an exclusively Chinese product for about 500 years until it appeared in Japan. It wasn't until another 500 years had passed that it found its way westward into Egypt, and finally Europe in the twelfth century. Its use for other than communication purposes was virtually nil until the mid-1800s when paper bags and boxes were invented. Around the turn of this century we find milk cartons, cups, plates, food wrappers, and all kinds of things being made of paper.

What is this wonderful stuff made of? Plants, of course! Actually, the fibers used in manufacturing paper come from rags, straw, grasses, old newspapers, and a number of other sources. But the ultimate source is always plant material, and most of it, better than 90%, is wood. *Cellulose*, the basic material for paper, is a material in the cell walls of secondary xylem, which is wood. You may have the feeling that paper manufacturing is very complicated. Actually the basic process is quite simple and hasn't changed much in its nearly 2000-year history. There are four steps: (1) make the pulp, (2) make the stock (optional), (3) make wet paper, and (4) dry it out.

Pulp is simply a suspension of pulverized wood cells. In the trade these cells are called fibers, but as plant biologists we know that most of them are really tracheids and vessels. This suspension of fibers is made in either of two ways. The logs with the bark removed may simply be ground against a grinding stone in the presence of water. Paper made from this kind of pulp is relatively inexpensive but weak and generally of poor quality. Newsprint, for example, is primarily of this type (Figure 7–7). Alternatively, the logs may be fragmented into chips and then pulverized, using hot sulfites, soda, or sulfates, depending on the various processes. These chemicals remove impurities

a.

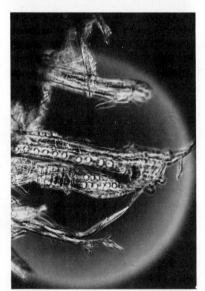

b.

FIGURE 7–7.

That paper is made from wood can be easily seen in these photographs. a. Tracheids prepared directly from pine wood. b. A torn edge of the local campus newspaper, which is the kind of paper called newsprint. The bordered pits and rays are evident in both photos.

from the chips, leaving only the cellulose; consequently the yield is only about 40–50% of the original wood, but the product is of much finer quality. The pulp is then bleached so that the paper will be white instead of wood colored. It can be used directly to make paper, but usually it passes through an intermediate step during which it is made into stock.

Pulp is refined into stock simply by further grinding. The purpose is to roughen and fray the extracted fiber fragments and to make them more uniform in size. As a result of this treatment, the fibers adhere to one another more readily, and the strength and quality of the paper are improved. Also dyes, strengtheners, and other additives may be introduced at this point.

The third step is to make wet paper out of the pulp or stock. This is accomplished by applying the pulp or stock to a wire-screen filter, draining off the liquid and thereby leaving only the wood fibers which form a thin, matted network on the surface of the screen. In times past this was done by simply immersing a screen in a vat which contains the stock, and then drawing the screen upward, causing the wood fibers to collect on it. This was done by hand, and one man could produce perhaps 750 sheets per day. Today you'll find manual operation replaced by a revolving wire-screen belt up to 30 feet wide which moves at rates up to 88 feet per second or about 60 miles per hour. This belt picks up the stock and drains off the water rapidly by using a vacuum on the underside (Figure 7–8). At such rates one machine can turn out in less than 2 seconds the equivalent of one man working all day using the techniques of the early 1800s.

FIGURE 7–8.

This is the wet end of a modern paper-making machine. In the foreground is a screen belt, carrying a film of wood fibers which it has just picked up from the stock beneath the camera. The belt is moving at about 60 mph toward the rollers in the dry end of the machine behind the workman. Photograph courtesy of Crown Zellerbach, by Alan Hicks Photography, Portland, Oregon.

	Percent of World's Wood Consumption	Percent of Wood Used in Industry				
United States	16.7	87	+	13	=	100%
Latin America	10.8	17	+	83	=	100%
	27.5					

TABLE 7–1. PERCENT OF WORLD'S WOOD PRODUCTION IN UNITED STATES AND LATIN AMERICA FOR 1960 TO 1962 AND PERCENT USED FOR INDUSTRY AND FUEL IN THOSE AREAS OF THE AMERICAS

The final step is to press the paper and dry it. This is accomplished by squeezing the paper between heated rotating drums. Finally, the paper is collected in giant rolls.

Unfortunately, our extremely high consumption of wood as paper is fast becoming something that we can no longer take great pride in (Table 7–1). As our use of paper has increased, two rather undesirable aspects of its presence in our society have become evident. First of all, and most noticeable, is paper pollution—in virtually every inhabited place in our country, from street corners to alpine meadows, we see evidence of careless disposal of paper. In addition to the litter, cities spend thousands of dollars to remove the trash of everyday living! Most of this is paper. Nearly everything purchased at our supermarkets comes packaged or labeled in paper, most of it used primarily for marketing purposes to make the product look appealing and immediately discarded when the product is used. We collect and throw away piles of newspapers, receive and throw away reams of superflous mail, carry home our purchases in easily disposable paper bags, get our milk in paper cartons. Clearly, the far-reaching result of this extreme dependence on paper is the rapid expansion of timber use which often means a decrease in the unspoiled lands needed for recreation and wilderness area. There has been some pressure applied recently through the media to re-evaluate our excessive use of paper and to recycle as much paper as possible. Some books and other printed materials are now being printed on recycled paper, and a few recycled-paper products, such as paper napkins and towels, are now appearing in a few supermarkets. This is a small beginning.

The Leaf: Its Architecture and Physiology

We have completed the major points concerning the anatomy of the stem and root and discussed the practical implications this has for paper production. Let us now look at several of the more important physiological processes in plants. We will start with leaves and the mechanism by which CO_2 and other molecules enter and leave the leaf.

Leaves exist in a wide variety of shapes, sizes, and thicknesses, and thus it is difficult to describe an average leaf. Nevertheless, it is necessary to start somewhere, and we begin with the internal structure of a "generalized" leaf which happens to be adapted to a reasonably moist climate. Later we will briefly review certain adaptions found in leaves of plants adapted to more extreme environments. Figure 7–9 shows that the leaf, like the stem, is composed of various tissue types. These tissues include the *palisade layer*, consisting of compact but elongated cells containing many chloroplasts, and the *spongy parenchyma* with irregularly shaped cells bounded by air spaces. These tissues are sandwiched between an upper and lower epidermis similar to that found on young stems. The epidermis is a tissue composed of tightly packed cells which produce and secrete a waxy material on their outer surface. This waxy layer (the cuticle) forms a rather impermeable barrier for water and gas exchange and would virtually seal off the leaf cells from their external environment if it were not for the discontinuous nature of the epidermis. That is, at various places in the epidermis there are small pores called stomates. These pores are usually more abundant on the lower surface of leaves, although there are exceptions (Figure 7–10). They provide a direct route for the exchange of CO_2, other gases, and water vapor between the leaf's interior and the surrounding environment (Figure 7–11).

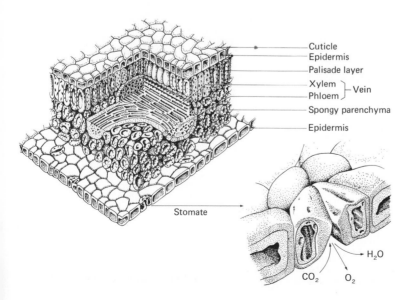

Cuticle
Epidermis
Palisade layer
Xylem ⎤
            ⎬ Vein
Phloem ⎦
Spongy parenchyma
Epidermis
Stomate

H_2O
CO_2 O_2

FIGURE 7–9. LEAF STRUCTURE

The leaf takes in carbon dioxide and gives off oxygen and water through the stomates.
Adapted from Biological Sciences Curriculum Study, *Biological Science: Molecules to Man*, Houghton Mifflin Company, Boston, 1968, Fig. 18.2, p. 467.

There are also bundles of vascular tissue extending into the leaf (veins) which provide for the internal transport of food and water out of and into the leaf from other areas of the plant.

It is important to view leaf anatomy not as just a static arrangement of cells which must be memorized for later regurgitation but in terms of an important physiological role of the leaf: photosynthesis. Recall from Chapter 2 that the CO_2 content of the atmosphere is relatively sparse (0.03%) and that such concentrations commonly limit photosynthesis. From this fact

FIGURE 7–10.

The leaves of water lilies float on the surface of the water and have stomates on the upper leaf surfaces rather than the lower. Why do you suppose there are no stomates on the lower leaf surfaces?

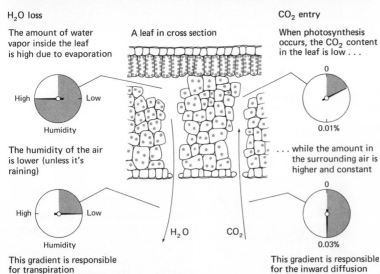

H$_2$O loss

The amount of water vapor inside the leaf is high due to evaporation

Humidity

The humidity of the air is lower (unless it's raining)

Humidity

This gradient is responsible for transpiration

A leaf in cross section

H_2O CO_2

CO_2 entry

When photosynthesis occurs, the CO_2 content in the leaf is low . . .

0.01%

. . . while the amount in the surrounding air is higher and constant

0.03%

This gradient is responsible for the inward diffusion of CO_2

FIGURE 7–11.

Leaf cross section and the mechanism of stomatal openings, according to the currently favored theory.

alone, one might speculate that leaves look the way they do because they have evolved anatomically so as to maximize their ability to capture CO_2 from the environment, and this does indeed seem to be the case. Leaves are typically thin and flat. This allows for a relatively large surface area to be exposed to incoming radiation. But more than that, it also means that CO_2 does not have to diffuse very far to reach the site of photosynthetic activity. This is not a trivial consideration since diffusion over distances larger than a few cells is painfully slow. Furthermore, there are many stomates on the lower epidermal surface of most leaves. Thus, CO_2 can enter the interior of the leaf directly *without* passing through the epidermal cells. This fact is itself a tremendous advantage to the plant because the diffusion of CO_2 (actually any gas) is more rapid in air than in water, and a cell is mostly water. Once inside the leaf, CO_2 still does not need to enter a cell immediately but rather can diffuse directly through the air passageways between the spongy parenchyma cells to the palisade layer where most photosynthetic activity occurs. The palisade cells in turn are constructed so that their chloroplasts are near an air space, again maximizing the efficiency of CO_2 transport and capture.

The exact area occupied by stomates is somewhat variable, depending on the species of plant, but from 1 % to 3 % of the surface area of the leaf is about average. CO_2 diffuses through the stomates into the leaf because photosynthesis depletes the internal concentration of CO_2 within the leaf, and therefore a concentration difference between the CO_2 in the outside environment and the inside of the leaf develops. In other words, CO_2, according to the laws of diffusion, will move from an area of high concentration (the outside environment) to an area of low concentration (the interior of the leaf) (see Figure 7–11).

One of the physiologically most interesting features of stomates is that they are not fixed pores within the epidermis but rather can be opened or closed, depending on the physiological state of the two specialized epidermal cells surrounding the pores called *guard cells*. Guard cells are unique in several respects, but the one anatomical feature that allows them to function in regulating the stomatal aperture is the peculiar thickenings in their cell walls (Figure 7–12). When the guard cells are filled with as much water as is physiologically possible, they become turgid and enlarge unevenly, producing a gap between two adjacent guard cells. We call this gap a stomate. However, when guard cells are not fully turgid, their geometry changes so that adjacent guard cells touch one another, abolishing the pore between them.

In order to appreciate fully the rather subtle physiology of the guard cells, you must understand the mechanism by which they take up and expel water and the environmental factors that trigger this response. In the past, botany texts such as this one simply stated that there were several theories, all as yet unproved, concerning the mechanism of water flux in guard cells. However, thanks to some rather elegant studies by Humble and Rasche at Michigan State University, we can now simplify the story considerably. It would appear that the water content of guard cells is regulated by the active (energy requiring) transport of potassium ions (K^+). Guard cells can take up K^+ from the surrounding medium or adjacent cells. This phenomenon increases the solute concentration of the cell, thereby causing water to be osmotically taken up (see Chapter 1) thus expanding the guard cells and producing a pore on the leaf surface. The pore ceases to exist when K^+ is pumped out of the guard cells, causing water to "follow" osmotically (Figure 7–13). Stomates will also close if the guard cells experience a direct water loss. Such situations are possible and similar to the conditions required for wilting—high temperatures and possibly low water availability in the soil.

We don't know precisely how the various environmental factors that induce stomatal opening or closing interact with the K^+ pumps. However, we can take light as an example and examine the information we do have. Typically stomates are closed during the night and open during the day. This makes sense in terms of overall plant strategy because it prevents unnecessary water loss through the stomates during the darkness when CO_2 is unnecessary. (Why?) One might therefore suspect that there is either some direct interaction between light and the K^+ pumps or an indirect effect of light, such as through the stimulation of photosynthesis, and the subsequent production of some product that would activate the ion pump.

FIGURE 7–12.   GUARD CELLS AND THE STOMATAL OPENING

There are many different types of guard cells throughout the plant kingdom. Each, however, exhibits some degree of differential thickenings in the cell wall which allow for opening and closing of the stomates as a result of turgor changes within the guard cells. a. This rather atypical style is found in the duckweed *Spirodela*. Note the cuticle (Cu), the nucleus (N), the vacuole (V), and the mitochondrion (M), and the substomatal chamber (SSC). Magnification 15,000×. b. The scanning electron microscope provides a different view of stomates. By looking through the stomate, one can see several mesophyll cells.

(a) From *Introduction to the Fine Structure of Plant Cells,* by M. C. Ledbetter and K. Porter, 1970, Springer-Verlag, New York, Inc., Fig. 7.5.2, p. 128. (b) From *Probing Plant Structure,* by J. H. Troughton and L. A. Donaldson, 1972, McGraw-Hill Book Company, Plate 29. Courtesy of the Department of Scientific and Industrial Research, Lower Hutt, New Zealand.

a.

b.

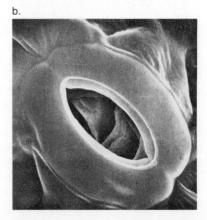

At the moment the latter proposal seems more likely, and in fact the energy harvest via the light reactions of photosynthesis may serve to prime the K^+ pump.

Stomatal opening is also regulated by the CO_2 content within the leaf's intracellular space, but that is another story which is even more poorly understood than the effect of light.

Stomates are necessary features of leaves in order to facilitate CO_2 entry. However, the necessity of having pores open to the environment also invariably results in the loss of water vapor from the leaves. This loss is called *transpiration*. Transpiration occurs because plant cells are composed in large part of water and the cell membrane of all cells is somewhat leaky with respect to water. This means that the cells (cell walls, really) bordering on the intracellular spaces within the leaf will lose water due to evaporation. As a consequence of evaporation the free space within the leaf is almost saturated with water vapor. On the other hand, the area outside the leaf (if it isn't raining) is far from saturated. A concentration difference exists between the water vapor inside and outside of the leaf, and water vapor will diffuse out of the leaf if the stomates are open

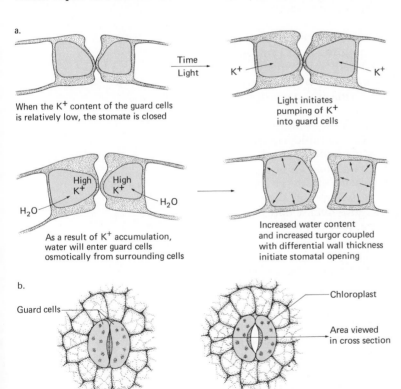

a.

When the K^+ content of the guard cells is relatively low, the stomate is closed

Light initiates pumping of K^+ into guard cells

As a result of K^+ accumulation, water will enter guard cells osmotically from surrounding cells

Increased water content and increased turgor coupled with differential wall thickness initiate stomatal opening

b.

Guard cells

Closed

Chloroplast

Area viewed in cross section

Open

FIGURE 7–13. MECHANISM OF STOMATAL OPENING

a. The steps in stomatal opening are shown in cross section. Note that it is the inner stomatal wall thickenings that are ultimately responsible for opening. b. The stomatal apparatus is shown in face view. Note the presence of chloroplasts in the guard cells—most epidermal cells lack chloroplasts. Presumably, directly or indirectly, the presence of chloroplasts allows the cell to produce the energy to activate the potassium (K^+) pumps.

Per day, midsummer		
Ragweed	6–7 quarts	
10 foot apple tree	10–20 quarts	
12 foot cactus	0.02 quarts	
Coconut palm	70–80 quarts	
Date palm	400–500 quarts	
Per growing season		
Tomato	100 days	30 gallons
Sunflower	90 days	125 gallons
Apple tree	188 days	1800 gallons
Coconut	365 days	4200 gallons
Date palm	365 days	35,000 gallons

TABLE 7–2.   TRANSPIRATION RATES

(see Figure 7–11). As water escapes, more will evaporate from the surface of the cells inside the leaf, thus maintaining the water-saturated condition of the intracellular space. Clearly this process of water loss and replacement will continue as long as the stomates are open, provided there exists a source of water to replace that lost to the environment. This ultimately means that water must be taken up from the soil and transported to the leaves.

Before continuing with water transport, however, it is instructive first to gain some idea of the magnitude of transpiration and to ask whether this process is advantageous to the plant or just a necessary evil. In addition, we will investigate the rather sophisticated means that plants living in a very dry environment employ to minimize water loss.

Perhaps we might first be inclined to doubt whether small microscopic pores on the surface of a leaf could contribute significantly to the amount of water lost by a plant, but they do indeed. In fact, well over 90% of the water lost from a plant occurs through stomatal evaporation. In some species of plants, this loss is 90% of a rather impressive volume of water. For example, a single adult corn plant may lose over a half gallon of water per day during the peak growing period. For an acre of corn, this amounts to a loss of something over 3,000 gallons per day. Equally impressive figures are also available for other species (Table 7–2). It is perhaps part of human nature to view such data and immediately ask, Just what "good" is transpiration? Of course we can only guess, but some possibilities seem reasonable. In the first place, as we will see shortly, transpiration is an important process in the chain of events that lead to water transport. That is, without transpiration it would be impossible for the plant to "pull" water from ground level to the highest leaves. This argument, however, becomes somewhat circular when it is realized that the transport of water itself is primarily necessary because transpiration occurs. Nevertheless, it turns out we cannot completely disregard this argument because the flow of water happens to be the mechanism by which most essential minerals (see Chapter 8) arrive at the growing regions of the plant. So perhaps we are on relatively safe ground if we say that transpiration appears to be indirectly responsible for the rapid delivery of mineral nutrients to developing cells.

There is another subtle benefit of transpiration, and this is *evaporative cooling*. This is a relatively easy concept to grasp in a general way but considerably more complex to understand completely. When sunlight strikes the leaf and is absorbed, only a small fraction is actually used in photosynthesis. The remainder of the light energy is transformed into heat, and therefore under most conditions a sunlit leaf will be warmer than the surrounding air. This phenomenon could be either useful or disastrous, depending on the tolerance and preference of the particular leaf in question. There are times, however, particularly in the tropics, when some mechanism to cool the leaf is essential to prevent injury or even death. The mechanism that has evolved is apparently transpiration since as water evaporates from any object, leaves included, that object is cooled. The magnitude of evaporative cooling is only in the range of 3° to 5° C when the atmospheric temperature is moderate (30° to 40° C) and thus of only marginal significance. But in very hot climates where air temperatures may reach 50° C, some plants are able to cool their leaves significantly by achieving very high rates of transpiration. In many cases this cooling effect is essential for survival because these plants could not otherwise tolerate the leaf temperatures that would develop.

Perhaps you are now convinced that transpiration can be a useful phenomenon. But just to place things in their proper perspective, we should consider the consequences if a plant cannot continually supply all the water lost by transpiration and also some of the adaptions that have evolved to minimize the chances of death by dehydration.

Plant Adaptations to Desert Living

The deserts in the United States are characterized by extremely high temperatures and a lack of moisture. Some of you may picture vast, rolling sand dunes, barren of life. There are sand dunes in most deserts, but they make up a relatively small part of the area as a whole, and even in dunes life abounds. Most of the desert consists of flat, brush-covered land interspersed with washes (which may change size and position dramatically from year to year) and small mountain ranges. The summers are long and hot, but the fall, winter, and spring can have very cold nights (below freezing) and warm days. Rainfall is infrequent but most commonly occurs in February. It usually measures less than 3 inches, but occasionally there are heavy cloudbursts in the summer months. Rainfall in the desert is exceedingly undependable, and there are numerous years on record when virtually no rain fell, emphasizing the fact that some truly remarkable adaptations to drought are needed for plant species to survive.

One approach to desert survival is to escape or avoid the dry conditions. Annual plants (plants that germinate, mature, and die within one year) do this by simply passing the driest part of the year as seeds. There are special adaptations built into these seeds that prevent germination until sufficient moisture is in the soil to allow them to develop to maturity. For example, some seeds contain a germination inhibitor in the seed coat, and approximately a centimeter of rain is required to leach it out before germination can occur. Other seeds will germinate with less rainfall but only when temperatures are cool. Thus, a heavy summer thundershower is not effective, but cool winter drizzles may bring spectacular results. Plants whose seeds do germinate with summer thundershowers have exceedingly rapid growth rates and can mature and set viable seeds in only a few weeks (Figure 7–14).

FIGURE 7–14.

Some plants, like this delicate annual *Eriophyllum,* may combine more than one strategy to face the dry desert conditions. Their small size and rapid growth rate allows them to complete their life cycle and produce seeds in the short time of the year when winter rains have made the desert more habitable. In addition, their stems and leaves are covered with a tangled mat of hairs which reduces the movement of air at the stem surface and decreases transpiration.

A number of other adaptations for coping with desert conditions are found among perennials (plants that live for several to many years). For example, some plants, instead of timing their exposure to the desert heat, grow only in a particular place (i.e., where there is adequate moisture). Palms, for example, and other plants that characterize an oasis need lots of water and therefore are found only in those desert areas where springs provide a plentiful water supply. Desert wash bottoms may also provide a suitable habitat for perennials due to subterranean springs.

Perennials growing away from oases and washes may escape the truly severe desert conditions in either of two ways. Herbaceous perennials, such as lilies and gourds, simply close up shop and pass through summer and fall dormant as bulbs or root stocks buried underground. With the cool winter rains they produce fresh shoots again and thrive in the exceedingly pleasant desert spring. Woody perennials cannot retreat underground in this way, but they can just as effectively close up shop by dropping their leaves. Ocotillo, for example, can produce a new set of leaves and lose them eight or ten times a year if necessary (Figure 7–15). Leaves are the organs of transpiration, and so leaf fall in response to drought is one way plants escape overexposure to desert conditions.

Other desert plants have developed different ways to deal

FIGURE 7–15.

Ocotillo can lose its leaves several times a year and grow a new set in response to alternating wet and dry conditions.

with adverse desert conditions. Faced with inadequate soil water, some plants have developed adaptations to maximize their ability to obtain all available water from the soil, while others, through specialized structures, retain within the plant body whatever meager water is available. Those desert perennials having an extensive root system are mostly shrubs, such as mesquite, occurring along desert washes and at the fringes of springs. They have roots which grow deep down into the soil where water supplies exist. In addition, many of the desert water supplies are extremely alkaline, so these plants must possess the ability to withstand high salt concentrations.

Elsewhere, the water table is too low to be reached, and the plants must absorb all the water they can during the short intervals when it rains. These plants generally have a substantial—wide but shallow—root system as well as some specialized adaptations for restricting water loss. One familiar adaptation is that exhibited by cacti (Figure 7–16). These plants have modified their growth and development to the point at which they no longer have recognizable foliage. Their leaves are represented by spines which have *no* stomates. The fleshy part of the plant is the stem which typically contains chlorophyll. The entire plant body above the ground is covered with a waxy cuticle which is impermeable to water. Other plants retain their leaves but have

FIGURE 7–16.

The leaves of cacti are represented by these spines and bristles which lack stomates and therefore conserve water. The fleshy part of cactus plants is the stem; it has a thick cuticle which restricts evaporation.

a special surface which greatly restricts water loss. For example, agave (Figure 7–17) has a thick, waxy coating overall which prevents water loss.

Another common adaptation is a fuzzy coating that covers the leaves and restricts surface air movements (see Figure 7–14). Reduced air movement means less evaporation and transpiration and thus greater retention of water. Finally, some plants conserve water by producing two different kinds of leaves. In the winter and spring these plants have large leaves whose properties are comparable to an average everyday plant. In the summer and fall, however, the large leaves fall off, and a new set of smaller ones appears. These smaller leaves have an exceedingly high osmotic concentration in their cells, about 10–15 times that of their spring leaves, and as a consequence the water is concentrated so that it evaporates only very slowly. It is not unlike the evaporation rates of pure water as compared with pancake syrup.

The big problem created by these various adaptations which restrict loss of water from plants is that they also restrict the entry of carbon dioxide. CO_2 is most often the limiting factor for photosynthesis and plant growth under natural conditions. In several hundred million years of the severest kind of natural selection, plants have been unable to solve this problem totally.

FIGURE 7–17.

Agave plants, unlike most cacti, have true leaves but reduce water loss by means of a thick protective cuticle and a few stomates.

We can see a clear causal relationship between the particular strategy a plant uses to survive the water problem on the one hand and its growth rate on the other. Annual plants and herbaceous perennials, which really escape the severest part of the year by going dormant in the form of seeds or underground roots, have rapid, even spectacular growth rates. On the other hand, the desert shrubs, which meet and adjust to the challenge of the desert environment, have very slow growth rates. Correlation is not necessarily causation, but in this case the relationship seems more than fortuitous.

A lesson for mankind becomes clear at this point. If adaptation to truly dry environments means slower growth rate, then agriculture under these conditions will be relatively unproductive. The reason is that agricultural plants must produce at a high rate, or they have little value to mankind. If plants are to maintain high productivity, they need to maximize CO_2 uptake, which in turn requires that the stomates remain open. But this allows high transpiration rates and requires a continuing water supply, and the whole point of it is lost.

These paragraphs were intended to give you an insight into some really interesting aspects of plant adaptations. Desert plants may not match the redwoods in their stature, but in character they reign supreme.

We have implied that virtually all the water (and it can be a lot) transpired from plant leaves ultimately comes from the soil. Therefore, if a plant is to survive, it must be able to extract the available soil water via its root hairs and to transport this water to all parts of the plant. At this time, then, it seems appropriate to begin a study of this long-distance water transport and the tissues involved.

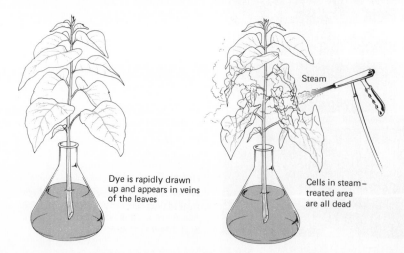

Dye is rapidly drawn up and appears in veins of the leaves

Steam

Cells in steam-treated area are all dead

FIGURE 7–18.

That water movement takes place in the xylem can be readily shown by the use of dyes and radioactive water. After the dye is drawn up into the plant, cross sections of the stem reveal that dye is found in the xylem. Furthermore, that living cells are not necessary for water movement is shown by the fact that transport will occur even after a portion of the stem has been treated with steam, thereby killing the cells in that area.

Plumbing, Water Pipes, and the Features of Water Transport in Plants

Before we delve directly into the old (1894) but still accepted theory of water transport, let's first assemble the major facts that any such theory must explain. First, relatively large volumes of water must be transported at times to compensate for the water lost as vapor to the environment. Second, some mechanism is necessary which will account for lifting water against the pull of gravity to a height of some 350 feet or so (some tall trees reach this height). Third, all the long-distance transport of water occurs in nonliving xylem elements, and therefore biological energy is not involved. This latter point was skillfully demonstrated years ago by various simple but definitive experiments (Figure 7–18). We must also account for the fact that the uptake and transport of water is correlated with the amount of transpiration, thus implying that there is some physiological connection between the two processes (Figure 7–19). Last, a mechanism is needed to account for the extremely rapid values that have been recorded for water transport in the xylem, up to 50–75 cm per minute under the most favorable conditions.

It is easy to look at all these requirements and then retreat in horror with the feeling that they would be impossible to fit into a single mechanism. However, there is one theory that does seem to explain all the facts and is easiest to understand if we begin by tracing in some detail the path of water from the leaf to the root.

When transpiration is occurring, water evaporates from the leaf cells bordering the intercellular spaces of the leaf. This loss of cellular water reduces the volume of the cells, and their turgor pressure falls. This means that these particular cells then have the *potential* to take up more water osmotically until they attain their original volume and turgor (review the appropriate parts in Chapter 1 if this is unclear). Plant physiologists call this potential the *water potential*, and define it as the energy status of water within the plant. By definition water will flow from an area of high water potential to an area of low water potential, and the water potential difference between the two areas can be considered to constitute the force behind water transport. All this might just seem to complicate your understanding of osmosis, but the terminology actually becomes very useful, if not essential, as we proceed. Our main point is that the cells facing the intercellular spaces are at a low water potential when transpiration is occurring. The adjacent interior cells haven't lost as much water, if any, via evaporation, and therefore water will flow osmotically from the area of high water potential (the interior leaf cells) to the area of lower water potential

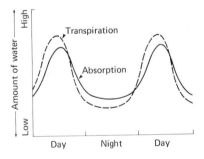

FIGURE 7–19.

The relationship between water uptake (absorption) and transpiration. Note the daily changes in the rate of both phenomena and that the peak rate of absorption lags slightly behind transpiration. Further reading in the text and study of Figure 7–20 should enable you to explain why this would be expected.

potential (the cells bordering the intercellular space). This flow rapidly results in a decrease in volume of the interior cells, and they will osmotically pull water from their neighbors. Rather rapidly a chain reaction or transpiration pull is initiated from leaf cell to leaf cell so that there will eventually exist a potential difference between the water-filled xylem cells in the veins of the leaf and those leaf cells adjacent to the xylem. The net result is a constant flow of water from the xylem into the leaf which is rapid enough to compensate for transpiration loss which ultimately initiated the flow (Figure 7-20).

At this point to simplify matters, consider the xylem to enclose long, continuous water columns which move upward as a unit. Of course, there must be a continual supply of water at the lower end of the pipe for the events discussed above to occur; this source is the root. As a result of water being removed from the uppermost xylem elements, a stress is placed on the entire water column within the xylem. We can correctly view this stress as equivalent to a decrease in the water potential of the xylem, and therefore a driving force is created so that water will flow into the xylem from the adjacent root cells. These root cells in turn draw water in from their neighbors to replace that which is lost, and this process continues until a root hair is reached (Figure 7-21). It is the function of a root hair to acquire, again via a water potential difference, water from the soil. This removal operation can be an easy, difficult, or impossible task,

FIGURE 7-20.

This is a very simple model showing the onset of water transport in plants. The gauges represent imaginary water potential sensors. The arrows represent turgor pressure (directed outward) or tension (directed inward). The long arrows indicate water flow. Note that evaporation initiates water transport, and that uptake from the root lags behind evaporation somewhat in time. Compare this with Figure 7-13.

From *The Living Plant,* 2d ed., by Peter Martin Ray. Copyright © 1963, 1972 by Holt, Rinehart and Winston, Inc. Reprinted by permission of Holt, Rinehart and Winston, Inc., Fig. 5.5, p. 82.

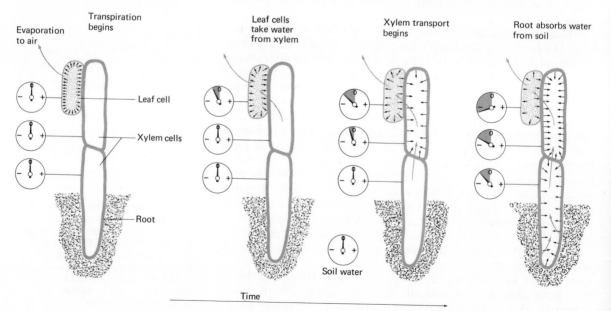

Evaporation to air — Transpiration begins — Leaf cell — Xylem cells — Root — Leaf cells take water from xylem — Soil water — Xylem transport begins — Root absorbs water from soil

Time

depending on the water content of the soil and how tightly the individual molecules are bound to the soil particles (Figure 7–22).

The Tension-Cohesion Principle, or How Much can Water be Stretched?

The basic theory presented above implies that there must be a large water potential difference between the soil water (highest water potential) and the air (lowest water potential). It further implies that all the plant tissues connecting the soil and air vapor must also have pressure differences in the correct direction. Unfortunately, it's a bit tricky to determine directly and simultaneously the differences in water potential in a single plant at various places along the route of transport, so we don't have very much data available. However, the data that are available suggest that the idea is correct. Furthermore, our model would predict that water flow would immediately stop if the water potential of any part of the system did not conform to the rules and regulations above. This in fact does happen regularly. For example, when the stomates close, transpiration ceases and so does water flow. This is perfectly logical because when the stomates close, the intercellular spaces within the leaf will rapidly become saturated with water vapor. This means that water will no longer evaporate from the cells bordering the intercellular spaces, and these cells will no longer have a water potential lower than their interior neighbors. Without a water potential difference between cells, there cannot be a flow of water (turn back to Figure 7–20 and try to determine the chain of events when transpiration ceases).

For a second example we can switch to the other end of the line and see what happens when the soil gets extremely dry. Water is bound to soil particles with varying degrees of tightness by attractive forces. As the water content of the soil decreases, it becomes more and more difficult for a plant to "extract" soil water (see Figure 7–22). We can express the degree to which the water is attracted to the soil particles in terms of the water potential difference (the units are called bars) necessary to cause a flow of water from the soil into an adjacent root hair. It should be clear from Figure 7–22 that this potential will vary greatly, depending on the water content of the soil. When the soil is very dry, the remaining water is bound extremely tightly, and it takes a very low water potential within the cell to initiate flow. Thus, at some point as the soil dries out, the root hairs will be unable to achieve a low enough water potential to initiate

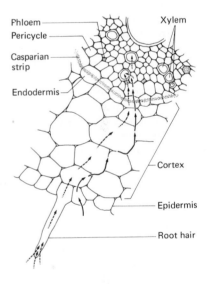

FIGURE 7–21.

This figure shows the two principal pathways of water transport within a root. The center arrows indicate a route that is strictly cell to cell. However, a certain portion of the water flow can take place along the cell walls, indicated by the other series of arrows. In either case, water flow is possible only because there is a water potential difference between the soil and root hair and between the root hair and the xylem.
From *Plant Anatomy,* 2d ed., by Katherine Esau, 1965, John Wiley and Sons, Fig. 17.14, p. 517.

water flow. When this happens, the entire transport process will cease, and since evaporation may continue from the leaf cells, dehydration and wilting will eventually result. In the extreme case, when dehydration becomes so severe that cellular function is limited, death of the plant will result.

One more feature of the transport system deserves special attention, and that is the nature and peculiar properties of the continuous water column within the xylem. This was an extremely troublesome point for early investigators for the simple reason that many plants are more than 33 feet tall. This figure is sort of a magic number for those who work in this area of research because a vacuum pump cannot pull water beyond this height. You might suggest that this is a trivial consideration which results from our inability to build a stronger pump, but it's not so. Rather, when a stronger vacuum is applied, the water in the column boils even at room temperatures (21°C), and the column breaks. Because plants can lift water to a height of 350 feet or so, something special is going on. The solution to this seemingly unsolvable problem is actually rather straightforward and is incorporated into the *tension-cohesion principle* of water transport.

This principle states that water is pulled up in the xylem directly, not by a vacuum. That is, we begin with a water column continuous from the roots to the leaves. Starting with the earliest growth of a germinating seed and continuing on to the end of its days, the cell-to-cell continuity of the water column is never broken. Now, add to this the fact that water molecules

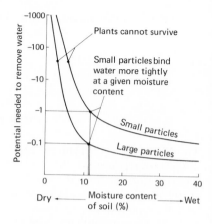

FIGURE 7–22.

Water is held in the soil by the cohesive properties of water molecules and by changes on the surface of the soil particles. The water potential needed to remove soil water varies according to the moisture content of the soil and the size of the particles (e.g., sand to clay) that make up the soil.

From "Soil moisture absorption curve for four Iowa soils," by M. B. Russel, in *Soil Science Society of America Proceedings* 4:51–54 (1939).

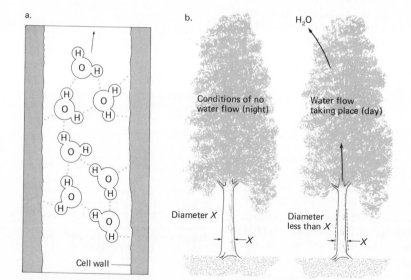

FIGURE 7–23.

a. Viewed on a microscale. Hydrogen bonding between water molecules occurs because of the attractive force created by a slight negative charge on the oxygen atom and by a slight positive charge on the hydrogen atoms of every water molecule. The attractive forces are indicated by the dotted lines. In a similar manner, hydrogen bonding can occur between water and the cell walls of the xylem elements. As the water column is pulled upward, it exerts an inward pull on the wall of the xylem elements. b. Viewed on a macroscale. The inward pull on the xylem elements can be detected by a decrease in trunk diameter during periods of rapid water flow.

behave differently in a very small diameter pipe (such as found in the xylem) than they do in a larger diameter pipe (such as is used to demonstrate water boiling under reduced pressures). The reason for this behavior isn't fully understood, but two points undoubtedly are important. First, we know that water molecules attract one another and form hydrogen bonds (Figure 7–23). Second, we know that in xylem elements, hydrogen bonding occurs between water and the molecules that make up the cell walls of the xylem. These two facts make up the cohesion part of the principle. Remember cohesion simply means that the water molecules in the xylem are held together in a continuous chain and that this column of water is also "attached" to the sides of the tracheids or vessels. The net result is a cohesive structured water column which resists being broken or pulled apart by virtue of internal and external hydrogen bonding. Thus the water columns within plants do not depend on a vacuum for support. They are already continuous (by virtue of continual synthesis at the top of prefilled cells) and are held in place by cohesive internal forces and also by a tendency of the column to adhere to the cell walls.

The tension aspect of this principle arises as a result of water being withdrawn from the top of the column. That is, as water is withdrawn, the remaining water molecules within the column become stressed and exert a pulling force on the lower molecules and on the cell walls of the xylem (see Figure 7–23). In some respects this force may be conveniently thought of as a suction force pulling inward on the cell walls and initiating a pressure drop in the xylem which will ultimately draw in water from the root cells adjacent to the lower xylem elements. The important differences in any direct analogy between a water pipe and a suction pump should be apparent at this point: (1) The entire xylem system is continuous, therefore there is no air trapped within it. (2) The narrowness of the xylem elements allows for a rigid cohesive structure in which water is not as easily vaporized as it is in a wider pipe, and therefore the water can withstand the large "suction" (better pulling) imposed on it during rapid transpiration. (3) There is a system of check valves within the xylem which prevents any air bubbles from moving elsewhere to destroy other continuous columns.

The third point so well illustrates once again the beautiful relationship between structure and function in plants that it is worth pursuing briefly. You know from this chapter and the previous chapter that tracheids are long, narrow cells that are dead at maturity. They differ most from vessels in that their ends are not open but rather have thin areas called pits which are points of discontinuity in the secondary wall (Figure 7–24). This relatively thin end covering offers very little resistance to water flow, and because tracheids overlap one another, we can consider the water column to be continuous. Furthermore, the pits in the end walls of tracheids prevent the passage of air bubbles. Therefore, should an air pocket form in a tracheid, it is trapped there and while this may destroy the cohesiveness of a single water column, it cannot affect others. This system has another system of checks and valves which also insures success. Vessels and tracheids may have small pores on their side walls so that in the event the end wall of one particular cell is blocked by an obstruction, water can move laterally into the next column and then upward. Again, however, should an air bubble form, it cannot move laterally for the same reasons.

Vessels, it should be noted, are somewhat more vulnerable than tracheids to serious disruptions of their water columns since they lack end walls, This means that if a bubble forms in a given vessel, the entire column is disrupted. Consequently, plants with a high ratio of vessels generally produce new functional vessels every year. Furthermore, this seeming disadvantage must be viewed in light of the functional advantages that vessels have over tracheids. That is, since they have a larger

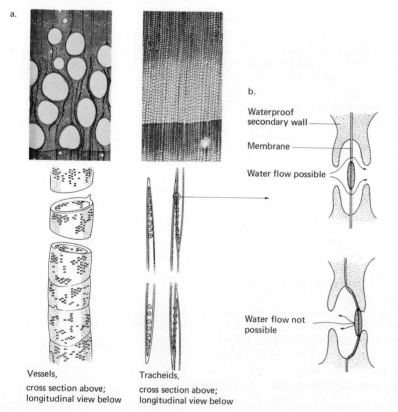

a.

b.

Waterproof secondary wall

Membrane

Water flow possible

Water flow not possible

Vessels,
cross section above;
longitudinal view below

Tracheids,
cross section above;
longitudinal view below

FIGURE 7–24.

a. The two principal types of *plumbing* in plants. Note that the individual tracheids are connected by pits. In vessels the end walls are almost completely eliminated, but pits are present on the sidewalls. It is important to realize that in both types the pits act as a sort of valve system to seal off individual elements if an air bubble should develop. b. The details of the *valve* system in xylem elements. The pit as it is normally (top). The pressure is almost equal in both sides (both xylem elements), and water can diffuse across the pit membrane. What happens when a bubble forms (bottom). Such a condition initiates an unequal pressure, and the pit membrane is forced to one side. The waterproof wall is therefore continuous, and the offending member of the chain is sealed off.
(a) From "How sap moves in trees," by Martin H. Zimmerman, Copyright © 1963 by Scientific American, Inc. All rights reserved. *Scientific American.* (b) From *Plant Anatomy*, 2d ed., by Katherine Esau, 1965, John Wiley and Sons, Fig. 3.3, p. 40.

diameter, they do not offer such a large frictional resistance to water flow and are thus more efficient. This consideration appears to be important for many plants; studies indicate that as plants evolve they favor production of very wide vessels and tend to eliminate the production of tracheids.

A summary of the water transport mechanism is found in Figure 7–25. Be sure to note the various forces involved in the system and to reread the appropriate text material if you do not fully understand the entire mechanism.

FIGURE 7–25.

Summary of the components of the water transport system. The imaginary gauges illustrate how water potential varies in different parts of the pathway. The gauges are marked in a logarithmic scale (except near 0) to illustrate the wide range between the soil, plant, and air.

(left) From *The Living Plant*, 2d ed., by Peter Martin Ray. Copyright © 1963, 1972 by Holt, Rinehart and Winston, Inc. Reprinted by permission of Holt, Rinehart and Winston, Inc., Fig. 5.2, p. 75. (right) Adapted from *Plant Physiology*, by F. B. Salisbury and C. Ross. © 1969 by Wadsworth Publishing Company, Inc., Belmont, California 94002. Reprinted by permission of the publisher, Fig. 7–11, p. 127.

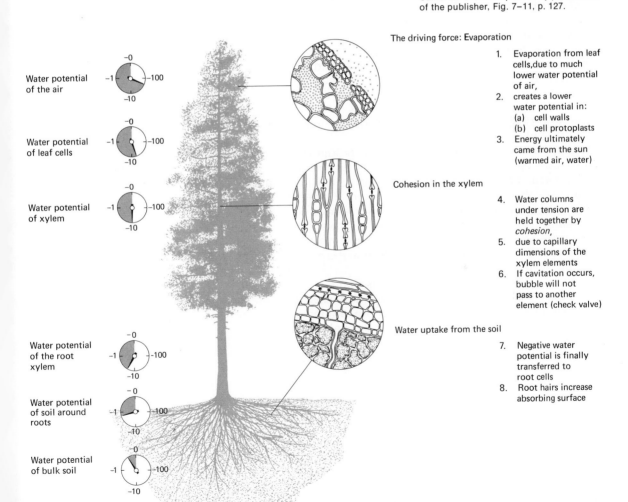

Water potential of the air

Water potential of leaf cells

Water potential of xylem

Water potential of the root xylem

Water potential of soil around roots

Water potential of bulk soil

The driving force: Evaporation

1. Evaporation from leaf cells, due to much lower water potential of air,
2. creates a lower water potential in:
   (a) cell walls
   (b) cell protoplasts
3. Energy ultimately came from the sun (warmed air, water)

Cohesion in the xylem

4. Water columns under tension are held together by *cohesion*,
5. due to capillary dimensions of the xylem elements
6. If cavitation occurs, bubble will not pass to another element (check valve)

Water uptake from the soil

7. Negative water potential is finally transferred to root cells
8. Root hairs increase absorbing surface

PART TWO

Food Distribution in Plants

For a moment let's think about animals instead of plants, and review the way in which food materials are distributed to the various cells of the body. First, animals ingest ready-made food material. This material is then gradually degraded as it moves through the digestive system, and at various points along the way the nutrients are absorbed into the circulatory system. Internal distribution and delivery are then accomplished via blood flow made possible by the action of a muscular pump, the heart. We don't mean to imply that this system is simple—far from it— but it *is* easy to visualize mechanistically.

Now contrast this with the problem of food distribution in plants. There is no heart nor even an analogous structure, yet plants, too, somehow distribute food materials over rather long distances. The photosynthetic machinery is located within the leaves, nevertheless all parts of the plant require and receive the products of photosynthesis.

At this point we will investigate the facts and theories concerned with the flow of nutrients in plants. Unfortunately, this phenomenon, called *translocation*, is not easy to visualize mechanistically nor is it fully understood. However, it is a rather intriguing and obviously necessary aspect of a plant's physiology. It seems most appropriate to begin our investigation of the translocation process by reviewing and studying in some depth the structural features of the cells that carry on translocation, the sieve tube elements.

Sieve tube elements are elongated cells that occur end to end within the phloem, forming a pipelike network where the flow of dissolved food materials takes place. An analogy to xylem elements is tempting, but any similarity ends abruptly with the pipelike arrangement of these cells. Sieve cells are living when functional and are incapable of transporting nutrients when even slightly disturbed—a fact of considerable annoyance for those who wish to study translocation! Second, sieve cells do not have the distinctive thick, lignified secondary walls that characterize xylem elements but are recognizable by some highly individual intracellular features. Perhaps the most remarkable feature is that during the development of a sieve cell, the nucleus disappears! In fact, sieve cells characteristically have a very low population density of virtually every common cellular organelle. This lack of subcellular machinery in cells that are alive and capable of carrying on work is somewhat surprising.

The explanation appears to be that sieve cells, for some un-known reason, depend on associated cells, companion cells, for much of their energy requirements and information-processing functions. The end walls between adjacent sieve tube elements contain porous areas, and it appears from electron micrographs that the plasma membranes from adjacent cells are continuous through these pores (Figure 7–26). This particular feature is of direct functional interest because it means there is no barrier to the transport of dissolved materials from sieve tube cell to sieve tube cell, and thus a series of open communication lines exist throughout the entire phloem. Obviously, sieve tubes present an entity far removed from what one "normally" pictures as a typical living, functioning cell.

As was previously mentioned, sieve cells are extremely sensi-tive to injury and plugging. In order to study even such basic questions about the translocation process as what nutrients move within sieve cells, investigators have had to be rather de-vious (Figure 7–27). It so happens that the aphid, a common garden pest, is a phloem feeder. Furthermore, when an aphid's stylet penetrates a leaf or branch, it taps a *single* sieve cell in such a delicate way that the cell remains functional. The bene-fit for the aphid is clear because phloem sap is rich in sugars. In fact, the source is so plentiful that often aphids exude excess food as honeydew which in turn supports the nutritional needs of certain species of ants and bees. But that is another story. Plant physiologists benefit from this fortunate circumstance in a subtle way. Aphids, a few or many, are allowed to tap the phloem of an experimental plant. They are then anesthetized, and the body carefully cut away from the stylet. When this operation is successful, the stylet remains in a sieve tube ele-ment and continues to exude the phloem contents, often for several days. It is now possible for the skilled, steady-handed investigator to collect the exudate from the stylet stump and analyze its contents chemically. Strange as it may seem, only

FIGURE 7–26.

Sieve tube elements are very fragile cells, and therefore it is difficult to make preparations that clearly show all the details wanted in a given prep-aration. To get around this problem, a drawing of two photomicrographs of phloem cells are shown (left). The larger "photomicrograph" gives some idea of the overall shape and ar-rangement of sieve tube cells and companion cells. The smaller "photo-micrograph" gives an idea of the porous end plates in sieve cells. For clarity, a diagrammatic sketch of several sieve tube cells and their as-sociated companion cells is also in-cluded (right). Note the nuclei in the latter cells and the lack of such organelles in the former cells.
(left) From *Anatomy of Seed Plants*, by Katherine Esau, 1960, John Wiley and Sons, Figs. 11.3, 11.4, pp. 126, 127.

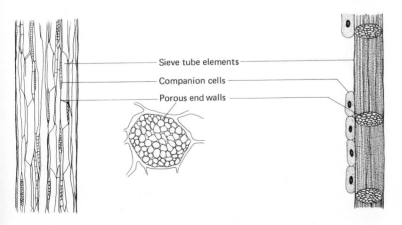

Sieve tube elements ——
Companion cells ——
Porous end walls ——

in this way is it possible to obtain pure samples of phloem exudate. It has been determined from such procedures that the phloem stream carries chiefly sucrose and also substantial amounts of other sugars containing 3 or 4 basic sugar units. Furthermore, the concentration of nutrients is a whopping 10–30% by volume. The same technique has been used to establish that there are also lesser, but detectable, amounts of amino acids and inorganic ions flowing within the sieve tube transport system.

The speed at which the nutrients flow within the sieve tubes can also be estimated via the aphid stylet technique; however, radioactive tracer methods are probably more suitable for quantitative studies. In a typical experimental situation radioactive CO_2 is fed to a leaf. The $^{14}CO_2$ becomes incorporated via photosynthesis into various compounds, and some of these are transported to other areas within the plant. The rate of translocation can then be determined by measuring how far down or up the stem the ^{14}C-label has traveled in some fixed time interval (Figure 7–28). Typically, rates on the order of 100–200 cm per hour are found, although higher rates have been occasionally reported.

Consider the following facts: (1) Nutrients flow within the sieve cells of the phloem, and such cells have a rather peculiar anatomy. (2) Concentrated levels of sugars appear to move. (3) The rate of flow is on the order of 100 to 200 cm per hour. How does the system operate? The question cannot be answered with any certainty, but most people favor a mechanism known as *mass flow*. In its most basic form the mass flow theory simply

FIGURE 7–27.

The aphid technique for analysis of phloem sap.
From *Plants, Viruses, and Insects,* by Katherine Esau, 1961, Harvard University Press, Fig. 15, p. 90. © 1961 by the President and Fellows of Harvard College.

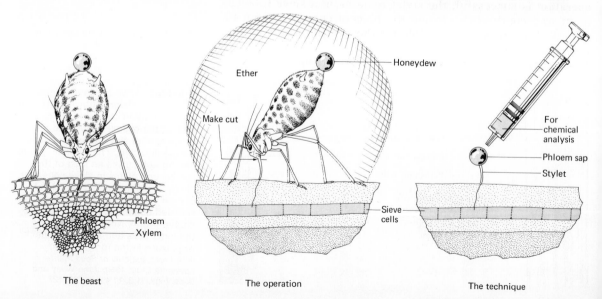

The beast   The operation   The technique

states that within the phloem there exists a pressure gradient and that the contents of the sieve tubes will respond to this gradient by flowing from an area of high physical pressure to an area of low pressure. Let's consider this proposal in a little more detail by first examining the end of the system that supplies the nutrients, the photosynthetic leaf.

As you know, a healthy leaf produces, via photosynthesis, more sugars than are necessary for its own cellular activities. The excess or overflow sugars, it is proposed, are actively pumped (via a membrane pump) into sieve tube elements within the leaf. In other words the cellular fluid within the leaf's sieve tubes is very rich in sugars. This necessarily means that, as the solute (osmotic) content increases, there will be a tendency for water to be drawn osmotically into these particular sieve tube cells from the nearby xylem. Carrying this a step further, we can get to the meat of the situation. Sieve tube elements are rather inelastic; therefore, as water is drawn into the elements located in the leaf, high turgor pressures are created. Now consider the situation in a nonphotosynthetic part of the plant. All the cells in a nonphotosynthetic area need nutrients; therefore, sugars will be continually removed from sieve tube cells. As this occurs, water will osmotically leave the sieve tubes. The net result is that sieve tubes in nonphotosynthetic parts of the plant will have a lower turgor pressure than do sieve cells at the sites of production. In short, it is possible to imagine that turgor pressure differences are created within the sieve tube network and that the highest pressures will exist in the areas of food production ("sources") and that the lowest pressures will exist in the areas of food utilization ("sinks"). The mass flow concept, then, states that the fluid within the sieve cells will respond to these pressure differences and flow from areas of high pressure to areas of low pressure (Figure 7-29).

As nice as this proposal might look on paper, you should be cautioned that mass flow is not proved. In fact, one would be

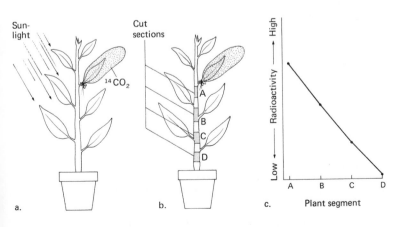

a.

b.

c. Plant segment

FIGURE 7-28. THE RATE OF TRANS-LOCATION

a. *Preparation.* Radioactive CO_2 is fed to a leaf where it is incorporated into organic molecules by photosynthesis. b. *Termination.* Plant sections are removed from below the point of feeding after 30 minutes to detect the radioactive compounds as they move down the stem. c. *Result.* The amount of radioactive CO_2 in each cut section is determined, and the rate of translocation or nutrient flow can be calculated.

hard pressed to think of another area in plant physiology that is such a hornet's nest of claims, counterclaims, and confusion. Yet, in spite of all the uncertainty and flux, there is perhaps a bit of reassurance in noting that of the many proposals put forth over the last 40 years to explain translocation, only the mass flow theory has remained viable for any significant length of time. A "time test" is a rather odd sort of justification for any scientific theory; nevertheless, it is a point. But there is more to mass flow than simply speculation and a long life. For example, physiologists have been able to demonstrate clearly that there is an osmotic gradient between the sieve cells in the area of a source and the sieve cells of a sink; the main postulate of mass flow has been experimentally verified. Unfortunately, this doesn't necessarily mean that the entire proposal is correct. It is still possible to argue, as many investigators do, that the magnitudes of the reported osmotic differences are insufficient to generate the turgor pressure differences that are believed necessary to achieve the known rates of nutrient flow (some 100–200 cm per hour).

Perhaps you can more fully appreciate the problem if you realize that a calculation of the pressure differences between sieve cells is very difficult to attempt and is made nearly meaningless by another separate but related controversy. Many people are skeptical about the mass flow concept because the end walls of the sieve tube elements have pores with a small diameter. It is

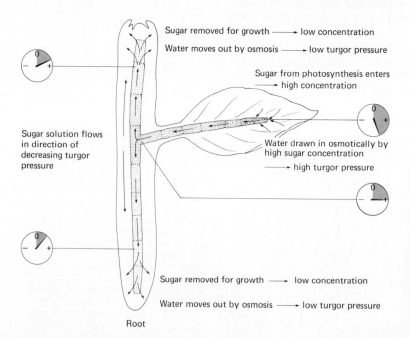

Sugar removed for growth ⟶ low concentration

Water moves out by osmosis ⟶ low turgor pressure

Sugar from photosynthesis enters ⟶ high concentration

Sugar solution flows in direction of decreasing turgor pressure

Water drawn in osmotically by high sugar concentration ⟶ high turgor pressure

Sugar removed for growth ⟶ low concentration

Water moves out by osmosis ⟶ low turgor pressure

Root

FIGURE 7–29.   THE MASS FLOW THEORY

All the components of mass flow are shown here. If you understand this, you now grasp the currently favored mechanism to explain the long-distance transport of nutrients in plants. The dots represent sugar concentration, and the gauges are imaginary turgor pressure sensors.
From *The Living Plant,* 2d ed., by Peter Martin Ray. Copyright © 1963, 1972 by Holt, Rinehart and Winston, Inc. Reprinted by permission of Holt, Rinehart and Winston, Inc., Fig. 7.5, p. 118.

claimed by the skeptics that the *resistance* to flow created by these small pores must be "very large," and a mass flow mechanism is not very appealing. Other scientists are quick to point out that the resistances "aren't too large." Unfortunately, it's a bit difficult to know which side has the advantage since neither can accurately define how much resistance is actually offered by the pores; therefore, the terms *very large* and *not too large* do not mean very much.

Until this question is settled, it is impossible to say what pressure differences are needed to give a particular flow rate. It would, however, be unfair to say that this is the sum total of the controversy. A few investigators claim that they can actually see globular materials move from sieve cell to sieve cell in strands, thus implying that there are really tubes within tubes— a concept which doesn't fit well with any known mechanism. Most workers discount these observations of globular flow entirely and suggest that they are experimental artifacts. Obviously this particular area of plant physiology is troublesome for those who insist on clear-cut answers. While the challenge may be stimulating for scientists, you may find it a bit confusing. At least the mass flow concept is sound in physical terms and will probably remain the choice of most investigators until there is evidence to prove it incorrect. Furthermore, you can now perhaps appreciate why it was originally stated that the distribution of nutrients in animals is reasonably straightforward but mechanistically subtler in plants. This also suggests that there is much good plant science yet to be done in translocation as well as in other areas—our next topic, for example.

The Flowering Response in Plants

We have seen that the plant body undergoes many changes as it progresses from a seedling to a mature individual. However, the most spectacular aspect of plant development is perhaps the flowering stage. This event is spectacular not only because it is pleasing to the eye (Figure 7–30) but also because it occurs with abruptness and reflects fundamental physiological changes. Plants can grow as strictly vegetative individuals and then, just as if some switch were thrown, the meristems cease to carry on their previous tasks and differentiate into flower buds. You already know something about the nature of this switch as a result of your direct observations of gardens and plants growing in the wild. Flowering is correlated with the seasonal changes. Of course, there are many environmental variables associated with these changes, so without further information, we can only guess what factor or combination of factors act as the flowering trigger. Fortunately, there is no need for guessing and hasn't been since 1918 when two workers at the U.S. Department of Agriculture discovered that the flowering of many plants is controlled by the length of day or *photoperiod*. As we briefly saw in Chapter 2, plants have some way of measuring the length of the day, and day length is related directly to the induction of flower formation. These pioneering experiments are simple enough and so clearly illustrate the points to be established that it is worthwhile to review briefly the observation that led to the photoperiod concept.

Maryland Mammoth tobacco is a mutant that arose by chance in an experimental plot in Maryland. This plant attracted attention primarily because it grew to a great height, but it was also of interest because it seemed that Maryland Mammoth would simply never flower (at that latitude). This latter feature was particularly frustrating because the experimenters were

FIGURE 7–30.

The flowering process is spectacular for a number of reasons, as illustrated here.

hoping to use the mutant in breeding experiments aimed at producing bigger and better commercial plants. So after a time, fearing that the oncoming winter would kill the new giant variety, the mutant plant was placed in a greenhouse and propagated by cuttings. Finally, in mid-December, flowering occurred, and seeds were eventually obtained. The next year the same thing happened. Maryland Mammoth grew beautifully in the field but did not flower during late summer as did the nonmutant plants. But the variety did flower again around December when snugly settled in the greenhouse. One obvious question is, What's so special about life in a Maryland greenhouse in mid-December? Many possibilities were investigated, and to make a rather long story somewhat shorter, it was finally discovered that this particular mutant tobacco plant required short days to trigger flowering. This observation was further verified and experimentally demonstrated by transferring the mutants during summer from the field to artificially lit growth chambers set on a short day (long night) regime. The plants flowered. When the winter days were lengthened via artificial lighting (or put another way, the nights shortened), no flowering was observed.

Many investigators quickly became interested in the photoperiodic concept, and rather soon it became apparent that the relationship between flowering and photoperiod wasn't a special feature of Maryland Mammoth but was common to many different plants. In fact, as it turned out, there is a general class of plants that flowers only if the *nights* are *longer* than some critical value: *short day plants*. In nature these plants would normally flower in late summer or fall. A few examples of common short day plants include soybean, chrysanthemum, strawberry, violet, and poinsettia. Another class of plants only flowers if *night* length is *shorter* than some critical value: *long day plants*. Such plants include red clover, oats, spinach, and dill.

In addition to these two types of flowering responses, there is also a third called *day neutral plants*. As the name implies, these plants are insensitive to daylength, and flowering seems to occur when a certain physiological maturity is reached. Perhaps it might at first seem that day neutral plants would be the best choice if you wanted to study the flowering process, but this isn't true. Flowering in photoperiodic plants can be precisely triggered by altering the lighting in a growth chamber. Thus, the investigator controls the switch. With day neutral plants, not only can't we precisely control the switch but we only have a general idea about what the switch might be. Thus, most of our knowledge about flowering has come from plants that respond to daylength, and we will emphasize only these types in our discussion.

Perhaps the most obvious question to ask is, What part of the plant is responsible for measuring the photoperiod? Logically, the plant apex is a prime suspect as the photoperiodic receptor because it is this region that will eventually develop into a flower. However, it turns out that logic (or at least this simple logic) fails to predict the right answer. The *leaves*, not the apex, are the receptors for the photoperiodic stimulus.

How can this be shown? Several individuals of a given species exhibiting photoperiodic behavior, for example Maryland Mammoth tobacco, are subjected to different photoperiodic experiences. Only the apex of one plant is exposed to short days, while the remainder of the plant is exposed to long days. What will happen? It won't flower. Or all the leaves of a given plant could be removed and the remaining axis exposed to short days. The plant may survive the surgery involved, but it still won't flower. However, if a *single* leaf is exposed to short days while the remainder of the plant body is kept under long days, flowering does occur (Figure 7–31). Thus, the pattern is clear; the leaf somehow perceives the stimulus, and the terminal bud, which is some distance away, flowers. Communication is therefore implied between the leaves (i.e., the receptors) and the bud (the site of the final event). Many people believe that this communication results from the production of a hormone called *florigen*. It would appear to work in the following way. The leaf responds to the proper photoperiod by producing florigen. Florigen is then somehow transported from the leaf to the bud where it initiates

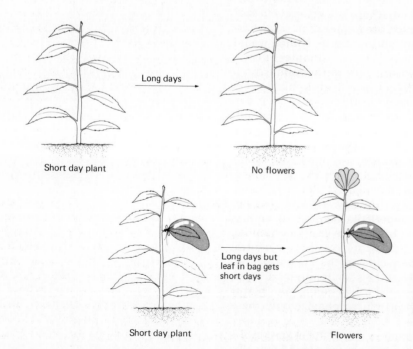

Long days

Short day plant

No flowers

Long days but
leaf in bag gets
short days

Short day plant

Flowers

FIGURE 7–31. LEAF PERCEPTION OF THE PHOTOPERIOD

It has been established that the leaf is the plant organ that perceives the photoperiod stimulus for flowering. There is then some means of communication between the leaf and the flower-forming area.

the chain of events that eventually causes a vegetative bud to become a flower. The existence of florigen is based on the kind of basic reasoning stated above as well as on more solid evidence based on grafting experiments.

Two individuals of the same species can be grafted together under lighting conditions that do not induce flowering. After the graft union joins the two plants, one of the individuals can be given an inductive photoperiod while the other is maintained under noninductive conditions. Both plants will flower, suggesting that a messenger molecule produced in the induced plant crossed the graft union and induced flowering in the plant kept under the noninductive photoperiod. All of these data, however, are indirect. What we really need is a plant extract containing florigen. However, such *direct* evidence is nil, for despite many attempts florigen has never been identified or its existence directly demonstrated. It's a rather disappointing story, to be sure, but a difficult one to appreciate fully unless you know how tricky such a search can be.

An example should give you some feeling for the problem. At one point in the history of florigen research (or perhaps it should be called search), it looked as though there was a chance that the flowering hormone was gibberellin. It was discovered that gibberellin could induce flowering in many long day plants, even if the plants were continuously maintained under short days. Thus, for a time it looked as if gibberellin was the florigen. Unfortunately, this did not prove to be the case because it was later discovered that gibberellin *cannot* induce flowering in short day plants. Some of you might be tempted to speculate that there are two kinds of florigens, one for long day plants (gibberellins) and one for short day plants (unknown). However, even this rationalization turns out to be incorrect because grafting experiments between long and short day plants have clearly shown that both long *and* short day plants respond to the same messenger hormone (Figure 7–32). It remains unknown just what florigen is chemically, and it is equally uncertain what relation, if any, there is between florigen and gibberellin. Other plant hormones have been tested for possible florigenlike activity, but these have proved rather uninteresting. The one exception is auxin which does have some activity in several species, but this finding is of interest only because these isolated species are of commercial importance. Auxin is definitely not *the* florigen.

When one discusses the concept of a photoperiod for short and long day plants, a mental picture emerges that the length of the light period is the critical factor for the induction of flowering. This isn't so. Rather, it is the length of the *dark* period that is the important factor. Put another way, short day plants require a dark treatment *longer* than some critical period in order to flower, while long day plants require a dark period *shorter* than some critical length in order to flower. The terminology is clumsy and somewhat unfortunate; however, the important point is to understand the concept. To this end, study Figure 7–33. Note that if nightlength is uninterrupted and exceeds a critical length (this varies somewhat from plant to plant), short day plants will flower. However, if the night is interrupted by even a flash of light, short day plants will not flower. Any interruption seems to prevent them from experiencing a night of sufficient length. Conversely, the long day plant will only flower if nightlength is shorter than some critical period.

The so-called night interruption experiments seen in Figures 7–33 and 7–34 lend themselves to an investigation of the pigment within the leaves which responds to a light interruption. In other words, it is a convenient way to approach the problem of what pigment is actually responsible for the photoperiodic behavior of flowering plants. The answer is *phytochrome* (see Chapter 6). How do we know? A night interrupted with red light promotes flowering in long day plants and inhibits flowering in short day plants. Light of other wavelengths is not effective. Furthermore, the red light break can be rendered ineffective if it is immediately followed by a far-red exposure (see Figure 7–34). This is by definition a phytochrome-controlled process.

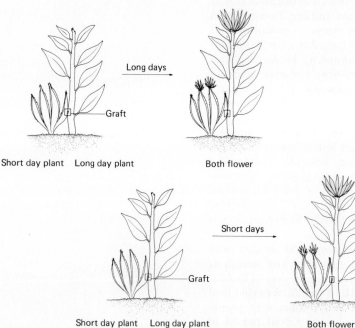

Short day plant   Long day plant   Both flower

Short day plant   Long day plant   Both flower

Long days

Short days

Graft

Graft

FIGURE 7–32.

Long days; short days? It makes no difference when a long day plant is grafted to a short day plant. As long as one plant is induced to flower, there can be a flow of florigen through the graft union, and the florigen will induce the partner plant to flower. Such experiments suggest that one type of florigen is common to both long and short day plants

Unfortunately, we do not know precisely how phytochrome interconversions serve to measure time intervals, but it would seem to be a rather sophisticated mechanism. We are also ignorant about how phytochrome ties into florigen production.

It is somewhat discouraging that this very important aspect of a plant's behavior has been so slow to yield its secrets, and because of the complexities of the system, there are no particularly good reasons to think that the remaining problems will be rapidly solved.

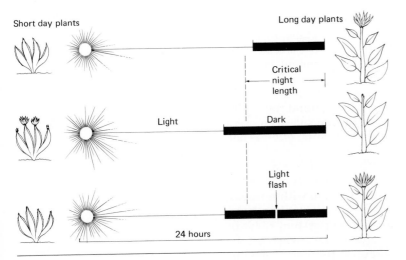

FIGURE 7–33.

The length of the night period determines whether or not long and short day plants will flower. If the night length is shorter than some critical period or if it is interrupted by a flash, long day plants will flower.
From Arthur W. Galston, *The Green Plant*, © 1968. By permission of Prentice-Hall, Inc., Englewood Cliffs, N.J., Fig. 5–12, p. 92.

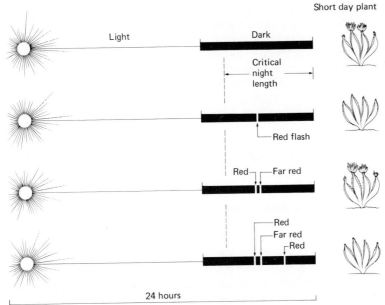

FIGURE 7–34.   NIGHT INTERRUPTION EXPERIMENTS AND THE ROLE OF PHYTOCHROME

A night interruption by red light prevents flowering; far-red light when used as an interruption does not. Furthermore, far-red light can reverse the effect of red light. Phytochrome is thus involved in the flowering response.
From Arthur W. Galston, *The Green Plant*, © 1968. By permission of Prentice-Hall, Inc., Englewood Cliffs, N.J., Fig. 5–13, p. 93.

Death in Plants

In this chapter and in the preceding one, an attempt has been made to organize the many facets of plant development and physiology into a roughly chronological chain of events. Clearly, we have almost reached the end of the story because flowering is the means for sexual reproduction in higher plants and for the eventual production of seeds, and we began our study with seed germination. The only remaining stage of development is deterioration and eventual death.

We can, and perhaps should, view the deteriorative processes that normally terminate the life of a plant or plant organs as being the continuation of the developmental process that leads to the mature individual; this phase of development is called *senescence*. You may question this statement because earlier we did emphasize that some plants can and do achieve great age and give the appearance, at least, of being immortal. Thus, it may seem a little contradictory to say now that senescence is a normal part of the developmental pattern in plants, *but* it is not.

Under the general term *senescence*, we recognize several different patterns, and every species falls into at least one of the categories (Figure 7–35). The most dramatic case is exhibited by annual grasses when an entire field crop will die within weeks. Another pattern is exhibited by species in which the above-ground portion of the plant dies, but the root system remains alive. The third case is typical of deciduous trees when all the leaves die in the autumn, but the rest of the plant remains alive. Last, we have the least dramatic case which includes even the redwoods and ancient pines. In this particular case the older leaves continually die but are also continually replaced by new ones as the plant grows. Thus, these "evergreens" have parts that regularly die but also parts that are continually reborn. Senescence, therefore, is a phenomenon that occurs in all plants, but in some cases it is more obvious and produces a different short-term result. Our discussion is divided into two parts, leaf senescence and total plant senescence.

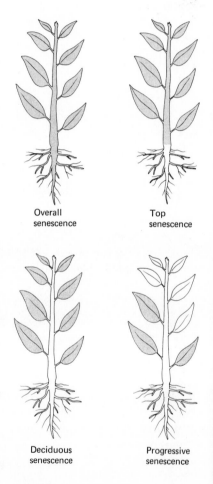

Overall senescence

Top senescence

Deciduous senescence

Progressive senescence

FIGURE 7–35.

Patterns of senescence. The senescent part of the plant is shaded. From *Plant Growth and Development*, by A. C. Leopold. Copyright © 1964 by McGraw-Hill, Inc. Used with permission of McGraw-Hill Book Company, Fig. 12–1, p. 195.

Senescence in Leaves

Almost everyone is familiar to some extent with the changes that take place in the leaves of deciduous trees. As autumn approaches, the leaves on such trees as the maple, aspen, sweet gum or liquidambar, and many oaks first become less green and subsequently take on yellow, orange, or red tinges. This color pattern intensifies as the season rolls along. Eventually, the leaves fall to the ground, and the tree remains dormant throughout the winter. These events represent visual verification of leaf senescence, obvious to even the most casual observer. But these are the terminal events, and there are earlier developments in the deteriorative process common to all leaves. Let's look at some of these.

The leaves of almost all plants are only at their prime photosynthetically when they are relatively young. That is, they are most efficient in fixing CO_2 just before or just after reaching full size. Thereafter they decline in photosynthetic and respiratory activity (Figure 7–36). As time passes, it can also be noted that there is a gradual decline in the total amount of protein and RNA found in the individual cells of a leaf. Apparently in the senescent leaf the biochemical pattern of cellular activity changes. The large macromolecules are gradually degraded, and their individual components mobilized so that they flow out of the leaf to be reutilized in younger regions or stored for future growth, a conservation project of sorts. Only after these events are well under way do we notice the loss of chlorophyll and the appearance of the yellow and red pigments associated with autumn. The terminal event in leaf senescence is the formation of the abscission zone which is illustrated and explained in Figure 7–37.

As you no doubt have noticed, our discussion thus far has dealt with leaf senescence in a purely descriptive manner, concentrating on the "what happens" aspect rather than on "how it works." As usual, this is because we don't have all the answers, but a few things are known. In the case of deciduous trees, and perhaps evergreens as well, some of the final events in the onset of dormancy and the final stages of senescence (such as the formation of the red and yellow pigments) are geared to seasonal changes. However, such environmental changes do not seem to trigger the earlier deteriorative process. Precise information pertaining to the nature of this control is virtually nonexistent. It is worth repeating at this point, however, that cytokinins prevent or at least retard senescence. As you recall from the previous chapter, cytokinin applications can initiate a flow of nutrients into treated areas of the leaf, and this flow appears to be correlated with a decrease in the rate at which the leaf ages. There exists, then, a known agent that seems to direct the flow of nutrients and to control the breakdown of already existing cell molecules. Now, let's suppose that normally the situation is reversed, and mobilization centers develop outside the fully expanded leaf. Would this not cause nutrients to flow out of the leaf, thus initiating senescence? It does seem at least possible that such a situation does exist, and as we will see, a similar case can be made for senescence of entire plants.

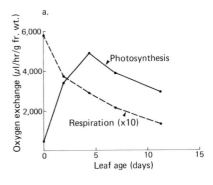

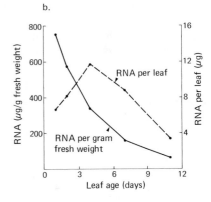

FIGURE 7–36.

Early events in leaf senescence. a. Actual data from a research publication showing the decline in both respiratory and photosynthetic activity in pea leaves with time. Pea leaves are fully expanded in about 4 days. b. Actual data, again with pea leaves, showing that the amount of RNA in a leaf declines with increasing age. From "Photosynthetic and respiratory activities of growing pea leaves," by R. M. Smillie, in *Plant Physiology* 37:717 (1962), Figs. 1, 2. (b) Reproduced by permission of the National Research Council of Canada, from the *Canadian Journal of Botany* 39:891–900 (1961).

Senescence in Plants

Picture yourself in a huge, green, lush corn field in Iowa in early August. The plants are not just healthy and growing but are doing so with a particular lust and vigor. Suddenly the time frame changes to October, and *every* plant is dead. What happened? Isn't it peculiar that some species die suddenly (as with corn), although others are capable of living for hundreds of years? If you examine this phenomenon in detail, you will soon discover that annuals appear to have evolved some physiological switch which, just after flower formation begins, triggers the various deteriorative processes discussed. Apparently other species lack or compensate somehow for this biochemical switchover, and flowering doesn't signal the end of their functional life. Again, we don't have an abundance of information on this rather striking phenomenon, but a few pertinent observations have been made. For example, if one removes the flower buds from certain species immediately after they are formed, senescence is significantly retarded (Figure 7–38). This again suggests some relationship between mobilization phenomena and aging. That is, the reproductive organs of a plant are known to require nutrients and thus act as mobilization centers. Nutrients, the argument would go, flow from the leaves into such centers; one could imagine the rest of the plant becoming so

FIGURE 7–37. THE ABSCISSION ZONE

In parts a, b, and c, the abscission zone is diagrammatically sketched. In part d, more anatomical details are shown. The actual formation of the abscission zone is triggered by environmental factors that influence the hormonal balance of the cells. Auxin and ethylene levels appear to be especially important in the regulation of the events leading to the formation of the separation layer.
From *Plants, Chemicals, and Growth,* by F. C. Steward and A. D. Krikorian, 1971, Academic Press, Inc., Fig. 4–10, p. 59. By permission of Dr. F. T. Addicott.

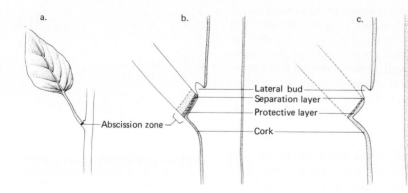

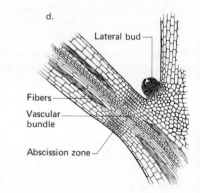

FIGURE 7–38.

Retardation of senescence in the soybean. As shown, senescence is most strongly deferred by removing the young flowers as they form, but partial retardation can also be achieved by removing the fruits at various stages of development.
From "Experimental modification of plant senescence," by A. C. Leopold, E. Neidergang-Kamien, and J. Janick, in *Plant Physiology* 34:571 (1959), Figs. 3, 4.

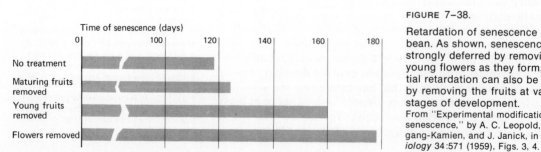

depleted that it actually starves to death. This sort of mechanism to explain senescence was suggested years ago and undoubtedly is partially accurate, but it still does not account for all the experimental facts. The major problem is that even when flowers are just developing and do not require large amounts of nutrients or act as mobilization centers, they still appear to trigger senescence in the leaves. Thus, we are led to speculate that in addition to acting as mobilization centers, the formation of flowers seems somehow to trigger independently the initial deteriorative process. It is indeed a curious, perhaps even mysterious, unsolved, and neglected phenomenon in the physiology of plants. You may find it interesting to speculate on whether the "planned" senescence so dramatically exhibited in grains has any advantages for a plant in terms of evolutionary progress. What about organ senescence?

Selected Readings

Bold, H. C. 1974. *Morphology of Plants.* 3d ed. New York: Harper & Row. Classic text on plant structure. Also good comparison of some schemes of classification.

Crafts, A. C., and C. E. Crisp. 1971. *Phloem Transport in Plants.* San Francisco: W. H. Freeman. Advanced reading on the transport of food in plants.

Esau, K. 1965. *Vascular Differentiation in Plants.* New York: Holt, Rinehart and Winston. Additional detail in plant anatomy.

Hillman, W. S. 1963. *The Physiology of Flowering.* New York: Holt, Rinehart and Winston. Advanced review of the flowering process.

Leopold, A. C. 1964. *Plant Growth and Development.* New York: McGraw-Hill. A very readable text in plant physiology and development.

Salisbury, F. B. and R. V. Parke. 1965. *Vascular Plants: Form and Function.* Belmont, Calif.: Wadsworth. A general text on plant anatomy as it relates to function.

Varner, J. E. 1961. Biochemistry of senescence. *Annual Review of Plant Physiology* 12:245–264.

Zimmermann, M. H. March 1963. How sap moves in trees. *Scientific American* 208(3):132–142. Nicely illustrated account of the physiology and anatomy of sap movement.

PART ONE

We live in a world that is becoming more and more mechanized, a world that sometimes seems to be dominated by plastic and synthetic goods. This trend has even spread to the agricultural products we eat, resulting in business practices that cater to our eyes while occasionally ignoring our stomachs. Man-made fertilizers are becoming extensively used in farming. Insects are controlled by spraying with man-made insecticides, some of which are now thought to harm the environment and give rise to ecological situations that may take years to rectify. This is all somewhat frightening, but we must not be too pessimistic or critical about modern agriculture in view of the fact that we are striving to feed a hungry world.

Perhaps as a result of our increased knowledge and use of mechanization in crop production, we are now experiencing among growing segments of American society an agricultural backlash, a return to more "natural" food products and farming techniques. "Organically" grown foods have enjoyed a remarkably profitable period lately, with people paying 10%–50% more for organically grown food which often looks less appealing than supermarket produce and perhaps isn't as nutritious either (Figure 8–1).

In this chapter we will investigate some of the underlying

FIGURE 8–1.

Health food stores featuring organically grown merchandise are enjoying increasing popularity. What do they really have to offer, and is it worth the price?

reasons for these current trends and critically examine current agricultural practices. We will explore such questions as: Should we really be concerned about the use of inorganic fertilizers? Can pesticides be used safely? What are the ecological implications of modern agricultural methods? What chances must we take to meet the demands of our rapidly expanding population? In order to answer any of these questions, we must first become acquainted with the medium through which some pesticides and most fertilizers reach the plant, namely the soil.

Soil: The Placenta of Life

Soil means different things to different people; but for now you are a plant biologist, and to such scientists soil is simply that part of the earth's surface in which the roots of a plant grow. Soil formation starts when rocks are fragmented into smaller particles. But to form a productive medium for plant life, soil must be much more than simply fragmented rock. A good soil is considered to be a complex mixture of (1) small mineral particles, (2) dead organic matter, (3) gases trapped between soil particles, (4) soil organisms, and (5) water with dissolved minerals. Let us look at the process of soil formation in a little more detail, and in doing so we will see how each of the components of a productive soil interacts to support plant and animal life.

As mentioned, soil formation begins with the weathering of rocks. This usually starts with small fractures in the parent rock caused by constant heating and cooling. Through the action of wind, water, and ice, smaller particles are gradually produced. Plants themselves can aid in this process by preventing the small particles from being blown or washed away, by facilitating fracturing, and by gradually contributing organic matter to the rock fragments (Figure 8-2). This is a slow process, and it may take thousands of years for an inch or two of soil to accumulate. For our considerations, however, we aren't too interested in the total thickness of the soil but rather are concerned with the thickness of the *topsoil*. Topsoil is the uppermost layer, the one that is usually rich in organic matter and the nutrients needed for plant growth (Figure 8-3). Of course, all soils are not the same. Some types are productive, and others are virtually incapable of supporting plant life. One of the questions that we should therefore address ourselves to is how can the quality of a poor soil be improved? This question can be answered at least partly if we understand the two basic terms that describe the characteristics of a particular soil: soil texture and soil structure.

The term *soil texture* refers to the size of the individual mineral particles that make up a soil. There are four main groups:

Soil Particle	*Diameter (mm)*
Coarse sand	2.0–0.20
Fine sand	0.20–0.02
Silt	0.02–0.002
Clay	Smaller than 0.002

b.

FIGURE 8–2. ECOLOGICAL SUCCESSION

a.

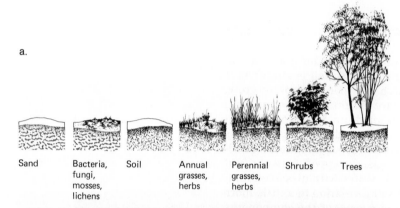

Sand — Bacteria, fungi, mosses, lichens — Soil — Annual grasses, herbs — Perennial grasses, herbs — Shrubs — Trees

a. Material such as sand and gravel cannot support large plants, but the smaller, more primitive plants can sometimes exist in such places. When these plants die they contribute organic matter to the developing soil, and it gradually becomes more productive as generations pass. Finally, such soil can be invaded by grasses and small shrubs. They in turn provide more organic matter until the topsoil becomes productive enough for trees. b. The seedling is facilitating the formation of soil and keeping the existing soil in place.

(b) From *Elements of Biology*, 3rd ed., by Paul B. Weisz, p. 81. Copyright © 1969 by McGraw-Hill, Inc. Used with permission of McGraw-Hill Book Company.

FIGURE 8–3.

A soil profile can be seen on many roadcuts and generally includes three layers or horizons. Horizon A (top) includes the fresh and decomposing litter recently dropped from the vegetation; water filters freely through, thereby leaching out various substances. Horizon B (middle) is the layer where those substances leached from Horizon A accumulate. Horizon C (bottom) is the underlying parent material.

Clay particles are the most important for our considerations because they have properties vastly different from those of the parent materials from which they were formed. Clay particles are not just finely divided rock but rather are particles that have undergone extensive physical and chemical changes. Most important, they are negatively charged and sheetlike in structure. This latter characteristic means that clay particles have a very large surface to volume ratio, and this in turn is important because it is the surface area of a particle that determines its ability to bind and store water and minerals. In contrast to the clay, sand particles are gigantic in size and do not possess this water- or mineral-holding ability. Can you now see why a clay soil has some advantages? It can bind and store large amounts of water and plant nutrients. But it also has some disadvantages, the most serious being that a clay soil does not possess many spaces or pores in which atmospheric gases can exist, and plant roots require such air spaces for proper growth. Sandy soils are just the opposite. They store little water or plant nutrients but have a good rate of water *percolation* and numerous air spaces. What, then, constitutes an ideal soil? The best soils combine the better characteristics of sand and clay, and this aspect brings soil structure and organic matter into consideration.

Soil structure refers to the arrangement of the individual soil particles into larger units called aggregates. We have seen that a soil containing much clay has some disadvantages, *but* if such soils contain the correct amount of organic matter, they can be greatly improved. We might view organic matter as a sort of stickum that will cause the small clay particles to clump together and form larger aggregates (Figure 8–4). Such particles have basically the same properties as individual clay particles with the added advantage of now possessing numerous pores or air spaces. A good agricultural soil will have aggregates about 1–5 mm in size and a crumbly or granulated texture. Such

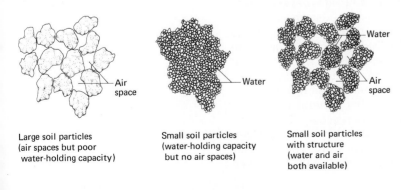

Large soil particles
(air spaces but poor
water-holding capacity)

Small soil particles
(water-holding capacity
but no air spaces)

Small soil particles
with structure
(water and air
both available)

FIGURE 8–4.

Proper soil structure is important for plant productivity because it provides for both air spaces and moisture. Organic matter is important in the formation of soil aggregates. Adapted from *The Green Revolution* by Maarten Chrispeels (unpublished manuscript).

Organism	Pounds per Acre
Bacteria	1,950
Fungi	1,750
Protozoa	300
Algae	250
Nematodes	35
Other worms and insects	900

TABLE 8–1. SOIL ORGANISMS
A soil doesn't become a fertile soil unless it is inhabited by organisms.
That's obvious, but the number of such organisms is what is so astonishing.
For example, the upper foot of fertile agricultural soil may contain 4000 to
6000 lbs of soil organisms per acre (excluding higher plants). The breakdown
in one particular study is shown above.

Source: From O. N. Allen. 1957. *Experiments in Soil Bacteriology*, Minne-
apolis: Burgess. By permission of Burgess Publishing Company.

soils optimize water- and mineral-holding abilities and at the
same time have enough pores so that plant roots will have
adequate aeration. Sandy soils already have enough large pores
so that organic matter isn't necessary for these holding and
aerating purposes. In sandy soils organic matter functions like
a sponge to increase the soil's water- and mineral-holding
capacity.

Organic Matter and the Lore of the Compost Pile

Just what do we mean by organic matter? How does it accumulate, and how is it recycled? It's really a fascinating story.
Organic matter is that fraction of the soil derived from living
organisms: leaves that have fallen, the dead bodies of soil
organisms and animals, dead roots, and more. Such matter is
slowly broken down by soil microorganisms which use it as an
energy source, much as we use sugars and fats. Larger organisms such as earthworms, ants, and insects also make use of
this matter as food, and in addition they aid in working it into
the soil and thus directly and indirectly facilitate aeration
(Table 8–1). The activities of soil organisms are important
because they not only digest and break down large molecules
but also release into the soil certain nutrients which can be used
by the plant. Actually, we have seen this general process in
earlier chapters, and it is called *chemical recycling*. The plant's
role in such a cycle is to obtain minerals from the soil which
have an essential function during the plant's life. These minerals become incorporated into the cells of the plant, and when

the plant dies, they are gradually released again as the plant body decomposes. It is indeed unfortunate that we don't know more about chemical recycling under field conditions; such information could be extremely valuable in understanding how to manage correctly new acreage coming into cultivation. For example, it would be helpful to know how much of a particular mineral is tied up in living organisms, how much is present in dead but not fully decomposed organic matter, and what percentage is readily available in the soil for supporting plant growth. Also, we would like to know at what rate this mineral is made available through weathering of the soil and how much is lost through run-off and leaching.

To summarize briefly what we have discussed thus far, we can say (1) that organic matter improves the aeration of the soil directly by acting as a sort of glue and indirectly by supporting various soil organisms such as insects and worms and (2) that the decomposition of organic matter slowly releases minerals that the plant needs.

Compost is a term heard frequently in connection with organic gardening. Oftentimes we get the impression that compost has some mystical and/or extremely special properties. Let's look at what compost really is and how to use it properly. Compost is a mixture of organic matter (including old lawn clippings, leaves, orange peels, and what have you) which we allow to decompose and work into the soil where it accomplishes the various functions mentioned above. All in all, compost production is a useful way to improve garden soil at low cost and at the same time to help recycle trash. However, the actual use of compost has many erroneous notions connected with it. Some of these may have developed because the improper use of compost can actually reduce the productivity of a garden. This occurs if dead plant material is worked into the soil before it is allowed to rot. Organic matter that is not fully decomposed when worked into the soil will stimulate the growth of soil microorganisms, and this can cause problems. Microorganisms need the same minerals as plants do, and thus a rapidly expanding population of soil organisms can actually deplete the soil's supply of minerals, leaving the plants, at least for the moment, with the short end of the stick. This situation is not permanent because eventually the soil organisms will decompose the compost or organic matter fully. As decomposition nears completion, the food source for many of these organisms is also depleted; a great many of them die and eventually the nutrients they have accumulated are returned to the soil. Nevertheless, this process takes time (perhaps 6 months or a year), and during this period the garden has suffered.

A properly tended compost pile, and subsequent use of the

compost, does not cause this problem because one first allows the organic matter to decay fully and *then* adds it to the soil. The full procedure goes something like this. One accumulates a large pile of grass clippings, leaves, and such. A little soil may be thrown in to provide a suitable "starting" population of microorganisms. The whole mass of material is then regularly turned and watered down. Basically, what is done is to set the scene for a population explosion of microorganisms. They rapidly multiply and begin to degrade the organic matter. This material is turned over from time to time to facilitate mixing and aeration, care being taken always to keep the compost pile moist. Under such favorable conditions organic matter is degraded to a usable and beneficial form in 3 weeks to 2 months. In the meantime, of course, a second compost pile has been started and so on (Figure 8–5). A lot of work is involved—especially considering that the same basic results in terms of nutrient availability could be achieved by working an inorganic fertilizer into the soil, although such inorganic fertilizers do not improve soil structure. Contrary to popular belief, there is absolutely *no* difference between the nutrients released from organic matter and those present in the inorganic fertilizers you can buy by the bottle or bag! This is easily seen if we consider the nutrients that a plant obtains from the soil. In doing so, we will see that as long as we supply these nutrients, neither soil nor organic matter is necessary for the production of normal, healthy, *and nutritious plants*.

FIGURE 8–5.

Note the steam rising from the warm, damp humus, indicating that decay organisms are hard at work in this compost pile which is being turned. The temperature within this compost was 71° C at the time the picture was taken; air temperature was 20° C.

Plants as Soil Eaters, or the Concept of Mineral Nutrition

You will recall from our discussion of photosynthesis in Chapter 2 the experiment performed by Van Helmont. Van Helmont showed that a willow twig could grow and obtain substantial weight while only extracting a few ounces of material from the soil in which it was planted. In the earlier account the importance of those missing few ounces of soil was neglected because we wished to emphasize the role of CO_2 in photosynthesis. Now we can begin to appreciate that they, too, are important in plant growth. That is, apart from their requirement for light, water, and CO_2, plants also need mineral elements which are normally provided in the soil and absorbed by the roots.

A method was established at the end of the last century to determine precisely which elements are essential and how much of each must be present for normal growth. This method is illustrated in Figure 8–6. Basically, the roots of a young seedling are placed in a bottle containing distilled water to which is added a multitude of various inorganic salts. Air is usually bubbled into the container to prevent the plant from "drowning" or asphyxiation. If we omit a particular element from our water culture and if this omission results in poor growth, disease, or death, the omitted element is considered to be *essential* for plant growth. By use of this method, it is rather easy to demonstrate that plants require relatively large amounts of nitrogen (N), potassium (K), calcium (Ca), phosphorus (P), magnesium (Mg), sulfur (S), and iron (Fe). We call these essential *macronutrients*. However, by taking special precautions and using very pure chemicals, clean glassware, and high-quality distilled water, you can demonstrate that plants also require

FIGURE 8–6.

Method used to determine whether or not a given nutrient is essential for a plant's nutrition. a. *Procedure.* Uniform young plants are transferred from sand to nutrient solutions. The control solution contains all known or suspected essential elements. The test plants are grown in a complete solution minus one particular element. b. *Results.*   After one or two weeks the test plants are compared with the control. If the absence of an element results in poor growth or conditions such as the loss of chlorophyll, we can consider the missing element essential.

a.

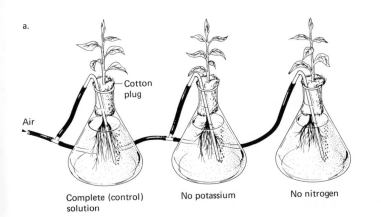

Complete (control) solution

No potassium

No nitrogen

b.

Complete

No potassium

	Compound	Element Supplied	Milligrams of Compound per Liter of Solution
Macronutrients	KNO_3	K; N	60.00
	$CaNO_3 \cdot 4H_2O$	Ca; N	94.00
	$NH_4H_2PO_4$	N; P	23.00
	$MgSO_4 \cdot 7H_2O$	Mg; S	24.00
	Fe-EDTA	Fe	0.69
Micronutrients	KCl	Cl	0.37
	H_3BO_3	B	0.15
	$MnSO_4 \cdot H_2O$	Mn	0.033
	$ZnSO_4 \cdot 7H_2O$	Zn	0.057
	$CuSO_4 \cdot 5H_2O$	Cu	0.012
	H_2MoO_4 (85% MoO_3)	Mo	0.0081

TABLE 8–2.   THE IDEAL NUTRIENT SOLUTION
Many years of work went into finding the ideal nutrient medium to use for the soilless culture of plants. Such a solution contains all the essential macronutrients and micronutrients at concentrations sufficient to produce maximal growth for several weeks, although in practice one generally irrigates with such a solution every few days. The recipe most commonly used is called Hoagland's solution. Its components are those listed above in the amounts indicated.

Source: Modified from E. Epstein. 1972. *Mineral Nutrition of Plants: Principles and Perspectives,* New York: John Wiley, Table 3–1. By permission of John Wiley & Sons, Inc.

minute quantities of other elements that we call *micronutrients.* These include chlorine (Cl), copper (Cu), manganese (Mn), zinc (Zn), molybdenum (Mo), and boron (B). Thus if all the necessary micronutrients and macronutrients are supplied to a plant in water culture, perfectly normal and healthy plants can be grown (Table 8–2). Note well that the addition of special organic fertilizers is unnecessary and does not improve growth. Potassium is potassium; the source is unimportant for mineral nutrition.

Hydroponics (the "soilless" cultivation of plants) is a term heard frequently. In fact, the technique of hydroponics is becoming commercially important. Its advantages are simply that it is possible to control carefully the "diet" of plants as well as to control more readily infections by disease-causing organisms which normally live in the soil and infect plants via roots. The chief drawback to the technique is cost; at present it appears that massive agricultural farming using this method is economically unfeasible. Nevertheless, there is increasing interest in hydroponics in many sectors of the economy (Figure 8–7). Even the United States government is in the "business" of hydroponic gardening on some islands in the Pacific. On these islands there is very poor soil or no soil at all, only rock. Thus hydroponic gardens are the *only* available means of growing fresh produce for government personnel. We might conclude that while soil is not essential for growing plants, it is presently the most economical medium in which to cultivate plants.

Fertilizers and an Introduction to the Nitrogen Problem

When we apply a fertilizer to soil, we are supplying the essential elements that a plant needs for growth. These nutrients are incorporated into plant cells where they are used as either structural components of the cell itself or as cofactors for the various enzymes that carry on metabolism. Since we continually remove plants and plant parts for our own uses, we also continually remove many of the essential elements originally in the soil. Run-off and leaching also contribute to this loss. Therefore, it is important periodically to add back those elements in short supply via the application of fertilizer.

It turns out that most soils are or become deficient most often in nitrogen (N), phosphorus (P), and potassium (K), and thus the replacement of these elements takes on a special significance in agriculture. Fertilizers supplying these three elements, the so-called NPK types, are the ones you will commonly find at

FIGURE 8–7.

At the San Diego Zoo—one of the largest in the world—grass seedlings are grown by using hydroponic methods. a. Seeds are placed on a screen within a growth chamber. From time to time a mineral nutrient spray from above irrigates the seedlings. After 3 to 4 days the seedlings with associated root mat are peeled from the screen. b. They are then fed to the animals.
San Diego Zoo photos.

a.

b.

your friendly neighborhood nursery or garden store. They are often referred to as complete or all-purpose fertilizers, and although this is by no means an accurate description, as you now know, they usually supply the minerals most likely to be in low supply.

Among the "big three" listed above, nitrogen deserves special consideration. Inadequate nitrogen in the soil is the most common mineral deficiency on a worldwide scale. There are really two reasons for this: (1) Plants require more nitrogen than any other element. (2) Unlike the other essential elements, little nitrogen is produced during the weathering process by which soils are formed. The occurrence of nitrogen in the soil is almost entirely dependent on rather special microorganisms that can capture nitrogen gas from the atmosphere and incorporate it into their cellular machinery. When these organisms die, they gradually decompose, and nitrogen is released in a form useful to plants.

Nitrogen is an element that can exist in various states. The nitrogen present in the atmosphere is an inert gas with the formula N_2. In this form it *cannot* be used by higher plants. For nitrogen to become useful, it must first be fixed. This means it must be combined with either oxygen or hydrogen, and only certain lower organisms are capable of performing this trick. It is worthwhile to examine in detail how nitrogen moves through the ecosystem, for this illustrates not only how plants get their nitrogen but also one of the most complex systems for recycling that is known.

Nitrogen Fixation: $N_2 \rightarrow NH_3$

Nitrogen fixation is the term given to the complex process by which atmospheric nitrogen (N_2) is converted to ammonia (NH_3). Ammonia represents the first biologically stable product of the fixation process, although plants usually do not store NH_3 as such. Ammonia is usually rapidly converted to some organic form directly useful to organisms. The process of nitrogen fixation is carried out by certain lower organisms, namely, some algae, bacteria, and fungi. Some of these organisms are free-living within the soil; however, others actually live *inside* the roots of certain green plants. This latter relationship is very important in agriculture and is mutually beneficial to both organisms involved. The plants provide the microorganisms with food in the form of organic materials, and the microorganisms supply the plant with fixed nitrogen. Such a give-and-take, mutual-benefit type of relationship is called *symbiosis*. Not all plants have such a handy relationship going for them. Some important crop plants that do have a symbiotic relationship are peas, beans, and peanuts, among others. These plants characteristically have small bumps on their roots called *nodules*, and it is here that a population of nitrogen-fixing organisms is located (Figure 8–8). The organisms within a nodule first fix nitrogen and then use it for assembling their own amino acids, some of which are secreted into the roots of the host plant. From here the secreted amino acids find their way to various parts of the plant and are used to make protein.

The nitrogen that has been fixed in symbiotic plants as well as the free-living microorganisms will eventually return to the soil. These organisms will die, decompose, and thus return the fixed nitrogen directly, or they will be eaten by animals whose bodies and excrements eventually return to the soil. It would be a mistake to say only that this process is important for plant growth; it is essential! All growth depends on fixed nitrogen, and its presence in the soil is almost entirely due, directly or indirectly, to nitrogen-fixing organisms.

Grasses have no symbiotic relationships with nitrogen-fixing organisms. Thus the accumulation of usable nitrogen in prairie soils, for example, is very slow and entirely dependent on the relatively inefficient, free-living soil microorganisms. In many plant communities there may be mixtures of plants, some with

FIGURE 8–8.

Root nodules and the invasion of nitrogen-fixing bacteria. a. Nitrogen-fixing bacteria normally found in the soil can enter plant roots and establish a symbiotic relationship. Remember that this phenomenon can only take place in the roots of certain specialized plants. Many species lack root nodules entirely and must rely on the nitrogen fixed extracellularly by free-living soil microorganisms. b. The photograph shows actual nodules on a broad bean plant.
(a) Adapted from *Plant Physiology*, by F.B. Salisbury and C. Ross. © 1969 by Wadsworth Publishing Company, Inc., Belmont, California 94002. Reprinted by permission of the publisher, Fig. 16-1, p. 334.

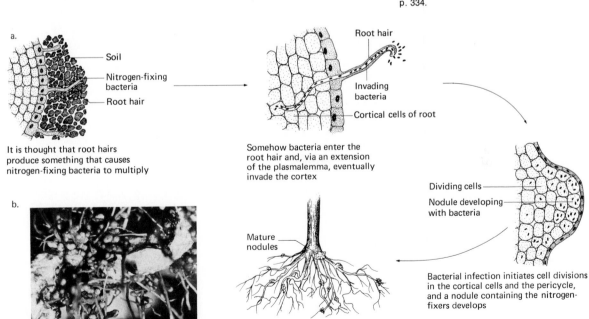

a.

— Soil
— Nitrogen-fixing bacteria
— Root hair

It is thought that root hairs produce something that causes nitrogen-fixing bacteria to multiply

Root hair

Invading bacteria

Cortical cells of root

Somehow bacteria enter the root hair and, via an extension of the plasmalemma, eventually invade the cortex

Dividing cells
Nodule developing with bacteria

Bacterial infection initiates cell divisions in the cortical cells and the pericycle, and a nodule containing the nitrogen-fixers develops

b.

Mature nodules

Finally mature nodules easily seen without magnification are apparent on the plant roots symbiotic bacteria and others without such a relationship. The nonfixers would of course benefit from the presence of the fixers. Farmers often alternate crops that lack symbiotic nitrogen fixers, such as corn, with crops that have a symbiotic relationship, such as soybean, in order to improve the useful nitrogen content of the soil.

Ammonification and Nitrification

Besides nitrogen fixation there is yet other activity beneath the soil surface, as was previously alluded to. There are decomposers present in the soil, specifically bacteria and fungi, which break down the proteins present in organic matter into amino acids and then into CO_2, ammonia, and water. This process is called *ammonification*. The free ammonia formed in this way can be taken up by plant roots and used directly. However, in most cases, free ammonia doesn't last very long within the soil because there are still other microorganisms that convert the ammonia to nitrate (NO_3). This process is called *nitrification*. Fortunately, NO_3 can also be used as a source of fixed nitrogen by higher plants.

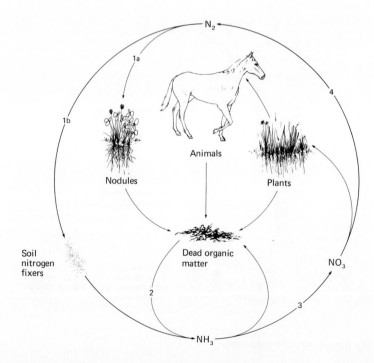

FIGURE 8–9.   THE NITROGEN CYCLE

The numbers indicate the basic stages in the cycle as discussed in the text. It is important to understand this cycle thoroughly.

Adapted from *The Green Revolution* by Maarten Chrispeels (unpublished manuscript).

Denitrification: $NO_3 \rightarrow N_2$

The nitrogen cycle is completed by bacteria which carry out the conversion of nitrate back to nitrogen gas. This process is called *denitrification*. At first consideration, denitrification might seem very wasteful because it converts potentially useful nitrogen into a form that can't be used directly by plants. However, it is an important and necessary process. Without such a process nitrogen in the air would gradually be converted into nitrate which would eventually find its way into the oceans and thus be lost for terrestrial purposes as well as be the cause of serious ecological problems in the oceans. Second, without denitrification the oxygen content of the atmosphere would also decline as you can see by considering the molecular character of nitrate (NO_3). Until recent times nitrogen fixation and denitrification balanced each other rather well, and thus on a global scale the amount of fixed nitrogen was fairly constant. This may not be true now. The need for nitrogen-containing fertilizers has increased greatly and is likely to continue. This need is being fulfilled in several ways. For example, industrial nitrogen fixation is becoming more common, and new ways are being found to exploit our remaining natural resources. In industrial fixation N_2 is taken from the air and reacted under carefully controlled conditions with H_2 to produce ammonia. Another new way is by introducing and expanding greatly the crops that contain internal nitrogen fixers. Thus the total amount of fixed nitrogen is increasing rather sharply. The ecological consequences of this are unclear at the present time.

Figure 8–9 summarizes and simplifies somewhat the major features of the nitrogen cycle. Can you amplify and explain the various numbered steps?

Water Pollution and the Aging of Lakes

Thus far we have concerned ourselves primarily with the relationship between land plants and their nutrient requirements. There is, however, another very interesting and important ecological problem that goes hand in hand with an increasing world population and our increased use of fertilizers: water pollution.

Perhaps the term *water pollution* first stirs up notions of oil spills, vast amounts of noxious industrial poisons flowing into our water supply, and so on. But this isn't all there is to it. In fact, as senseless as these examples are, we are polluting our water in yet another way that is subtler but perhaps potentially more dangerous. We are polluting our water with nutrients. To illustrate this problem, we need only to look as far as the nearest freshwater lake or pond. As close to home and as important as this problem is, let us take a moment to look at it in more detail.

It is important to realize that lakes are not the static, unchanging bodies of water that they may appear to be on casual examination. In virtually every body of water, some form of life is present. Fish, of course, but also various other animals, bacteria, and plants make up the dynamic, interacting community that is a lake. For our purposes we will be primarily concerned with one member of this community, the blue-green algae. This concern is not because we are interested only in plants but rather because the entire lake community is directly or indirectly dependent on the relative abundance and control of these small plants.

The blue-green algae are a rather diverse and primitive group of organisms ranging in size from single-celled species to others about 1 m long (see Chapter 9). However, the blue-greens do have some very distinguishing characteristics. Like bacteria they lack an organized nucleus and other membrane-bounded organelles such as mitochondria and chloroplasts. Their cell walls are more similar in structure to bacterial walls than to those of higher plants. Another characteristic is the presence of a group of pigments. These pigments, together with chlorophyll, give these algae their characteristic color—usually blue-green.

Probably the most important factor regulating the growth of blue-greens, as well as other algae, is the concentration of available nutrients, and this is the variable we are most interested in. When a lake is first formed, it usually contains only limited amounts of organic matter and nutrients. As time passes, the lake is said to age, which means, among other things, that it gradually accumulates organic matter and nutrients. This is a natural and virtually unavoidable phenomenon. Such a development isn't necessarily bad because it may directly or indirectly provide more food for fish and other animals, and herein lies one of the secrets of good fishing water. Unfortunately, too much of a good thing may get out of control, and such a situation is termed *eutrophication*. This is what seems to be happening today. Man and his modifications of the environment are greatly increasing the rate at which many of our lakes are aging. Let's examine how this comes about and some of the results we can expect.

As you probably have noticed, we dump all sorts of materials into our waters: garbage, old automobiles—you name it and you can probably find it at the bottom of some lake. Sewage, however, is probably our most serious pollutant. Raw sewage is still dumped into many of our water sources at an alarming rate, contributing not only nutrients but potential sources of disease. Contrary to popular belief, however, the problem doesn't end when sewage is fully processed. Processed sewage still contributes large amounts of organic matter, phosphorus, nitrogen, and other nutrients to our water supplies. Fertilizers are also important sources of nutrient pollution. It has already been mentioned that most fertilizers contain nitrogen, phosphorus, and potassium (the NPK type). Such fertilizers are beneficial to the soil, but unfortunately they don't always stay there. Rainfall and the resulting run-off move downhill and eventually end up in a river, reservoir, well, or some such water source, bringing the dissolved N, P, and K leached from the surrounding farmland. Laundry detergents are another rich source of phosphates and greatly contribute to increased algal populations in our water sources.

By now, the main point should be clear. Man is greatly increasing the available nutrient supply, particularly nitrogen and phosphate, and these nutrients flow into lakes. The net result is that our lakes are aging more rapidy than they would naturally. When the situation gets out of control, one indication is greatly increased algal populations, particularly blue-greens. When an algal "bloom" results, the real trouble is only just beginning. Some blue-green algae produce toxic compounds which poison other organisms in and around the lake, most of them cause the water to have a strange and unpleasant taste. But there are still more problems. As the bloom begins to decay and sink, there is a population explosion of bacteria because their food source, the dead algae, greatly increases. This in turn results in the depletion of oxygen within the water and can cause the death of many fish species. The dead fish decay, further increasing the nutrient supply and thus setting the stage for an even greater algal bloom the next summer. Once begun, this cycle is repeated until the lake is virtually dead.

What can be done? One approach to the problem of eutrophication can be summarized by examining the progress made by one forward-looking community. About 10 years ago Lake Washington, located in the heart of urban Seattle, was plagued with recurring and worsening algal blooms. Most of the damage accrued from poor sewage-disposal practices resulting in a dramatic increase in the nutrients within the lake. This problem was recognized and new disposal techniques were rapidly adopted. The lake is returning to a healthy condition; swimming is safe, and salmon continue to return in good numbers. What did it take to achieve this? Money, of course, but also planning and forward-looking laws which were properly enforced. Unfortunately, there are many other examples without such a happy ending. We can hope that with increased awareness of these pollution problems there will be increased action toward finding badly needed solutions.

PART TWO

Have you ever noticed that natural communities such as forests, grasslands, and swamps manage to flourish without man's help? They do not require the continued application of insecticides to control insect damage nor do they require yearly fertilization. In short, natural communities are in balance with their environment. Built-in mechanisms exist which tend to stabilize such natural populations and assure their continued existence. We have already discussed one of these mechanisms in regard to the nitrogen cycle. The complexity of ecosystems should now be emphasized and related to some of our agricultural practices and problems.

To suggest how complex natural ecosystems are, let us say that you wish to establish a balanced community by making a *terrarium*. First, you will need a large glass container. An aquarium is best, but you can make do with a large fishbowl or something similar. Add a layer of gravel or coarse sand to facilitate drainage. Some people also like to add a little charcoal to maintain the proper water chemistry. Next, you will have to decide what sort of community you wish to establish: desert, bog, or maybe woodland. It's your choice, but don't try to mix communities because it doesn't work. Let's say you select the bog environment. Your first task is to prepare the soil. This can be accomplished by blending some organic matter like sphagnum moss and some topsoil in the ratio of about 1:2. Place this mixture on top of the gravel or sand base so that you have a soil base of at least 2 inches. Now you're set to begin planting. If you live near a bog-type environment, you might try collecting a few species of plants that flourish there, or you could try more exotic species such as the sundew plant (Figure 8–10), obtainable from some nurseries and mail-order houses. Once planted, the terrarium should be kept moist.

You may find that preparing a woodland terrarium is considerably less trouble. Your planting mixture should consist of mostly topsoil. It also should be kept moist but not wet. Undoubtedly moss will begin growing rapidly because the topsoil will contain spores from which these plants arise (see Chapter 9); however, small ferns and other seedlings will require planting.

If you are lucky and have carefully planned your terrarium, it may flourish without too much outside manipulation. However, be prepared for problems. You may find that it is very difficult to duplicate a given environment exactly. One species

FIGURE 8–10.

The sundew plant (*Drosera*) has viscid, tentacled glands on its leaves. Insects, perhaps mistaking them for honey, alight and adhere to these glands. (Note the white fly caught on the leaf at the right.) The tentacles then bend downward, carrying the insect to the leaf surface where enzymes secreted by the leaf digest it. The products are absorbed by the leaf, thereby providing nutrients and enabling such plants to live on poor soils, particularly those soils that are low in nitrogen.

may dominate all the others, and it will be difficult to determine why. Your plants may fail to reproduce, or they may develop abnormal appearances. The lesson is that natural communities are extremely complex in their regulatory mechanisms, and although you may think that you have successfully simulated a given environment, it is easy to overlook an inconspicuous but important functional part. In any case, this exercise, real or imaginary, should make it easier to visualize some of the problems that occur when man displaces a naturally regulated community. Let's examine some of these problems, and in doing so, you should understand better the very serious ecological predicament we find ourselves in today.

One of the first things man does when he settles in a new location is to clear the land of its native vegetation and replace it with the particular crops he wishes to grow. In other words, he clears away a vegetation that consisted of *many* species and in its place substitutes one or two species of commercially important crop plants. This, of course, results in rather fundamental ecological malfunctions because the intrinsic stability of the natural community was closely linked to the very diversity of organisms that composed it. That is, in a natural environment made up of many interacting species, there are built-in checks and balances which tend to lessen the chances of any given species getting out of control. This is so for the simple reason that each member is somehow dependent on some other member. A complex community also has built-in "alternatives" to cope with changes in the environment. In short, complexity *means* checks, balances, alternatives, and therefore stability. Simplification on the other hand almost always leads to instability, a fact that you may have discovered in attempting to duplicate a particular environment in a terrarium. Of course, in addition to simply reducing the stability of the system, the very act of clearing the land may have direct effects on its subsequent fertility.

A brief yet interesting example of this phenomenon concerns one of the changes that takes place when one tries to establish agricultural land in the tropics. When tropical land is cleared, the soil temperature will rise as much as 20–30°. This increase in soil temperature greatly accelerates the rate at which organic matter breaks down, and in turn decreases the capacity of the soil to absorb water and retain nutrients. Within a few years the soil becomes too hard for agricultural use. Fortunately, not all tropical soils behave in this way, and even for those that do, there are some sensible alternatives. For example, it is possible to remove some of the forest but leave enough to provide shade. Crops can then be planted under the canopy provided by the trees. Under these conditions it is possible to grow crops such as tea, coffee, cacao, and rubber.

The Need for Control: To Provide for our Needs

The problem of developing good agricultural land is more complex than indicated by the simple examples given above. Once a commercially valuable crop is planted, regardless of the location, continued maintenance and interference by man are required. How long would a garden last if it was left unattended? One, maybe two years at most. This need for constant attention stems from the fact that crop plants are selected for their ability to produce a specific type of food, and they function rather well in this regard. However, such selection has also led to a strong dependence on man. Modern-day corn, for example, has seeds that are so firmly attached to the cob and enclosed in the husk that man is a necessary agent for seed dispersal. Without any interference by man, many commercially *useless* plants will outcompete crop plants for nutrients, air, water, and space and thus take over. Such nuisance plants by definition are called *weeds*. Therefore, one of the problems arising in agriculture is the control of weeds. Perhaps the situation wouldn't be too serious if we had to worry only about weeds, but there are also the other more serious and complex problems involving insects and plant pathogens. When we disrupt and simplify a natural community, we affect the balance not only of plant life but of animal life as well. Animal populations become less diverse and therefore unstable. The general trend is toward a vastly increased population of certain species which must in turn be eliminated by man. In other words, man must attempt to stabilize artificially a situation that is intrinsically unstable. Thus we have the reasons underlying our development of *pesticides*. One variety, called a *herbicide*, is used to control weeds; the other type, called an *insecticide*, is used to control insects. As we discuss each type, take care to remember the rationale for using them, but at the same time critically evaluate their limitations and potential dangers.

Herbicides: The Plant Killers

Herbicide is the name given to a wide variety of chemical compounds with one characteristic in common: They kill plants. The strategy in using them is to kill some plants while sparing others. To facilitate our discussion of herbicides, let's divide them into two groups: (1) synthetic hormones and (2) non-hormone-related herbicides. Let's first consider the hormone-related herbicides.

The first of the hormonelike herbicides to be discovered was 2,4-dichlorophenoxyacetic acid (2,4-D). This compound (Figure

8–11) exhibits auxinlike activity when applied at very low (hormonal) concentrations. That is, it stimulates cell elongation, promotes rooting, and so on. When 2,4-D is applied at higher concentrations or herbicidal levels, it rapidly and effectively causes plant death. However, it has been found that the level of 2,4-D required to kill varies, depending on the type of plant involved. Broad-leaf plants, such as the common weeds mustard and dandelion, are more sensitive to 2,4-D than are many so-called narrow-leaf plants, such as corn and wheat (Figure 8–12). Thus, spraying a field of corn, for example, with a certain level of this synthetic hormone will leave the corn unaffected, but most weeds will be killed. Our scientific understanding of this particular phenomenon which allows us chemically to weed a field is rather sketchy despite the fact that vast amounts of money and experimentation have been expended on the effects of 2,4-D. In general terms, however, it would appear that the sensitivity of broad-leaf plants is due to the fact that they absorb 2,4-D into their cells and/or are unable to readily metabolize 2,4-D to some harmless product. The mechanism by which 2,4-D actually kills plants is also obscure. In fact it may be that many of the effects elicited by 2,4-D are not due to the synthetic hormone itself but rather to ethylene. Herbicidal amounts of 2,4-D induce the formation of herbicidal amounts of ethylene which can inhibit or kill small seedlings and also cause defoliation. You might say that this is just begging the question, and you would indeed be correct. Whether the killing agent is 2,4-D or ethylene makes little difference. The question is, How are living plants killed by an application of 2,4-D? The answer to this question in precise terms is not known. However, it would appear that the herbicidal levels of 2,4-D change the metabolism of plants so that their production of energy is adversely affected. In some cases this may result in outright kill or may affect the plant so that it is more easily invaded by bacteria, fungi, and insects which then ultimately cause disease and death.

The discovery of 2,4-D as a broad-leaf weed killer probably occurred in the early 1940s. We say probably because much of the early research on the action of this compound was performed under military secrecy during World War II. In addition to 2,4-D, hundreds of other chemically similar compounds have been synthesized and tested. One such compound, 2,4,5-trichlorophenoxyacetic acid (2,4,5-T) was also found to be a very potent herbicide (see Figure 8–11). In fact, you may recognize 2,4-D and 2,4,5-T as the compounds used to defoliate trees in the forests of South Vietnam. Leaf abscission, that is, leaf fall, is a normal process that occurs in deciduous trees in autumn. In nature it is controlled by a rather complex interaction between daylength, temperature, and several plant-growth hormones.

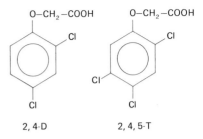

FIGURE 8–11.

Structures of 2,4-D and 2,4,5-T. Both compounds are hormone-like herbicides. That is, they mimic many of the actions of indoleacetic acid (IAA) when used at low concentrations, but at higher levels they kill plants.

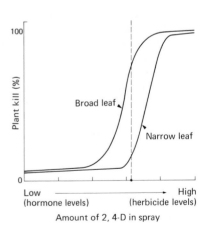

FIGURE 8–12.

Selective killing action of 2,4-D. By spraying a field with 2,4-D at the concentration indicated at the point on the graph one can selectively kill a high percentage of broad-leaf plants, while narrow-leaf plants will be relatively unaffected.

Ethylene is one of these hormones, and artificial treatment using it greatly accelerates abscission. The strategy in Vietnam was to apply massive doses of 2,4,5-T and 2,4-D to the leaves of forest trees; abscission resulted. Such treatment of an entire forest may eventually cause irreversible damage and severe ecological disturbances. In the case of the mangrove stands, outright killing of the trees has been reported. Only in time will we know the full extent of such activities, but at least it serves to illustrate the potential dangers that can arise from questionable use of scientific knowledge.

Besides direct ecological damage, there are other disturbing aspects of the indiscriminate use of herbicides. For example, let's consider 2,4-D which is actually one of the better herbicides. It is apparently harmless to humans. It is rapidly degraded in the soil, and therefore its effects are temporary. However, it is hard to keep any herbicide in one place. We spray herbicides on fields to control weeds, yet they can drift into parks and forests or even into a neighbor's field containing a broad-leaf crop. Such incidents are not uncommon. Other herbicides, however, appear to be considerably more dangerous; 2,4,5-T and various contaminants produced during its synthesis may affect human health. It has been suggested that the contaminants sometimes found in 2,4,5-T preparations may even possess cancer-inducing properties.

The Non-Hormone-Related Herbicides. The non-hormone-related herbicides are really so diverse in structure and action that they virtually defy any sort of systematic classification. For example, chlorinated acids such as trichloroacetic acid and dalapon are very effective in killing grasses but relatively ineffective against broad-leaf plants (Figure 8–13). Therefore, they can be used to control the growth of wild grasses in such crops as beans and

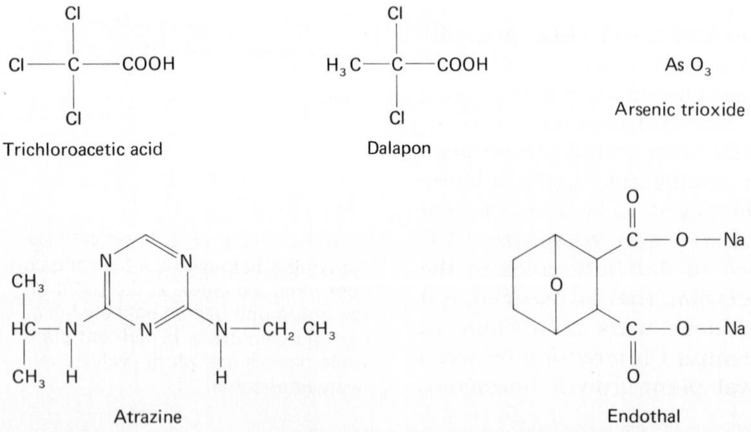

FIGURE 8–13.   THE NON-HORMONE-RELATED HERBICIDES

The point here is not to learn these structures but simply to look at the variety. Systematic classification is impossible. The discovery of the herbicidal activity of a given compound was made simply by trial-and-error testing of thousands of compounds. The examples here were chosen at random from perhaps 50 or so active herbicides.

sugar beets. Other compounds such as triazines and substituted ureas are rather indiscriminate plant killers since they affect a critical step in photosynthesis. Often non-hormone-related herbicides may be incorporated into the soil prior to planting. In some cases great care must be taken to place the seeds of the crop plant where the plant killer isn't. Tordon, for example, is a herbicide that remains in the soil for years and is capable of virtually sterilizing a given parcel of land. What happens when this washes into productive areas? Fortunately, the use of Tordon is now carefully controlled.

Insecticides and Insect Populations: If We Can Beat 'em, We Can Eat 'em!

As we have seen, agricultural use of land causes many profound changes in the natural ecological balance. We are forced to increase our input in an attempt to stabilize the artificial condition we have created. Certainly one of the most serious threats to agriculture is the insect, and it is easy to see why man is interested in insecticides. *Insecticides* are chemical compounds that *kill insects.* They are tremendously important in our modern economy yet potentially very dangerous since they may have lasting side effects on various nontarget organisms. The net result is that insecticides may modify the very nature of insect populations.

Let us first consider the group of insecticides that we call chlorinated hydrocarbons. The prime example is DDT (Figure 8–14), but you may also have heard of dieldrin, endrin, and aldrin. These compounds, in some as yet unknown way, interfere with the central nervous system of insects, preventing normal impulse transmission. The chlorinated hydrocarbons are very stable compounds and not easily degraded by either enzymes or chemicals. This means that once they are sprayed on the land or the water, they stay around a long time. Exactly how long DDT remains active isn't altogether clear; however, estimates suggest that more than a decade is required for 50% of the applied DDT to be degraded. This fact coupled with the mobility of DDT means that its total concentration in the environment has increased during every year of its use. This increase occurred not only on our farmlands but also on every square foot of the earth's surface. DDT has been detected in the Arctic and Antarctic, in the air over Barbados, and high in the Sierra Nevada. Indeed, even our upper atmosphere is contaminated since a certain amount of DDT evaporates along with water from the surface of the earth.

The rather alarming content of DDT in the environment has

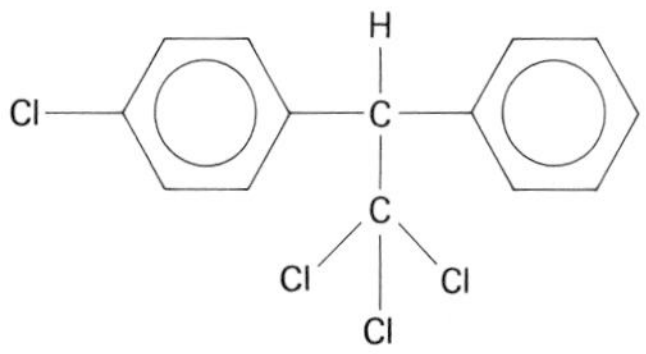

FIGURE 8–14.   THE STRUCTURE OF DDT

Dichlorodiphenyltrichloroethane (DDT) is a very stable compound and remains active within the environment for years. It is not precisely clear how DDT kills insects, but it is known that its general mode of action is through the nervous system.

really two consequences that should concern us: (1) an effect on insect populations as a whole and (2) an effect on other living members of the world community including man himself. Let us first briefly examine the second effect and then return to the first.

The effects of DDT and other chlorinated insecticides on noninsect species have only recently attracted serious attention and legislative action. In fact, DDT has been banned from all but emergency use in the United States. Nevertheless, we do not yet know all the ramifications of the DDT already used, and thus some consideration of its possible effects seems in order.

Chlorinated hydrocarbons are nearly insoluble in water but very soluble in fats. As a result they are not easily excreted from animals but instead tend to accumulate in the fatty tissues. The tendency for animals to accumulate these chemicals becomes more and more apparent as we progress along the food chain. At each successive level more DDT is concentrated in the body of the consumer (Figure 8–15). This phenomenon is called biological magnification, and it is of great practical importance to man since he is at the top of some food pyramids. It is not surprising, then, that serious questions are now being asked about the side effects that these chlorinated hydrocarbons may have on higher organisms.

Fish are very sensitive to chlorinated hydrocarbons. In the early 1960s endrin, a chlorinated hydrocarbon, was used extensively in the lower Mississippi Valley on cotton and cane crops. Coincident with its use was the loss of 10–15 million fish, including many commercially valuable species, in the lower Mississippi and its tributaries. The livelihood of many individuals was severely affected. In 1968 approximately 700,000 young coho salmon died as a result of DDT toxicity in waters of Lake Michigan and its tributaries. Other commercially valuable fish such as tuna have been shown to have rising DDT levels in their bodies. Indeed, the recorded loss of nontarget organisms as a result of DDT use is increasing at an alarming rate.

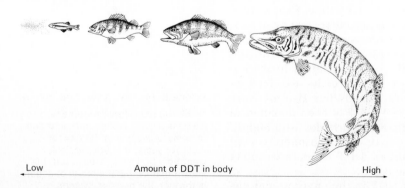

Low   Amount of DDT in body   High

FIGURE 8–15.

As one moves up the food chain from producer to consumer to secondary consumer, and so on, the amount of DDT in the tissue of the consumer increases. This occurs because the DDT is not readily eliminated from the body. Therefore, let's say a small fish has 1 part DDT. A larger fish may have 10 parts DDT because he gets and maintains a certain dose from every small fish he consumes, and so the process continues with biological magnification.

An especially well documented correlation exists between DDT concentrations and the reproductive failure of aquatic carnivorous birds. This particular phenomenon is apparently due to DDT's effect on calcium metabolism in birds which results in eggs with thin shells. Such shells are more prone to damage and crushing by the weight of the nesting parent. The problem has become severe enough that the survival of entire species has now become questionable. For example, nesting failures among bald eagles have depleted the population to dangerous levels, and the species could become extinct. Other species declining at an alarming rate include the brown pelican, the peregrine falcon, and the Bermuda petrel. What about man and DDT? As yet we have no reason to be particularly alarmed about DDT directly affecting our physical health. We emphasize "directly" and "as yet" because many facts are simply unknown. Indirectly DDT may prove to be a serious danger because any agent that affects the food chain obviously affects us.

Let us now turn to the direct effects of chlorinated hydrocarbons on insects themselves. In order to fully appreciate the problem, we must be aware of how insecticides affect insect populations, and indeed there are some unusual and unexpected results. The first problem encountered is *resistance.* When an insecticide is first used on a particular species, it kills a very high proportion of the individuals of that species, perhaps as many as 99.9%. However, because of biological variation, a small number of individuals will for one reason or another normally be resistant to a particular man-made insecticide. Following treatment the resistant individuals will often multiply at a much faster rate than is normal because competition with other members of their species for available resources has been greatly reduced. Furthermore, many of the natural enemies of the target insect may also have been killed by the insecticide and this, too, contributes to the rapid reproduction of the resistant few. The pattern here can be understood if one realizes that in nature predators are always less numerous than prey (such is the nature of a food pyramid) and that it will thus take the predators longer than the prey to recover. The net result is that the original population of largely susceptible insects will eventually be replaced by a resistant population. This concept is dramatically illustrated by the number of insect species which have become resistant to DDT since its widespread use began in the mid-1940s (Table 8–3).

As you may have already guessed, there is another problem associated with the use of insecticides. Much evidence suggests that in direct response to insecticides, the population density of certain insects *increases* so drastically that they become pests, while before such applications they were in sufficiently low concentrations so as to cause only minor or insignificant

Year	Number of Species Resistant to DDT
1946	0
1948	12
1954	25
1957	76
1960	137
1967	165

TABLE 8–3. INSECT RESISTANCE TO DDT
Due to variation and selection more species are now resistant to DDT than in the past.

Source: Data from Richard A. Wagner. 1971. *Environment and Man*, New York: W. W. Norton, pp. 228–229. Copyright © 1971 by W. W. Norton & Company, Inc.

crop damage. You might actually go so far as to say that certain pests are the *creation* of the pesticide industry. For example, the European red mite is now a serious pest in apple orchards. This situation is the direct result of eliminating the particular moth that was the natural predator of the red mite. In other words, it is possible with the use of chlorinated insecticides to eliminate one pest but create another. Do you see why? Probably our best hope for coping with the myriad of problems we have created via agricultural use of land is not the use of more and more chlorinated insecticides but is rather the initiation of integrated pest-control programs.

Integrated Pest Control

In order to understand integrated pest control, it must first be realized that the population density of any insect population is not static; rather it rises and then falls, rises again and falls. In other words, it fluctuates around some equilibrium position (Figure 8–16). We don't always know exactly what causes these fluctuations; however, such factors as food supply, predator level, and weather are certainly involved. Before a pest-control program is initiated, it only makes sense to determine the "economic threshold level" of insects which a particular crop can tolerate before the farmer is actually losing money. This level is reached when the value of the additional crop obtained by killing the pests exceeds the cost of the pest-control program. Such reasoning seems so obvious and sound that it is rather staggering to realize how many pesticides are applied only in the belief that the economic injury inflicted by the pests will exceed the cost of treatment. Pesticides themselves are very profitable products, and the companies that sell them benefit regardless of whether or not they are needed. Indeed, it is a self-perpetuating business. Excessive, nonessential use promotes the number of resistant species. Consequently other insect species become pests. This calls for greater pesticide applications and of course greater profits for the manufacturers. It is pointless to single out specific examples, but there are many, many examples of irresponsible advertising and false claims of increased production yields in the popular pesticide literature.

Returning to Figure 8–16, we should note that pesticides like DDT do not change the long-term equilibrium level in an insect population. Rather they usually bring only temporary relief by drastically reducing the immediate population of a given insect. However, the insect population will gradually return to the original level. Biological regulators (see below) on the other hand lower the equilibrium level (Figure 8–17). This is an important difference, therefore, between chemical pest control and biological pest control because by lowering the equilibrium level we make it less likely that fluctuations in insect den-

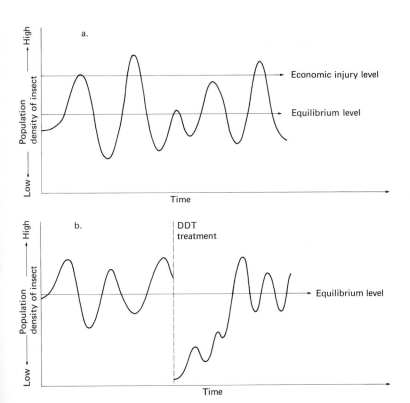

FIGURE 8–16.   INSECT POPULATION DYNAMICS

a. The population of a given insect fluctuates with time around some equilibrium level. The economic injury level to crops is determined by that population of insects which causes sufficient damage to offset the cost of the insecticide used to lower their density. Unfortunately, this distinction is not always made clear, and very often insecticides are used more frequently than it is economically profitable to do so. b. A pesticide like DDT gives only temporary relief. It quickly lowers the density of an insect population, but with time the population increases so that the same equilibrium level is again reached.

Adapted from "The history of a pest species," by Robert L. Rudd, in *Environment: Resources, Pollution, and Society,* ed. William W. Murdoch, 1971, Stamford, Conn.: Sinauer Associates, Inc., Figs. 1, 2, pp. 281, 284. By permission of Robert L. Rudd.

sity will exceed the economic threshold level. Let us examine some of the biological controls that are now becoming more popular and that seem to be much safer than chemical control.

The introduction of predators and parasites is undoubtedly the biological program that has met with the greatest success. In general, this approach is most successful if the pest itself is imported from somewhere else (see the example below). This is because the imported plant or animal usually has natural enemies only in its place of origin. Thus, one has only to find a predator that will not cause ecological disturbances of its own, and introduce it in the desired area. An endemic pest (not imported) usually has predators already, and thus one must either find a way to increase the predators' population or find species that do a better job in lowering the equilibrium density of the pest species.

An example of this type of biological control is the prickly pear cactus, a plant species introduced in Australia before the turn of the century. Due to an absence of natural enemies, its population exploded, and it threatened to become a serious pest. In 1925 a natural enemy, a moth, *Cactoblastis cactorum*, was introduced from Argentina. It might seem strange that a cactus and moth are enemies, but it's not too weird for the biological world. The larva of the moth bores into the interior of the cactus. This causes dehydration and also provides "open wounds" for various kinds of infections. The cactus dies. After the introduction of C. *cactorum* in Australia, the entire cactus population was destroyed in about 15 years.

Interestingly, pest control via the introduction of predators has reached the point that it is even being practiced by casual, part-time gardeners. Sears now sells ladybugs in bulk, and by mail order no less! The strategy here is to release these ambi-

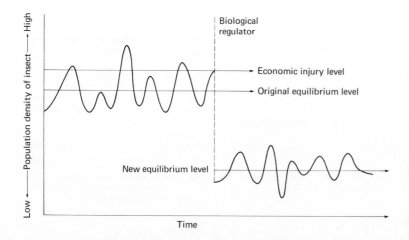

FIGURE 8–17.

Biological pest-control measures lower the equilibrium level of insect populations. Chemical pest control does not do this (see Figure 8–16). The advantage of biological pest control is that with a lower equilibrium level it is less likely that an insect population will ever reach the economic injury level.

tious creatures in your garden in order to lower the aphid population. Ladybugs do not eat plants, but they do effectively reduce the number of pests that do. Unfortunately, it is difficult to convince the ladybugs that they should take up permanent residence in your yard, and thus it is necessary to replenish the supply periodically. Nevertheless, this method is fun, relatively inexpensive, and more "natural" than insecticidal sprays or powders—assuming, of course, that supplying ladybugs by mail is natural.

Perhaps a potentially more useful, and certainly subtler, means of biological pest control is the introduction of sterile mating partners into a population of insects. The most important condition to be met in order for this technique to be useful is that the female mates only once. If this mating occurs with a sterile male, the result is sterile eggs. An example of the value of this technique is the eradication of the screwworm fly from the southeastern United States. A program was initiated by the U. S. Department of Agriculture to capture and sterilize the male screwworm fly. Sterilization was accomplished by irradiation, and the males were then released in large numbers in the infested areas. The screwworm fly was virtually eliminated in these areas within a matter of weeks.

Insect hormones have also attracted considerable attention as substitutes for chemical insecticides; however, their use on a large scale still appears to be something for the future. One hormone that shows exceptional promise is the *juvenile* hormone. This naturally occurring hormone must be present at certain stages of development and absent during other stages in order for insects to mature normally. If insects are artificially treated with the hormone during the "wrong" developmental stage, death will result. Such compounds are very specific for insects and have no known side effects on other animals. Furthermore, since these are naturally occurring compounds, insects cannot become resistant. The possibilities are present, therefore, for an almost ideal insecticide. Unfortunately, this hormone is too potent and kills all species of insects. Its widespread use would have serious ecological implications. Other hormones and sex attractants appear to have greater specificity and can possibly be designed to affect only certain pests. Undoubtedly, as our understanding of insect physiology increases, so will the possibilities of using the insects' own hormones as insecticides.

Possibilities also exist for the introduction of special strains of crop plants that are resistant to certain pests. These strains often take the form of so-called nonpreference plants, which means that the plants lack the necessary attractants and the insects feed elsewhere. Certain plants can also be bred which have compounds making them unpalatable to insects. Unfortunately, after 10–15 years insects that can attack the so-called resistant species usually evolve.

Drawing Conclusions

In summary there is much that can be said, but hard and fast conclusions are difficult. As our discussion indicates, the very nature of modern agriculture causes an ecologically unstable situation. Weeds and insects become pests that must be controlled if we are to continue to increase agricultural output. The situation becomes sticky when we try to decide the least harmful way to cope with this unpleasant but unavoidable task.

Unfortunately, the percentage of crop losses to insects has remained almost the same despite enormously increased use of the chemical pesticides. The chlorinated hydrocarbons such as DDT appear to have outlived their usefulness except in rare instances. Furthermore, it has now been shown that DDT affects not just insects but also many other nontarget organisms. Its persistence in the soil and atmosphere is causing and will continue to cause ecological malfunctions for years to come. Truly, it is a worldwide problem and one that is gradually becoming known. Legislation and even total bans on the use of DDT are being enacted, and the total picture is not an altogether pessimistic one. Substitutes for chlorinated hydrocarbons including the organophosphates, carbonates, and naturally occurring toxins such as rotenone are becoming more widely used. Rotenone is especially popular with home gardeners. All these compounds are much less dangerous ecologically and are not persistent; they are rather rapidly degraded in the environment. Let us hope that with more research and the use of biological controls, we will totally eliminate the use of the persistent and dangerous chlorinated hydrocarbons. There is really no reason that the transition should lead to serious consequences for human health and nutrition. But even if it did, the continued use of persistent insecticides would sooner or later result in even more havoc and devastation. Let us therefore proceed with care, and if necessary, man those hoes!

Selected Readings

Aaronson, T. 1971. Gamble. *Environment* 13:20–29. Discussion of possible harmful effects of herbicides and a cost-benefit analysis of their use.

Carson, R. 1962. *Silent Spring*. Boston: Houghton Mifflin. A classic in popular ecology.

Epstein, E. 1972. *Mineral Nutrition of Plants: Principles and Perspectives*. New York: John Wiley. Advanced text on mineral nutrition.

Hasler, A. D. 1969. Cultural eutrophication is reversible. *BioScience* 19:425–431. Community action and how it improved lake quality in Germany.

Orians, G. H., and E. W. Pfeiffer. 1970. Ecological effects of the war in Vietnam. *Science* 168:544–554. An early assessment of the effects of deforestation.

Powers, C. F., and A. Robinson. November 1966. The aging Great Lakes. *Scientific American* 215(5):94–100. (Offprint no. 1056.) Formation of the Great Lakes and man's impact.

Wigglesworth, V. B. 1970. *Insect Hormones*. San Francisco: W. H. Freeman. Good discussion of insect hormone physiology and possibilities for pest control.

Williams, C. M. July 1967. Third generation pesticides. *Scientific American* 217(1):13–17. (Offprint no. 1078.) Describes how hormones may aid in the control of insects.

Woodwell, G. M., P. P. Craig, and H. A. Johnson. 1971. DDT in the biosphere: Where does it go? *Science* 174(4014):1101–1107.

9 *Diversity Among Organisms: Classification and Life Cycles*

PART ONE · THE PLANT KINGDOM

Nearly all living organisms share certain unifying features: metabolic activity, genetic phenomena, adaptation. sexual reproduction, and, most important perhaps, cellular organization. Thus, in the previous chapters dealing with anatomy and physiology, it was possible to deal almost entirely with a "generalized" plant, a hypothetical organism defined by the characteristics common to most plants. At this point, however, this approach will be abandoned in order to study and examine the *diversity* that exists in the plant world. Take a quick survey of the plants around you. You may observe flowering plants occurring in many shapes and sizes, each using more or less divergent strategies to cope with its terrestrial or aquatic environment. And, a careful observation indicates that not all plants have flowers. There are cone-bearing plants, ferns, mosses, and algae. In short, the noun *plant* can mean a great deal more than the kinds of organisms we have examined up to now.

Classification Sometimes Involves Arbitrary Decisions

Classification systems do not exist in nature but are man-made attempts to organize; they reflect our perception of the diversity in plants. Nevertheless, now as in the past, a system of classification is an efficient way to organize and transmit information regarding a particular plant or group of plants. It is instructive, therefore, for us to begin this discussion by considering how classification systems were developed and to understand how and why they have changed.

The basis for the earliest forms of identification and classification was undoubtedly utilitarian. This is suggested by studies of aboriginal tribes living today in remote corners of the earth far removed from literary and other contemporary influences. These people are able to distinguish with precision various kinds of plants and even occasionally particular individual plants used for food, building materials, weapons, and so on. Other plants of little utility are also recognized as different kinds but with considerably less discrimination. That being the case today, it is a good guess that it was also the case thousands of years ago before man left a written record of his work.

The earliest botanical work for which we have a written record is that of Theophrastus (about 372 to 287 B.C.), the so-called father of botany. Theophrastus began as a student of Plato and worked in Athens during the fourth and third centuries B.C. During those early days and for over a thousand years thereafter, the strong emphasis in plant studies was again utilitarian and primarily medicinally oriented. This was probably the case because the early botanists were also physicians who required certain kinds of plants in their work. Thus, plants were classified with respect to their medicinal properties, with no attempt to group the various kinds according to any other relationship (Figure 9–1). Beginning with the Renaissance, the emphasis of botanical studies gradually changed from primarily utilitarian toward an interest in the intrinsic properties of plants themselves. Many new kinds of plants were described, and plant structures were studied in some detail so as to discover their function.

The culmination of this phase in the history of plant classification is probably the work of Carolus Linnaeus (1707–1778), a Swedish botanist, who classified plants primarily according to numbers of stamens, pistils, and other flower parts (see Figure 9–6). This scheme offered a simple and direct aid in identifying an unknown plant. It quickly became very popular and stimulated interest in botanical studies. During and after this period botanists began to recognize that some groups of plants were more similar to one another than to other groups on the basis of their structure, morphology, and mode of reproduction. And so the first notions developed that there might be some natural relationships among these similar groups. The idea of evolution was unknown, however, and so the basis for these natural groupings was generally thought to be some sort of divine pattern which man might discover by continued and diligent study.

The studies of Charles Darwin (1809–1882) provided the breakthrough to understanding the basis for natural relationships among plants and animals: common descent. His facts

FIGURE 9–1.

Early botanical studies often emphasized the medicinal values of plants, as suggested in this photograph taken in Kew Gardens near London, England.

and reasoning can be summarized in five points, although you must understand that the arguments and supporting evidence are far more extensive. (1) *Overproduction*: All organisms tend to produce more offspring than can survive in the course of time. (2) *Constant population size*: Minor short-term fluctuations aside, in the long run the number of individuals of any kind of plant or animal tends to remain fairly constant. These first two points were verified by observation. (3) *Competition*: This idea was a deduction based on points 1 and 2. Clearly, there is not adequate space for every seed and seedling to grow to maturity and then itself produce a number of offspring which do the same. Given overproduction and yet constant population size, it necessarily follows that there will be competition for the limited available resources. (4) *Variation*: No two individuals of sexually reproducing organisms are identical. Slight variations in many traits from individual to individual are always found when they are examined closely, even among members of the same species. This point was also made as a result of direct observation. (5) *Preferential survival and reproduction*: This idea is sometimes called survival of the fittest but is really the preferential survival and reproduction of the fitter as compared with the less fit in a group. Basing this point on points 3 and 4, Darwin reasoned that since no two individuals are identical, some plants are probably better adapted to the particular environment in which they find themselves than are some of the others. The survivors are those that grow to reproductive maturity, and these are more often from the better adapted variants than from the more poorly adapted. Today we know that these adaptive traits are heritable because they result from genetic mutation and recombination (see Chapter 4).

Now let us carry this reasoning one step further in order to see how these ideas might lead to the evolution of related but distinctive plant groups. As a basis for discussion, consider the hypothetical case of a plant that grows at an elevation of 1000–4000 feet in the foothills of a mountain range in western North America. Many environmental conditions within this general location will vary, but this is an important aspect of the example. Temperatures will be generally higher and the precipitation less at lower elevations than at higher ones. Because of these and other differences in the environment, the particular variants produced in each generation will respond accordingly. Perhaps, for example, larger leaves are the superior adaptation at high elevations where water stress is less important and they provide greater photosynthetic capacity; smaller leaves are useful at lower elevations because water stress is important there and smaller leaves reduce the amount of transpired water. It follows that at higher elevations those individuals with larger leaves will more often be the larger, vigorous plants which produce more offspring. At lower elevations the same type of plant will often die from water stress, while their smaller-leafed neighbors will develop and grow vigorously. Clearly, if this sort of selection carries on very long, we will find two plant types corresponding to the two habitats: the large-leaf race at higher elevations and the small-leaf race lower down. In the long course of time, perhaps these two types would become so different in these and other traits that we would recognize two separate species instead of one species that includes two types (Figure 9–2). Two species that have arisen in that way are related by descent, that is, they are derived from the same ancestor.

Just as all cells come from pre-existing cells and all individuals come from their parents, so it is that all species of plants come from pre-existing species. Being related by common descent is the natural relationship that results in the recognizable groupings of organisms.

Thus, you can perhaps now understand why the purpose of modern-day classification is to group similar kinds of plants together—to reflect their genetic relationships. While this goal is rather straightforward, the actual decisions involved in grouping plants are not so easy as they may appear at first blush. Life has been on earth a long time, and biological evolution has been proceeding as long as life has existed. Clearly, some kinds of plants became distinct from their parental types (diverged) long ago and now are only very distantly related, whereas others are more closely related because they have diverged more recently. The difficulty encountered when classifying plants is that we have only a limited amount of information to work with. We weren't on the site watching and recording evolutionary events during the past several hundred million years. Our evidence therefore includes only plants living today and the fossil record which is largely incomplete, particularly so for the early evolution of life forms that are of concern to us here. It follows then that as new fossils are found and new techniques reflecting relatedness are discovered, classification systems may be re-evaluated and changed. A reasonable example of this process of re-evaluation concerns the establishment of the broad categories called *kingdoms*.

FIGURE 9–2.

Shown here are three closely related plants (note the flower and leaf patterns) which illustrate the kind of adaptive evolution outlined in the text. a. *Delphinium recurvatum* grows in dry, warm, and often alkaline flats along the western margin of the San Joaquin Valley, California. b. *Delphinium gypsophilum* grows in the intermediate grassland habitat above the valley floor and below the woodland. c. *Delphinium hesperium* grows at higher elevations in the generally cooler and damper oak woodland above the valley.

From "*Delphinium gypsophilum,* a diploid species of hybrid origin," by Harlan Lewis and Carl Epling, in *Evolution* 13:515, Fig. 2 (December 1959).

a.

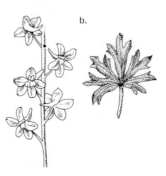

b.

c.

Kingdoms: An Example of Re-evaluation

Only several years ago all organisms were placed in the plant kingdom or animal kingdom, using some of the characteristics listed in Table 9–1. Although such a system functioned rather well in serving to separate most organisms, there were a number of problem cases such as euglena (*Euglena*) (Figure 9–3). This organism consists of a single cell, lives in fresh water, and is capable of swimming via the motion of its *flagellum* (a whiplike appendage at one end). Some kinds of euglena contain chlorophyll in chloroplasts and are photosynthetic, whereas others lack chlorophyll and require a source of pre-made food. Beyond that, some green euglenas lose their chlorophyll after being cultured in darkness for a time, and they do not recover and turn green again when returned to the light. These organisms have no cell wall, only a cell membrane. Their stored food is neither starch (like plants) nor glycogen (like animals) but is a similar carbohydrate called paramylum. Finally, there is a light-sensitive eyespot at the anterior end of the organism. So what do we say —are euglenas plants or animals or both?

Our purpose here is to point out some kinds of data used by biologists in investigating relationships among organisms and to mention a few of the problems that arise. In relation to the plant-animal problem with euglena, the argument as seen today goes something like this: (1) All euglenas and other single-celled organisms are more closely related to one another than they are to multicellular organisms and therefore should be classified together in the same group. (2) The sum of all their characteristics indicate that they are somewhere between the plant kingdom and animal kingdom and therefore cannot unambiguously be classified in either one. (3) We are left with the option of either (a) grouping all living organisms, plants, and animals together in a single kingdom, or alternatively (b) recognizing at least three kingdoms. The first alternative defeats the purpose of classification which, as viewed today, is to communicate degrees of genetic relationship. There is no question that oak trees and whales belong in different kingdoms; the only problem is where to draw the lines. Consequently, biologists generally follow the second alternative (b) and recognize the need for the creation of other kingdoms.

Summing up all the available evidence and taking all living things into consideration, biologists now recognize five kingdoms. The animal kingdom (kingdom Animalia) includes all those things we normally think of as animals, such as sponges, jellyfish, worms, snails, insects, starfish, and, of course, the vertebrates. The plant kingdom (kingdom Plantae) includes all the seed plants, ferns, mosses, and algae. The fungi are now considered different enough from other organisms to be placed all alone in a separate kingdom (kingdom Fungi). All the re-

Plants	*Animals*
Possess chlorophyll	No chlorophyll
Photoautotrophic	Heterotrophic
Nonmotile	Motile
Cell wall; cellulose	No cell wall; no cellulose
Continuous growth	Limited growth in higher forms
Totipotency	Irreversible differentiation
Starch as storage carbohydrate	Glycogen as storage carbohydrate
No centriole or astral rays	Possess centrioles and astral rays
Lack nervous system	Possess nervous system

TABLE 9–1. CHARACTERISTICS THAT DISTINGUISH HIGHER PLANTS FROM MORE ADVANCED ANIMALS

maining organisms have been assigned either to the kingdom Protista, where a true nucleus with a membrane has been identified, as in diatoms, euglenas, and protozoa; or to the kingdom Monera, including blue-green algae, bacteria, and other organisms that lack a true nucleus. Viruses are also typically placed in this kingdom although, as you will see, they present special problems.

Such a classification system emphasizes the point that *within* each kingdom the organisms are thought to be more or less closely related to each other, whereas the genetic relationships *between* kingdoms are generally more distant (Table 9–2, Figure 9–4).

FIGURE 9–3. *EUGLENA*

Note how the organisms change shape as they move about, indicating a flexible outer covering rather than a rigid cell wall (left). Various organelles mentioned in the text are shown here (right).

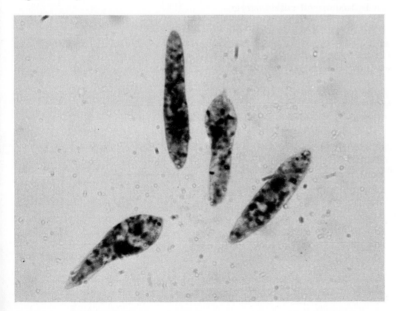

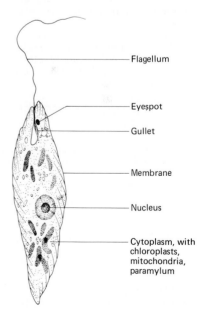

- Flagellum
- Eyespot
- Gullet
- Membrane
- Nucleus
- Cytoplasm, with chloroplasts, mitochondria, paramylum

Kingdom Monera: Procaryotic cells (lacking nuclear membrane), most unicellular but some colonial, lacking plastids, mitochondria, and other intranuclear organelles. Nutrition mostly absorption, some photosynthetic, or chemosynthetic. Nonmotile or with simple flagellum or gliding.

> *Subkingdom Myxomonera*: Lacking flagella. Included here are division Cyanophyta (blue-green algae) and some so-called sliding bacteria in division Myxobacteriae.
>
> *Subkingdom Mastigomonera*: With simple flagella. Includes most bacteria in the three divisions Eubacteriae (true bacteria), Actinomycota (mycelial bacteria), and Spirochaetae (spirochetes).

Kingdom Protista: Eucaryotic cells (with nuclear membrane), mostly unicellular but some colonial (some with multicellular stages in life cycles), with plastids, mitochondria, and other intranuclear organelles. Nutrition by photosynthesis, absorption, or ingestion. Nonmotile or with advanced flagella, some motile by other means.

Included here are five phyla* of protozoa (single-celled animal-types), one of which probably gave rise to the animal kingdom, and division Euglenophyta which may have given rise to plants. There are four other divisions as well.

Kingdom Fungi: Eucaryotic, generally with many nuclei in each cell. Cells generally organized into septate filaments, which may be further aggregated, as in mushrooms, by lacking plastids and photosynthesis pigments. Nutrition absorptive. Mostly nonmotile. Eight divisions, the larger of which are the following:

> *Division Myxomycota, plasmodial slime molds*: Lack cellular organization, consisting of acellular multinucleate mass of protoplasm.
>
> *Division Oomycota, oosphere fungi*: Cellulose cell walls, biflagellate zoospores present.
>
> *Division Zygomycota, conjugation fungi*: Chitin cell walls, spores not flagellate, meiotic spores produced in fusion cell called a zygospore.
>
> *Division Ascomycota, sac fungi*: Chitin cell walls, spores not flagellate, meiotic spores produced with a sac.
>
> *Division Basidiomycota, club fungi*: Chitin cell walls, spores not flagellate, meiotic spores produced on club-shaped basidium; mushrooms.

TABLE 9–2. MAIN CLASSIFICATIONS OF ORGANISMS FORMERLY INCLUDED WITHIN THE PLANT KINGDOM. (THEIR PRESUMED RELATIONSHIPS ARE SHOWN IN FIGURE 9–4.)

Source: From R. H. Whittaker, "New concepts of kingdoms of organisms," *Science* 163(10):150–160, Table 1, (January 10, 1969). Copyright 1969 by the American Association for the Advancement of Science. By permission of R. H. Whittaker.

TABLE 9–2 (*cont.*)

Kingdom Plantae: Eucaryotic cells, multicellular (some algae unicellular), with cell walls, photosynthetic pigments in plastids, mitochondria, other organelles. Nutrition mostly by photosynthesis. Mostly nonmotile. Nonvascular plants lack vascular tissues (xylem and phloem), may be aquatic or require moist habitat, and include the algae and bryophytes; vascular plants have vascular tissues distributed throughout the plant body, and include ferns, gymnosperms, and flowering plants. A few of the major divisions are listed below.

Division Rhodophyta, red algae: Contain chlorophylls a and (in some) d, also phycocyanin and phycoerythrin, floridean starch as food storage, and no flagella. No marked somatic cell differentiation.
Division Phaeophyta, brown algae: Contain chlorophylls a and c, also fucoxanthine, with laminarin and mannitol as food storage, and zoospores with two flagella. No marked somatic cell differentiation.
Division Chlorophyta, green algae: Contain chlorophylls a and b, and starch in plastids as food storage. No marked somatic cell differentiation.
Division Bryophyta, mosses and liverworts: Terrestrial (in moist habitats) with some somatic cell differentiation but not vascular tissue. Contain chlorophylls a and b and starch, as in Chlorophyta.
Division Tracheophyta, vascular plants most with true leaves, stems, and roots.

Subdivision Pteridophyta, ferns and fern allies: Vascular plants that have a distinct alternation of generations, reproducing by spores but no seeds. Ferns and fern allies include several classes, the largest and most common today being the true ferns. True ferns are plants lacking seeds with both gametophyte and sporophyte generations green and photoautotrophic.
Seed plants include a number of groups which reproduce by seeds. The gametophyte generation is much reduced and parasitic upon sporophyte. The largest and most common representatives today are the following subdivisions and classes within the division Tracheophyta.
Subdivision Coniferae, gymnosperms: The seeds develop from ovules borne on the surface of a scale, usually aggregated into cones; no flowers present.
Subdivision Anthophyta, the flowering plants:

Class Dicotyledonae: Embryo with two cotyledons.
Class Monocotyledonae: Embryo with one cotyledon

*The International Code of Botanical Nomenclature stipulates that within kingdoms the largest categories in the classification of plants will be called divisions, rather than phyla as in zoological nomenclature. This presents a problem as to the proper term to use when the divisions of the old plant kingdom and phyla of the old animal kingdom are reorganized according to a five-kingdom scheme. The answer to this difficulty here is to use the term *division* throughout except when phylum is clearly indicated, as for the single-celled animals in kingdom Protista. The reasons for this are, first, that in this book the rules of the botanists, rather than the zoologists, are generally followed, and, second, that most of the groups in question have traditionally been included within the plant kingdom in the two-kingdom scheme.

In Part One of this chapter, discussions will deal exclusively with those organisms placed in the new plant kingdom. Part Two will briefly survey several of the other kingdoms. As you make your way through this chapter, you will have the opportunity to observe directly some of the organisms mentioned. In fact, you can use your text as a kind of informal guide through a local park, greenhouse, forest, or even your own yard. On such a walk you should see representatives of most of the organisms mentioned, get an overview of the material presented here, and at the same time enjoy one of the very real pleasantries of plant biology.

Beginning with the Most Advanced: Angiosperms, the Flowering Plants

The principle characteristic that distinguishes angiosperms from the other plants possessing a vascular system (Table 9–2) is that their ovules and seeds are *enclosed* within a pistil which later develops into a seed-bearing fruit (Figure 9–5). This somewhat technical description of this group is used instead of the

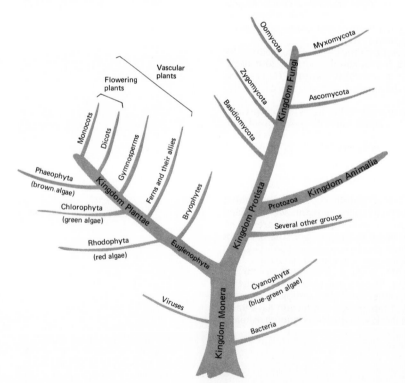

FIGURE 9–4.

Phylogenetic relationships of organisms according to the classification system of five kingdoms. In the kingdom Plantae, dicots and monocots are flowering plants; dicots, monocots, gymnosperms, and ferns are vascular plants. See Table 9–2 and the text for further details.
From "New concepts of kingdoms of organisms," by R.H. Whittaker, in *Science* 163(10):157, Fig. 3 (January 1969). Copyright 1969 by American Association for the Advancement of Science.

more obvious characteristic, flowers, simply because the flower is very difficult to define accurately enough to distinguish it unambiguously from the cones of gymnosperms.

A detailed and complete life cycle for a typical angiosperm is presented in Figure 9-6. In addition, the life cycles for other groups within the plant kingdom and selected examples from other kingdoms will be presented later in the chapter. Since detailed life cycles represent a summary of many important points, one can conveniently compare the various groups through these illustrations. However, there are several points to be made now in order for you to understand more readily how to "read" a life cycle and make use of the kinds of information that can be extracted from one.

Most important, sexual reproduction requires the production of haploid (n) cells via meiosis, and from these cells the sex cells are eventually produced. This haploid phase of the plant's life cycle is called the *gametophyte* generation (see Chapter 4). In many plants, angiosperms and gymnosperms in particular, the gametophyte generation consists of only a few cells and is rather inconspicuous. In other plants (ferns, mosses, and algae), however, the gametophyte is more conspicuous. Therefore, one of the important features to be noted about a life

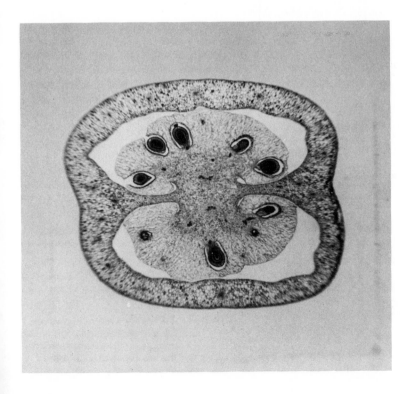

FIGURE 9-5.

Ovules and seeds of angiosperms are enclosed within the ovary wall. Compare this photograph of a young tomato with those of gymnosperms in Figure 9-9.

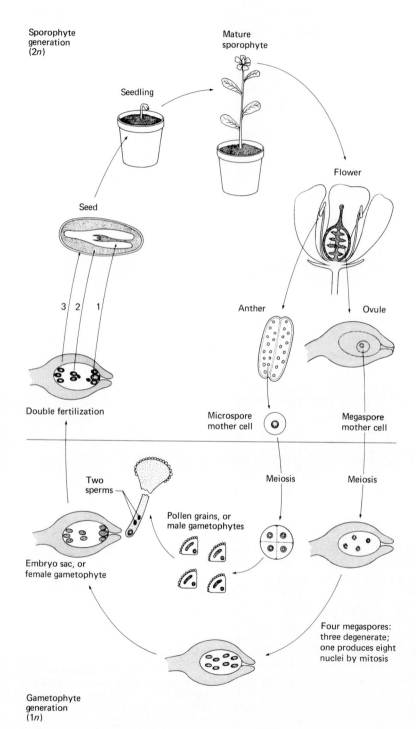

Sporophyte generation (2n)

Mature sporophyte

Seedling

Flower

Seed

Anther

Ovule

3 2 1

Double fertilization

Microspore mother cell

Megaspore mother cell

Meiosis

Meiosis

Two sperms

Pollen grains, or male gametophytes

Embryo sac, or female gametophyte

Four megaspores: three degenerate; one produces eight nuclei by mitosis

Gametophyte generation (1n)

FIGURE 9–6.

Life cycle of the flowering plant, *angiosperm*. As a result of double fertilization, (1) an embryo develops from the zygote (egg and sperm), and (2) an endosperm develops from the polar nuclei and sperm. (3) A seed coat develops from the ovule wall.

cycle is the relative size and complexity of the gametophyte. Following fertilization the diploid $(2n)$ or *sporophyte* generation is produced. Again, the relative size and complexity of this phase of the life cycle is important. Where and how meiosis occurs (giving rise to the gametophyte and thus completing the cycle) should also be noted. In other words all plants' life cycles involve an *alternation* between the diploid and haploid conditions. The relative importance of each condition is one of the features that separates various groups of plants from one another. In addition some plants can reproduce asexually, and therefore some life cycles may have an asexual "loop" indicating the particular features of this type of reproduction.

Using the angiosperm life cycle as an example, we can see that the sporophyte is the conspicuous part of the life cycle and that both male and female gametophyte plants are produced *within* flowers (see Figure 9–6). The gametophytes consist of only a few cells which give rise to the egg and sperm. When the egg and sperm eventually fuse, a new sporophyte generation is produced. In the case of angiosperms, the early development of the new sporophyte involves seed formation while it is enclosed within the pistil of the parent sporophyte plant.

The angiosperms exceed all other groups of plants in their range and diversity of habitats. However, amid this seemingly wide spectrum of sizes, shapes, and life strategies, two rather distinct classes can be recognized: the Monocotyledonae (monocots) and the Dicotyledonae (dicots). Let us first consider the monocots and the characteristics that set them apart.

Grasses are representative of the Monocotyledonae. One characteristic of a grass that quickly tells you it is a monocot is the *parallel* venation in the leaves. If you pick a blade and look at it closely, you will see that the veins run parallel down the length of the leaf.

Other characteristics that distinguish monocots from dicots are shown in Figure 9–7. Among these characteristics are the arrangement of flower parts: threes or multiples of three. This is an especially easy and reliable rule of thumb to use in tandem with the parallel venation rule. Put it to the test by locating a plant that has parallel venation and is flowering. Count the petals and the stamens in each flower. If it's a monocot, you will find three or six parts almost every time. Even the pistil has three carpels. Occasionally, you'll find in cultivation especially showy monocot flowers called *doubles*. In such flowers the stamens develop as extra petals, and there are few or no stamens. These are rather special genetic strains established in the horticultural trade because of their striking appearance and are exceptions to our general rules for quickly distinguishing between monocots and dicots. An additional problem is likely to be encountered in the grasses and certain other monocots

FIGURE 9–7.

Comparison of monocot (left) and dicot (right) characteristics.

a. Parallel venation, palm leaf

b. Reticulate venation, sumac

c. 3-merous flower, day lily

d. 4-merous flower, *Fuchsia;* 5-merous flower, *Hibiscus*

FIGURE 9–7 (*cont.*)

Comparison of monocot (left) and dicot (right) characteristics.

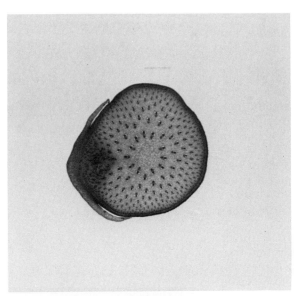

e. Stem cross section, asparagus

f. Herbaceous stem cross section, *Helianthus;* woody stem cross section, magnolia

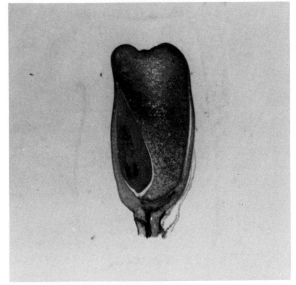

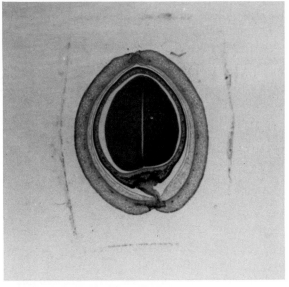

g. One cotyledon, corn

h. Two cotyledons, lima bean

that have lost flower parts in their evolutionary history. But on the whole you will find that the rules hold true.

Economically, the Monocotyledonae is a very important group. All the world's grains and cereals, dates, coconuts, and bananas are monocots, as are onions and garlic. Many ornamentals are in this group, too, including *freesia*, calla lilies, canna lilies, orchids, and aloes. If we had to select one single group of plants as the most important in human history and civilization, it would probably be the grass family in the Monocotyledonae.

The second of the two classes of angiosperms is the Dicotyledonae. Most of the large conspicuous plants you see are probably dicots. In contrast to the monocots, dicots typically have *reticulate* rather than parallel venation, and the other general features shown in Figure 9–7. The leaves of a dicot may be either *compound* or *simple*. If you look at the leaves of a rose bush, you will most often find five leaflets per leaf (Figure 9–8). How do we know that those five things are leaflets of a single leaf rather than five separate leaves? A leaf has a bud (auxiliary or lateral bud) in its axil which has the potential to grow into a branch or flower, and leaflets do not.

Flowers are another readily distinguishable feature of the dicots. If you count the parts of a dicot flower, you'll find they occur in fours, fives, or many but almost never in threes. For example, most ornamental hibiscus plants have five sepals, five petals, and many stamens, whereas fuchsias have four sepals, four petals, and eight stamens (see Figure 9–7). Again, except for doubles and other very showy ornamentals, you will find this pattern to be a remarkably accurate way of identification.

Economically the dicots are important for a number of reasons. All so-called hardwoods are dicots, including not only the rosewoods, mahoganies, walnuts, and oaks used in making fine furniture but also the very soft balsa wood popular with model builders. Many food plants, too, are dicots, such as peas and beans, potatoes, tomatoes, carrots, and lettuce.

Scientists have been unable to total accurately the number of different kinds of angiosperms. A fairly good estimate, however, would be on the order of 200,000 species with perhaps a fifth of these monocots and the rest dicots.

Gymnosperms: Seeds Not Enclosed

The second group of plants to draw our attention is the gymnosperms. Members of this group include pines, firs, junipers, cedars, yews, and ginkgoes. They are considered separate and

FIGURE 9–8.

Compound leaves are those that have two or more leaflets. For example, each leaf on a rose shrub consists of several, usually five, leaflets.

distinct from the angiosperms because their seeds develop on the surface of a reproductive appendage (usually on a scale of a cone) and are *not structurally enclosed* (Figure 9–9). Perhaps a more workable definition would be to say that gymnosperms are seed-producing plants *without flowers*. Most commonly, gymnosperms produce cones that consist of a central axis which bears numerous scales. Pines and many other gymnosperms produce both male and female cones (Figure 9–10). In specialized areas on the male and female cones, cells undergo meiosis giving rise to haploid cells from which the male and female sex cells will eventually arise. The details concerning the life cycle of pines are presented in Figure 9–11. In short, however, the cycle is very similar in principle to that exhibited by the angiosperms. That is, the dominant part of the life cycle is the sporo-

a.

b.

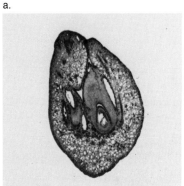

FIGURE 9–9.

Young developing seeds of gymnosperms: (a) juniper and (b) pine. Note that the seeds are not structurally enclosed. Compare this with Figure 9–5.

FIGURE 9–10.

Shown here are (a) male and (b) female cones of the commonly cultivated Deodar cedar (c).

a.

b.

c.

phyte, and the male and female gametophytes consisting of only a few cells, are not free living but rather produced in localized areas on the diploid cones.

Since almost every garden or park of any size has some gymnosperms, you can hardly miss them on your strolls. If possible, pick up a cone and examine it closely. Can you find any seeds? In most cases the leaves of the gymnosperms are needle-like (see Figure 9–10) or perhaps scalelike and may cover the young stems (Figure 9–12).

There are several gymnosperms of historical interest, but perhaps the most curious story involves the Dawn redwood (Figure 9–13). In order to appreciate this story fully, it is important to know that in the course of evolution many species have appeared and disappeared from the earth, leaving only their fossilized remains as evidence of their existence. Such was assumed to be the case with the Dawn redwood. This plant was named and studied from fossilized structures and presumed to be extinct until the early 1940s. The initial discovery of *living* Dawn redwoods resulted from explorations of the forests of western China initiated by the upper-class Chinese who were anxious to flee China after the Japanese invasion. After World War II a rather large expedition, organized with financial

FIGURE 9–11.
Life cycle of the pine.

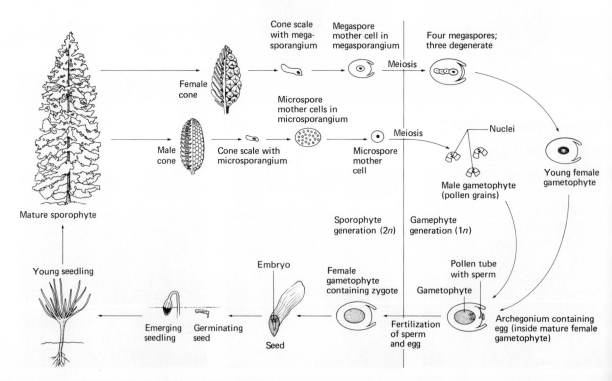

support from the United States, confirmed the original find by discovering 1000 of these trees living in the Shui-Sha Valley. Seeds were collected from these specimens and distributed throughout the world so that today the Dawn redwood is found in many botanical gardens and nurseries throughout the world.

Ferns and Their Relatives: Vascular Tissue but No Seeds

Ferns and their relatives are the third group of plants to be considered. This group has true vascular tissue (as do angiosperms and gymnosperms) but a very different and more primitive mode of reproduction which does not involve the production of seeds. A detailed outline of the life cycle of ferns can be found in Figure 9–14 and should be referred to from time to time as the characteristics of this group are described.

Most ferns have large to very large *compound* leaves called fronds which expand in a circinate manner, uncoiling as they grow (Figure 9–15). On the backs (underside) of mature fronds you may notice small brown or rusty-colored areas. These areas are *sporangia* (Figure 9–16), and it is within sporangia that certain cells undergo meiosis to produce thick-walled haploid cells called *spores*. When the spores are released and fall onto a suitably moist substrate, they germinate to produce tiny *independent* haploid plants—free-living gametophytes. Because of the size, usually only a few millimeters in diameter, the gametophyte stage is seldom seen by the casual observer. When the gametophyte is mature, both male and female sex organs form and produce via mitosis sperm and eggs. The sperm cells produced by the gametophyte are motile and can fertilize the egg only if the surface of the gametophyte is sufficiently moist to permit the sperm to swim to the egg. Thus, fertilization is dependent on surface water, and water then becomes an important limiting factor in the potential growing range of the gametophyte. Once the egg is fertilized, the *diploid zygote* (the first cell of the sporophytic fern plant) begins growing directly on the gametophyte. Eventually the gametophyte withers and dies (Figure 9–17).

The life cycle of the fern is the first opportunity you have had to see a plant in which there is a well-developed *alternation of generations* with both a large spore-producing sporophyte ($2n$) and a small photoautotrophic, independent gametophyte ($1n$). For comparison, it would be useful to review the life cycles of angiosperms, gymnosperms, and ferns with special attention to the structure and independence of the gametophyte generation.

FIGURE 9–12.

Scalelike leaves of *Thuja* (shown here) as well as junipers, cypresses, and some other gymnosperms may be closely appressed to the branches and conceal the young stems.

FIGURE 9–13.

A small specimen of Dawn redwood (*Metasequoia*).

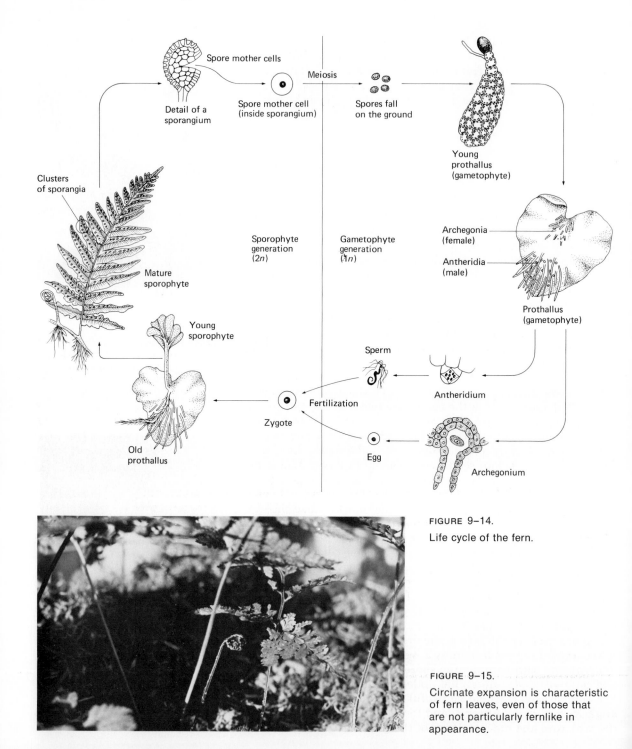

FIGURE 9–14.
Life cycle of the fern.

FIGURE 9–15.
Circinate expansion is characteristic of fern leaves, even of those that are not particularly fernlike in appearance.

FIGURE 9–16.

In ferns sporangia are typically found on the underside of the leaf. The size and shapes of these structures vary.

With this background information on ferns, we can now get back to our informal tour. When you locate a fern, you obviously will want to observe the pattern of sporangia formation, but in addition observe if possible the mode of leaf or frond growth. You should also note that the stem is quite reduced and at ground level (except in the case of tree ferns); that is, of a "typical" fern virtually all you can see are the leaves. Usually the stem grows horizontally underground and is called a *rhizome*. True roots grow directly from this underground stem.

Relatives of the fern, including horsetail (*Equisetum*), *Psilotum, Selaginella,* and others, have an external morphology that is quite variable and not at all like that of ferns (Figure 9–18). Many of these plants have small leaves called *microphylls*, which are narrow and often scalelike with a single, unbranched vein. The stems also vary from erect to prostrate and are usually less than half a meter tall. Lacking large leaves, the stems are often green themselves and carry the primary burden of photosynthesis. Since these plants are so different in external appearance from the ferns, you might well wonder why they are considered to be relatives. The answer is that they contain vascular tissue as do ferns, the life cycles are very similar to

FIGURE 9–17.

A young fern sporophyte, with tiny branched veins just visible in the leaf, is seen here still attached to the gametophyte, which consists of two thin, flat lobes.

FIGURE 9–18.

Relatives of ferns include (a) horsetail (*Equisetum*) and (b) *Psilotum*. They have vascular tissue, and their life cycles are similar to that of ferns.

a.

b.

that of ferns in broad outline, and from the fossil record it would appear that both groups arose from a common ancestor.

Ferns and their close relatives were the dominate form of vegetation during the late Paleozoic era (200 million–350 million years ago) and were among the primary contributors to the great coal-forming forests of the Carboniferous period (see Table 10–1). Ferns can still be found today in a number of habitats ranging from desert to tropical zones.

The Bryophytes:
An Introduction to Nonvascular Plants

With the introduction of the bryophytes, we come to a very important division in the Kingdom Plantae: the separation of the vascular plants—angiosperms, gymnosperms, ferns, and the fern relatives—from the nonvascular plants such as the bryophytes and all other groups discussed in the following pages. The most conspicuous and easily found member of the bryophyte group is the moss, and therefore we will use it as the basis for our discussion. If you can spend a few moments looking around in almost any garden, especially along damp, northern exposures or under shrubs, you will undoubtedly encounter this member of the bryophytes. Mosses usually appear as a dense green carpeting of plant material; however, on close examination you can see that this carpet is made up of many small plants growing close together. Each possesses a short stalk from which small, green, scalelike structures called *phyllidia* grow in a spiral pattern. These structures are rather leaf-like in appearance but are not true leaves because they lack vascular tissue.

The other group of bryophytes, the liverworts, are even less conspicuous. They consist of a simple, flattened green piece of plant material, sometimes with wavy or leafy margins, growing prostrate on damp soil or forest litter (Figure 9–19). You probably won't see liverworts growing in a cultivated garden unless someone has planted them there. To spot them in the wild, you must have good vision, even though they are quite common. Again, there is no vascular tissue present in liverworts.

We would like to emphasize the lack of vascular tissue in the bryophytes for several reasons. First, it allows us to rationalize why bryophytes in general are so small. Lacking specialized tissues for water and food transport, the nonvascular plants must rely on capillary action, diffusion, and cytoplasmic streaming for the distribution of materials and water. Since these methods are relatively inefficient over long distances, nonvas-

FIGURE 9–19.

The liverworts, so called because the plant body resembles the lobes of a liver (wort means plant), is an example of the division Bryophyta. Each of these lobes is about a centimeter wide.

cular plants are severely restricted in size. Second, the lack of a vascular system severely limits the habitat of these plants; they can only survive in a place where surface moisture is abundant.

To put this into a different perspective, imagine the landscape of the early Paleozoic era before ferns dominated. There were no large land plants at all. The vegetation was limited to very damp places or to water. Invasion of the terrestrial habitat by larger plants was only possible after vascular systems evolved, allowing plants to draw water and minerals out of the soil and to move these materials over long distances to the aerial parts of the plant. Thus the earliest plants were nonvascular, and from these, vascular plants evolved, This evolution represents a great turning point in the history of life on earth. The moss you observe today may represent the modern-day counterpart of an early attempt to move from a relatively moist habitat to a more typical terrestrial one.

The life history of mosses may be observed on your tour. Perhaps you can see, growing from the tip of a tiny, green moss plant, a thin, greenish to brown *stalk* terminating in a small *capsule* (Figure 9–20). This structure (stalk and capsule) is the sporophyte, and the more conspicuous green plant is the gametophyte. Therefore, in mosses we find a continuation of the trend pointed out in ferns. That is, a distinct alternation of sporophyte and gametophyte generations. In bryophytes, however, the roles are reversed: the sporophyte is nongreen at maturity and therefore is totally dependent on and physically attached to the gametophyte.

Haploid spores are produced by meiosis in the capsule at the tip of the sporophyte. When these spores are released and later germinate, they give rise to the green photoautotrophic and haploid moss plant. When this gametophyte matures, male (antheridia) and female (archegonia) sex organs are produced at the tips of the plant. The male structure produces motile sperm which swim to an archegonium and fertilize the egg contained within. Water is again, as with ferns, a necessary factor for fertilization to occur. After fertilization the diploid sporophyte begins to grow on the gametophyte and eventually produces a capsule and spores, and the moss life cycle continues (Figure 9–21).

The story is essentially the same for liverworts. In this case the flat plant body represents the gametophyte.

Economically, the bryophytes are of little importance and play a very insignificant role in most of our lives. Peat moss, called *sphagnum*, is used commercially as fuel in some parts of the world. It is also used as a packing material in shipping plants and sometimes in making planting mixes for container plants (Table 9–3).

FIGURE 9–20.

Moss sporophytes form a tiny forest only 1 to 2 cm tall. The dense mat below is the gametophytes.

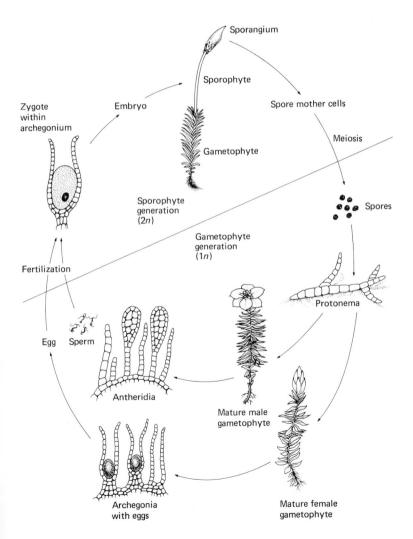

FIGURE 9–21.

Life cycle of moss.

Soil mix	INGREDIENTS, PER CENT BY VOLUME		WEIGHT, POUNDS PER CUBIC FOOT		MAXIMUM WATER CONTENT†		Comments and suggested uses
	Fine sand	Peat moss*	At max. water content†	Oven-dry	Percent by volume	Percent by weight	
A	100	0	117	89	43	30	Seldom used; densest and least retentive of nutrients; for cans, flats, beds
B	75	25	105	76	46	38	Commonly used; excellent physical properties; for cans, flats, beds
C	50	50	94	63	48	48	Commonly used; excellent physical properties; for pots and beds
D	25	75	66	34	51	94	Light weight, excellent aeration; for pots and beds
E	0	100*	43	7	59	530	Very light weight; used for azaleas, sometimes gardenias and camellias

TABLE 9–3.   THE FIVE BASIC U.C. SOIL MIXES
Scientists at the University of California have extensively tested various soil mixes for container-grown plants. Here are their recommendations.

*Redwood shavings may be used for part of the peat in mix E to improve aeration and reduce cost. Redwood shavings or sawdust or rice hulls may also be used for some or all of the peat in other mixes.
†Maximum water content, and weight at the moisture level, are typical for a 6-inch column of a mixture of fine sand of the Oakley series and Canadian peat moss.

Source: Modified from Kenneth F. Baker, ed. 1900. Manual 23, *The U.C. System of Producing Healthy Container-grown Plants.* Berkeley: Agricultural Publications, University of California.

Algae, the Most Primitive Group in the Plant Kingdom

The last major group in the plant kingdom is the algae. These plants are characterized by a relatively undifferentiated plant body (no true stems, leaves, or roots) and no vascular tissue. Another major feature that sets them apart from groups discussed previously is their mode of reproduction. In the algae the zygote does not develop into an embryo while enclosed within the female sex organ. This characteristic of the algal life cycle is thought to be a primitive feature and is quite important from an evolutionary standpoint. In fact, the algae exhibit many evolutionary trends which help us to visualize some of the events preceding the evolution of land plants, and these points will be emphasized as we discuss the algae.

It would be easy to overlook the algae on your botanical tour because they are usually small and inconspicuous. Nevertheless, if you know where to look and what to look for, representatives of this group can usually be found. The algae are predominantly aquatic, so if you chance to walk by a pond, lake, or river, you could probably find several species: slimy, filamentous, clumpy, green masses or tiny, single-celled forms which turn the water green. If you live in a very damp climate, you will probably notice that the north side of many trees is a bright green; this is another alga, not a moss as many mistakenly suppose. And if you are near the coastline, at your disposal are a wide number of algal species growing in the tidal zones. Marine algae can be quite large with flat, leaf-like structures, while others are small and quite delicate. In short, the growth forms of the algae are exceedingly variable as are their life histories.

You will recall that the generalized life history of plants is a diploid sporophyte generation alternating with a haploid gametophyte generation. This is also true among the algae except that virtually all possible variations are superimposed on the basic pattern. Thus, in some algae meiosis takes place immediately after fertilization, so that the diploid sporophyte is limited to the zygote cell itself and nothing more. In other algae we find well-developed sporophyte and gametophyte generations, but even within this life strategy, all variations are possible. The range extends from plants with gametophytes essentially indistinguishable from the sporophytes to those in which the two generations bear no morphological similarity. The degree of variation even extends to the gametes: sometimes male and female gametes are indistinguishable and sometimes they are very different. Life is indeed complex and variable among the algae, and you will see selected examples of most types as we proceed.

The algae can be separated into three major groups depending on pigmentation, characteristics of life cycles, and certain other considerations. The groups are the green algae, Chlorophyta; the brown algae, Phaeophyta; and the red algae, Rhodophyta (other groups do exist, but we will not consider them here). The three groups to be examined are usually readily distinguishable in seaweeds, but in freshwater forms the differences are not always obvious.

The Chlorophyta perhaps deserve special attention because they are considered to be on the main evolutionary line leading to the emergence of higher plants. Even among the green algae today, we can see in some species various features that may be regarded as primitive and in others more complex or advanced features. Thus, although the true ancestors of the vascular plants probably no longer exist, the present trends may mimic the events that gave rise to the first land plants. Several of these trends can be illustrated with the genera *Ulothrix*, *Ulva*, and *Oedogonium* (Figure 9–22).

Ulothrix is an alga often found in streams, although marine forms also exist. It is filamentous with the basal cell modified for attachment to some substrate. *Ulothrix* produces haploid gametes by mitotic divisions in the cells of the filament (Figure 9–23). All the gametes are structurally similar in size and shape (i.e., no obvious female gamete or male gamete) and therefore are called *isogametes*. When two of these isogametes fuse, a thick-walled zygote develops which can remain dormant for a period of time. Eventually though, the zygote undergoes meiosis to produce four *meiospores* which eventually develop into new filaments, and the cycle begins anew. The points to remember about *Ulothrix* are: (1) gametophyte dominates,

FIGURE 9–22.

a. *Ulothrix* is a green filamentous alga, like *Oedogonium,* with cells arranged end to end without branching. b. *Oedogonium* is a filamentous alga with cells always arranged end to end. The enlarged cells are forming zoospores; the other cells are strictly vegetative. c. *Ulva,* another green alga, consists of a thin flat sheet two cells thick, as can be seen near the torn edge shown here.

a.

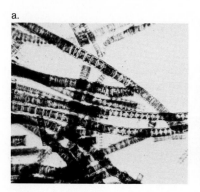

b.

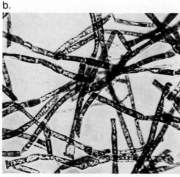

c.

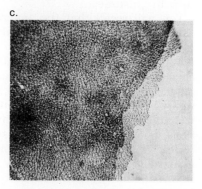

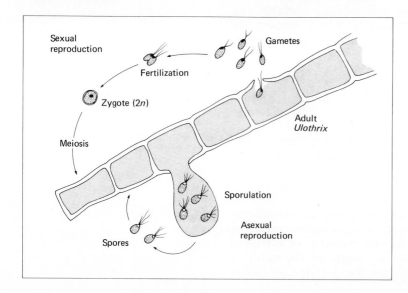

FIGURE 9–23.
Life cycle of *Ulothrix.*

(2) filamentous growth form, and (3) reproduction by the fusion of isogametes.

If we now compare the structure and reproductive mode of *Ulothrix* with that of *Ulva*, a green alga found in the subtidal zone, you can see changes which may reflect the trends that took place in the evolution of land plants. The zygote of *Ulva* germinates into a filament (rather than four meiospores) and via continued mitotic divisions produces a "leafy" sporophyte plant perhaps 10–15 inches long but only two cells thick. This sporophyte is free living and eventually produces meiospores by meiosis in cells on the "leaf" margins. These haploid cells germinate, producing a filament which also develops into a leafy gametophyte plant which looks exactly like the sporophyte. The gametophyte plant eventually produces gametes which fuse again, giving rise to the zygote (Figure 9–24). While gametes produced by *Ulva* look identical, we now know there are physiologically two types involved, called plus (+) and minus (−) gametes, and only the fusion of a plus and a minus will give rise to a viable zygote.

In *Ulva*, as compared with *Ulothrix*, we can then see certain advances: (1) an alternation of generations with the sporophyte assuming more importance, and (2) a divergence from the primitive isogametes to the more advanced form of reproduction where gametes are of two distinct types. The continuation of the later trend can be seen in *Oedogonium*.

Oedogonium is a freshwater, haploid filamentous green alga that differs from *Ulothrix* and *Ulva* in that it produces gametes

FIGURE 9–24.
Life cycle of *Ulva*.

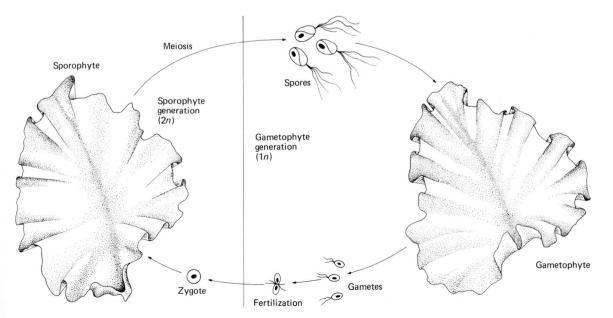

Sporophyte

Meiosis

Spores

Sporophyte generation (2n)

Gametophyte generation (1n)

Gametophyte

Zygote

Fertilization

Gametes

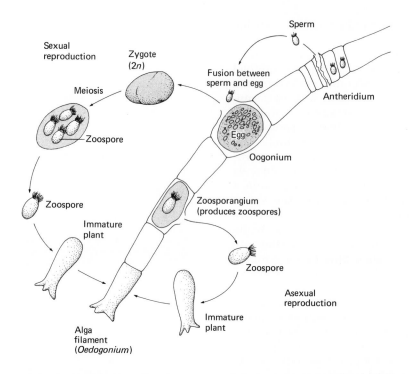

FIGURE 9–25.

Life cycle of *Oedogonium.*

FIGURE 9–26.

Macrocystis, a large brown kelp, is harvested off the coast of Southern California. The harvesting machinery is located at the ship's stern, therefore the ship must move backward through the kelp bed.
Photograph courtesy of Kelco Company, San Diego, California.

which are clearly differentiated into eggs and sperm. Fertilization takes place when the mobile sperm swims to the nonmotile egg and fuses with it. Once formed, the zygote then undergoes meiosis to produce meiospores which develop into haploid filaments (Figure 9–25).

The trend or advance exhibited in *Oedogonium* is, of course, the development of anatomically different gametes. Perhaps the most significant factor, however, is the production of a large, nonmotile egg. In terms of selective advantages, this seems to be important because a large egg can contain stored food while energy is expended only by the tiny sperm cell. Furthermore, the development of egg and sperm would be a step toward the reproductive strategies of higher plants.

Economically, the algae are more important than you might guess. In a world view of life, algae are often at the base of the food chain, and much of the earth's photosynthesis occurs in algae. In fact biological productivity (including photosynthesis and growth) is as high in the seaweeds along the shore as in the most productive of all terrestrial ecosystems. Algae are also eaten in large quantities in some parts of the world, serve as fertilizers, and are important in a number of industrial processes.

Kelp, the large brown seaweeds, often attain a length of over 100 feet and are harvested regularly for the colloidlike material alginate, obtained from their cell walls (Figure 9–26). This material serves as a stabilizer in ice cream and gives it a smooth consistency. These colloids are also used to retain moisture or provide structure in toothpastes, lipsticks, cake frostings, paints, chocolate milk, syrups, salad dressings, polishes, insecticides, and a few other items. Kelp is also used as a food for certain farm animals and as a fertilizer, since it has a high nitrogen content.

PART TWO • A BRIEF SURVEY OF THE MONERAN, PROTISTAN, AND FUNGUS KINGDOMS

When Leeuwenhoek reported seeing "animalculae" and "wee beasties" in 1676, little did he realize that these two groups would one day be recognized as two kingdoms of organisms, distinct from one another and from other kingdoms as well. His so-called animalculae were in fact single-celled organisms which today we know contain a nucleus and many of the cytoplasmic organelles described in Chapter 1. They are considered to represent a major advance over the simpler organisms which lack such intracellular differentiation. Here, in fact, is the distinction between the Protistan kingdom, comprising single-celled organisms with a nucleus, and the Moneran kingdom, which includes all those organisms without a true membraned nucleus. This latter group—Leeuwenhoek's wee beasties—are nevertheless very important to our way of life.

We will have a more detailed look at these two kingdoms, as well as the Fungus kingdom, in the following pages. As you proceed through this chapter, ask yourself if these organisms are more like plants or more like animals. You may eventually agree that the best answer is neither and that the preferred solution to an overall classification scheme is the five-kingdom approach.

The Moneran Kingdom

Within the Moneran kingdom are all the unicellular or colonial organisms which lack a true nucleus. We will consider individually the two main groups: bacteria and blue-green algae. Viruses, although not cellular in structure, will also be considered here.

Bacteria. A "typical" bacterium is diagrammatically illustrated in Figure 9–27. These organisms are of a relatively small size. Note the cell wall and the presence of a single circular chromosome not contained within a nuclear membrane. A bacterial cell also differs from the typical plant or animal cell in that there are no distinctive organelles such as mitochondria or an endoplasmic reticulum. Some examples illustrating the variations in bacteria are given in Figure 9–27b.

The most common reproductive method in bacteria is simple, asexual fission. In fission the genetic material doubles, and the cell subsequently divides into two individuals. Sexual reproduction, however, does sometimes occur when two cells come together and exchange parts of their chromosomes.

Under favorable conditions a bacterial population can grow extremely rapidly with each cell dividing as often as every 20–30 minutes (Figure 9–28). This rapid growth potential makes bacteria ideal subjects for certain types of scientific investigations in which large populations are necessary for statistical analysis and the isolation of mutants. Bacterial populations rarely, however, approach the population densities theoretically possible because depletion of nutrients or the accumulation of toxic wastes eventually limit growth.

Bacteria can be found in virtually every environment on earth. We have already suggested the role these organisms play in the decomposition of dead organic matter and in the nitrogen cycle. Such functions are indispensable in recycling and thus in maintaining a balance in our ecosystem. Bacteria are also found in our digestive tracts; they produce and release several vitamins that we cannot produce ourselves. Of course, disease-producing bacteria also exist and take a heavy toll of both animal and plant life. Bacteria may in fact indirectly limit the population density other organisms can achieve. Can you suggest some examples?

You may be able to observe directly bacterial infections on higher plants, but viral infections are easier to recognize. The gall-like structures on some stone fruits and rose bushes are indications of crown gall disease which is thought to be caused by a bacterium (see Figure 5–12, Chapter 5, for a more detailed

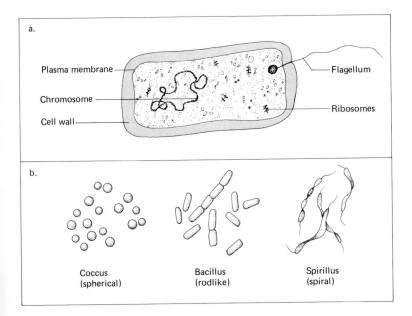

a.

Plasma membrane

Chromosome

Cell wall

Flagellum

Ribosomes

b.

Coccus
(spherical)

Bacillus
(rodlike)

Spirillus
(spiral)

FIGURE 9–27.   BACTERIA

a. In the "typical" bacterial cell in the diagrammatic sketch, note the absence of a nuclear membrane and many of the organelles found in the cells of higher organisms. b. Diversity of bacterial form is shown here.

discussion of this problem). Soft rot of fruits and vegetables may be caused by bacterial infections. That is, if spoilage of fruits and vegetables occurs and the affected area is soft and mushy, the causal agent is either a fungus or bacterium. Fire blight is another bacterial disease which is common among fruit trees (Figure 9–29) in general and pears in particular. As the name implies, the outward appearance of an infected tree suggests that it had been burned by fire.

Blue-Green Algae. In spite of the name, these organisms are probably more closely related to bacteria than to the algal groups discussed in Part One of this chapter. The genetic material in blue-greens consists of short filaments localized within the cellular matrix (Figure 9–30). Blue-green algae, unlike most bacteria, do contain chlorophyll and thus are photosynthetic. The common name of these organisms arises from the presence of a blue pigment, phycocyanin. The blue-greens typically reproduce asexually by fission and may exist as isolated cells or filaments surrounded by an external gelatinous sheath (see Figure 9–30). As previously discussed, blue-green algae are important elements in nitrogen fixation, and eutrophication.

Viruses. In Figure 9–31 the typical structure of a virus is illustrated. A virus particle presents some fundamental and interesting problems concerning our definitions of living organisms. That is, should viruses be considered living organisms or simply a highly organized collection of macromolecules? This difficulty stems from the fact that a virus particle existing as an isolated unit does not grow, metabolize, reproduce, or possess any of the characteristics usually associated with living organisms. However, when a virus infects a plant, animal, or bacterial cell, it becomes active and begins to multiply, utilizing the molecular machinery within the host cell. Eventually, the host cell becomes filled with the viral particles and bursts, and the newly released virus can infect other cells; in this way the population expands. Viral particles seem, therefore, to fall somewhere between the living and nonliving world. They can reproduce themselves but only by taking over the biochemical machinery of a true cell.

Many diseases are caused by viruses, and you should have no trouble finding symptoms of viral infections in some of the angiosperms in your own garden or local park. If possible, locate a tomato, squash, or, best of all, a tobacco plant. If the leaves appear mottled with alternating light-green or yellow and dark-green areas (Figure 9–32), a viral infection is strongly suggested. The other possibility for such a condition is a mineral deficiency, but if your plant is in a well-fertilized area, you can be fairly certain virus is involved.

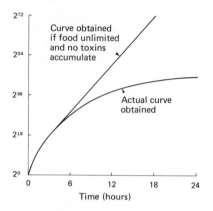

Number of bacteria (logarithmic scale)

Curve obtained if food unlimited and no toxins accumulate

Actual curve obtained

Time (hours)

FIGURE 9–28.

The geometric growth curve for a bacterial population dividing every 20 minutes. If reproduction is unchecked for 24 hours, 2000 tons of bacteria would exist from the starting population of one individual cell. From *Contemporary Biology: Concepts and Implications,* by Mary E. Clark, 1973, W. B. Saunders Company, Fig. 4.2, p. 104.

FIGURE 9–29.   FIRE BLIGHT

Photograph from Photo Science, Cornell University, Ithaca, New York.

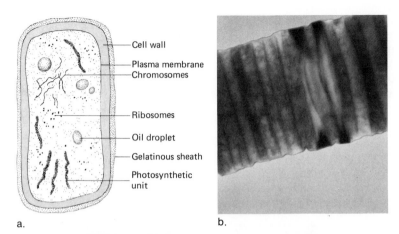

a.

b.

FIGURE 9–30.

a. Typical blue-green algal cell.
b. Filamentous blue-green alga.

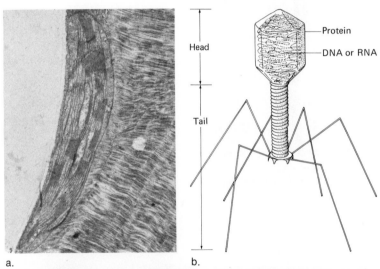

a.

b.

FIGURE 9–31.

a. Infection of a host cell. Tobacco mosaic virus (rod-shaped units) within a tobacco cell. b. Basic viral structure. A typical virus that attacks bacterial cells is called a bacteriophage. Note that the head portion contains DNA or RNA (as in the case of most plant viruses) and that the remainder of the structure is protein.
(a) From *Contemporary Biology: Concepts and Implications,* by Mary E. Clark, 1973, W. B. Saunders Company, p. 109. (b) Adapted from *Botany, An Ecological Approach,* by William A. Jensen and Frank B. Salisbury, p. 246. © 1972 by Wadsworth Publishing Company, Inc., Belmont, California 94002. Reprinted by permission of the publisher.

FIGURE 9–32.

Virus infection in cucumber.
Photographs from Photo Science, Cornell University, Ithaca, New York.

Bumps will often occur on squash fruit, and this condition also may be caused by a virus. However, many plant viruses are relatively symptomless in their many hosts and result in such a gradual decline in yield that their presence may escape notice for some time. For example, potato virus *X* is readily detectable only if the juice from a potato is rubbed on another susceptible plant such as tobacco. The leaves of the tobacco will become mottled, and the vigor of the plant will generally decline.

Plant viruses have no known effects on humans, and their ingestion is, as far as we know, harmless.

a.

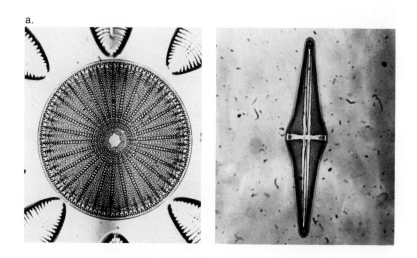

b.

c.

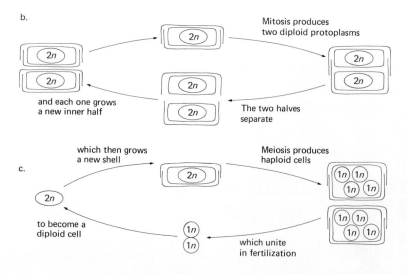

FIGURE 9–33.

Diatom structure and reproduction is summarized as an example of the kingdom Protista. Each individual diatom consists of a protoplast contained within a shell made of two concave halves—one slightly larger—which fit together in the manner of a pillbox. a. There are two types of diatoms, some with *radial* symmetry (left) and some with *bilateral* symmetry (right). b. Asexual reproduction: *mitosis.* c. Sexual reproduction: *meiosis.*
Photographs from C. David McIntire, Oregon State University.

The Protistan Kingdom: Single-celled or Colonial Organisms with a Defined Nucleus

The Protistan kingdom is perhaps one of the most diverse of all the kingdoms. It includes all single-celled organisms that have a well-defined nucleus bounded by a membrane. Within this kingdom are the protozoa of the old animal kingdom, *Euglena* and its relatives, and the diatoms (Figure 9–33).

Members of this kingdom are rather easy to find. Look in the same types of places that you did for mosses—on the soil along the north side of a building, the north side of a tree trunk, wooden or clay pots in the greenhouse, around a leaky water spigot, and essentially in any place that has been damp for some time. The green, filmlike material coating such damp surfaces is likely to be single-celled organisms and, if not an alga, a member of the Protistan kingdom. You could check to be sure what you have by examining the material under a microscope. While you are looking, you might as well go ahead and check some pond water, a drop of water from an old puddle, or perhaps the water from a flower vase which has been sitting around for a few days. You should find an assemblage of small organisms cavorting about together or floating listlessly in the water. Many of these organisms are protistans, single-celled organisms or sometimes clusters of cells (colonies), each with a nucleus enclosed by a membrane. Some will have specialized structures allowing for mobility (flagellae or cilia); others will be immobile. Some will be heterotrophic, while others will be green and photoautotrophic. Clearly, some seem more plantlike than others. Most of the time these organisms reproduce asexually by simple cell division. However, some also have a sexual stage when two cells fuse and exchange genetic material in a manner corresponding to fertilization and meiosis in higher organisms.

The organisms belonging to the Kingdom Protista together with some tiny multicellular animals account for most of the free-floating biota in our oceans, lakes, and rivers; they are called plankton. They are responsible for the bulk of the photosynthesis in our aquatic ecosystems and form the base of the food pyramid there.

The Fungus Kingdom

Fungi and most bacteria do not carry on photosynthesis. They are therefore called heterotrophs and obtain their food by digesting dead organic matter or by parasitically feeding on

1. *Non-photosynthetic*: Fungi obtain their nourishment from premade food, oftentimes first partially digesting organic matter outside their cells via the secretion of digestive enzymes.

2. *Cell Wall of Chitin*: The cell walls of many but not all fungi are composed of the polysaccharide chitin, which is the same material found in the hard shells of crustaceans such as the lobster.

3. *Filamentous Structure with Tip Growth*: Fungi do not have specialized tissues such as one finds in higher animals; however, the aggregation of filaments can produce large and rather complex forms.

4. *Heterokaryosis*: This term means that genetically different nuclei exist within the cytoplasm of a single fungal organism. The mechanism leading to this condition is briefly illustrated in our discussion of the field mushroom later in the text.

TABLE 9–4.   CHARACTERISTICS OF THE FUNGI

living matter. Decomposition of organic matter is of vital importance in world ecology because it is the primary mechanism leading to the release of CO_2 into the atmosphere and the recycling of nitrogen and other minerals. Therefore, in their role as decomposers, the fungi are beneficial and important components of any ecosystem. However, the fungi are not particularly selective in their sources of organic matter and thus can also do great damage to crop plants as well as other items of commerce. In the tropics where temperature and climate are most suitable for fungal growth, they will attack and at least partially decompose virtually anything that has some nutritive value: books, crops, wood, leather, clothes, crude oil, and so on. Even in milder climates fungi cause many plant diseases and destroy much potentially valuable food. Thus, like many organisms, they are both beneficial and harmful in our man-altered ecosystems.

There are presently about 80,000 species of fungi recognized and probably many more not yet discovered. The fungi do not seem to have any direct evolutionary connection with plants; at the present time they are thought to have evolved from some member of the Protista.

It is rather difficult to characterize adequately all members of the very diverse Fungus kingdom with a few simple criteria. Nevertheless, there are some generalizations that can be made (Table 9–4). Considering the fact that the fungi are no longer classified as true members of the plant kingdom, we will examine only three representative members in detail. Each of these examples represents a separate division in the fungal kingdom and will serve to give you an overview of the diversity of growth and reproductive patterns found.

Zygomycota.   Members of this group generally live in the soil and exist on organic matter. Perhaps the genus of this group with which you are most familiar is *Rhizopus*, black bread mold (Figure 9–34). *Rhizopus* consists of tubelike, multinucleate, haploid filaments called *hyphae*. Hyphae grow prostrate sometimes with branches and produce erect structures called *sporangiophores* (see Figure 9–34). Black-walled *spores* are produced at the end of the sporangiophore when the terminal section is cut off by a complete cell wall. The thick-walled spores which then develop are responsible for the black color of the fungus. When these spores are released and germinate, they produce more multinucleate, haploid hyphae and continue the process of fungal proliferation. Spore production in *Rhizopus* is a means of asexual reproduction common to this group. However, a sexual mode of reproduction is known, and the details are presented in Figure 9–35.

Oomycota.   Although many members of this group are aquatic, we will consider a terrestrial species, *Phytophthora infestans,* because of its economic importance. This species is responsible for a disease called *late blight* which is particularly destructive to potatoes and tomatoes (Figure 9–36). In fact, this organism was responsible for the Irish potato famine of 1845 to 1851.

The drastic consequences of the infestation of the Irish potato crop with this fungus are well known. The Irish economy previous to the famine was based almost entirely on potatoes, and the peasantry literally depended on it for their food. Estimates indicate that before the famine the average individual ate 8 pounds of potatoes per day! Therefore, when the fungal infestation occurred drastically reducing the harvest, there

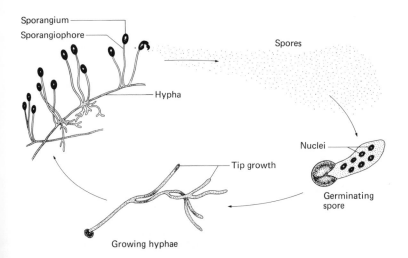

FIGURE 9–34.

The structure and asexual life history of *Rhizopus,* or black bread mold, is summarized as an example of the kingdom Zygomycota (see Figure 9–35).

were many deaths due to starvation and a subsequent typhus epidemic as well. Mass migrations to other countries including the United States occurred. These factors are reflected in the decline in the population of Ireland: 8.5 million in 1845, 6.5 million in 1851 and 4.7 million in 1891. The potato blight fam-

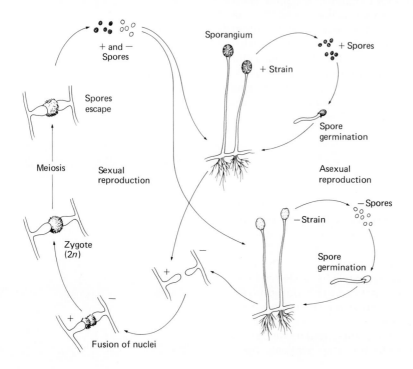

FIGURE 9–35.

Life cycle of bread mold, *Rhizopus.*

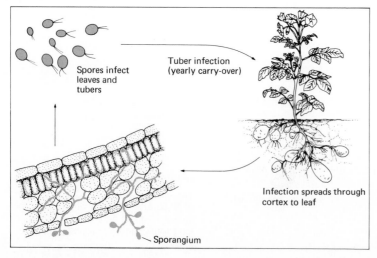

FIGURE 9–36.

The life history of *Phytophthora infestans,* which causes late blight of potatoes and was the basis for the Irish potato famine of the mid-1800s.

ine also had other far-reaching effects. For example, it was partly responsible for the fall of Prime Minister Robert Peel's government; it was a factor in influencing freer trade laws in England; it indirectly allowed the Irish to win independence; and it changed the social and voting structure of parts of America because of the mass migrations which ensued.

Development of the late blight disease in potato begins when the filaments of *Phytophthora* invade a potato tuber (see Figure 9–36). The fungus usually grows up the stem through the cortex and eventually causes stem collapse. When the spreading filaments reach the aerial parts of the plant, sporangiophores are usually produced through the stomata of the leaves. Spores are produced on these terminal structures which, when dispersed by wind or rain, cause additional infections on germination. Under favorable conditions (warm and damp) the time from infection to spore production can be as short as four days. Thus, from this form of asexual reproduction, many generations and new infections can occur in one growing season.

Control of late blight is typical of the control of most fungal plant diseases and involves sanitary measures and spraying with fungicides. An important preventive measure is to seed a new potato crop with infection-free tubers selected because they have a relatively high degree of resistance to late blight. Further care must be taken to insure that diseased plants and tubers from the previous crop are destroyed.

Basidiomycota. Members of this group are probably the most familiar of all fungus species and include the mushrooms, puffballs, and shelf-fungi as well as the commercially harmful crop pathogens the rusts and the smuts (Figure 9–37). We will examine only one of the 25,000 or so species that make up this class, the common field mushroom *Agaricus.*

The life cycle of the field mushroom (Figure 9–38) begins when a haploid *basidiospore* falls to the ground and germinates to produce many filamentous hyphae. Initially, this filamentous mat, a *mycelium*, consists of cells which contain one haploid nucleus. The mycelium spreads underground, forming a ring sometimes as large as 120 feet in diameter. All the materials necessary for mycelium growth must come from the soil. As this growth is occurring, some of the cells of the filaments undergo a change so that in the sexually mature mycelium many cells contain two haploid nuclei. These cells are called *dikaryotic*. This developmental event can occur either with the fusion of two mating types (Figure 9–39) or with the absence of cell wall formation following nuclear division. These mating types are called plus and minus strains rather than male and female because they are morphologically indistinguishable.

FIGURE 9–37.

Members of the class basidiomycetes can produce fruiting bodies (basidio-carps) of variable shape and size. Photographs by Avery H. Gallup, San Diego State University, California.

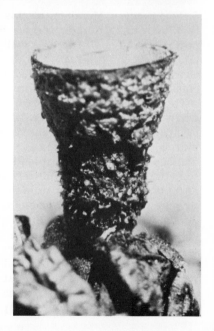

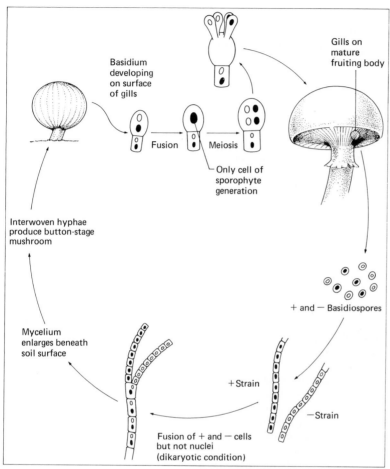

Basidium developing on surface of gills

Fusion

Meiosis

Only cell of sporophyte generation

Gills on mature fruiting body

Interwoven hyphae produce button-stage mushroom

Mycelium enlarges beneath soil surface

+ and − Basidiospores

+ Strain

− Strain

Fusion of + and − cells but not nuclei (dikaryotic condition)

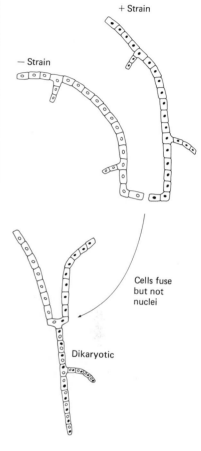

+ Strain

− Strain

Cells fuse but not nuclei

Dikaryotic

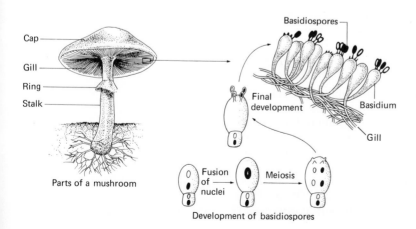

Cap

Gill

Ring

Stalk

Parts of a mushroom

Basidiospores

Final development

Basidium

Gill

Fusion of nuclei

Meiosis

Development of basidiospores

FIGURE 9–38.

Life cycle of the field mushroom, *Agaricus* (top left).

FIGURE 9–39.

Fusion mating types in the basidiomycetes (top right).

FIGURE 9–40.

Basidiospore development in *Agaricus* (see Figure 9–38).

The dikaryotic mycelium begins to bud on its outer edges, producing in time a ball of interwoven filaments which form the button stage of mushroom development. The hyphae composing the button stage absorb large quantities of water and expand rapidly into what you recognize and call a mushroom. Botanically, this structure is called the *basidiocarp* (Figure 9–40).

Within a basidiocarp are structures called gills. On the surface of the gills, specialized cells are formed in which the two nuclei of the dikaryotic stage fuse (see Figure 9–40). This is the only true diploid stage in the life cycle of the field mushroom. After fusion meiosis occurs producing four haploid nuclei which are enclosed in four terminal cells, each of which is called a *basidiospore*. When released, the basidiospores germinate to produce a mycelium, and the cycle begins anew.

The cultivation of *Agaricus bisporus*, a species very similar to our generalized example, is an important commercial enterprise. Blocks of mycelium called seed or spawn are placed in especially prepared beds containing partially degraded compost. In mushroom production it is important to keep the spawn beds moist and at the correct temperature in order to insure rapid growth. Great care must also be taken to insure that the starting mycelium actually is *A. bisporus* and that no foreign mycelium is growing in the starting compost mix. These precautions are necessary because this group contains some of the most highly poisonous of all the fungi.

Edible and poisonous mushrooms can only be separated by specific botanical features, and unfortunately there are no short cuts to distinguish the edible types. Characteristics such as tarnishing of a silver spoon by poisonous fungi are totally unreliable as are observations based on the palatability of mushrooms to insects and animals. Therefore, great care should always be taken in collecting and sampling wild mushrooms, and in fact, the only truly safe way to indulge in this sport is to take along an expert.

Reflections

This ends our brief introduction to the plant and plantlike kingdoms. We can now reflect on the question of relationships among the various groups. It should be remembered that evolution has continued to give rise to new life forms ever since life began more than 3 billion years ago. Consequently the relationships among organisms are complex.

We can draw an analogy to our own families. Every single one of us has parents, but not everyone has offspring. Of those who have offspring, some have one or very few, while others have many. So it is with evolution and a phylogenetic tree. All organisms have ancestors, and we can often trace that ancestry back in the fossil record for millions of years. But it does *not* follow that all organisms in the fossil record have left descendants. Applying this idea to the phylogeny of plant groups, we see from the fossil record that flowering plants were the latest major group to appear. Gymnosperms were prominent before flowering plants, ferns and fern allies earlier yet, but the algae were the earliest of all.

Does this mean then that the algae gave rise to the ferns, which evolved to gymnosperms and thence flowering plants? Not necessarily, and it is undoubtedly a simplified view, but this general pattern is one that most plant biologists favor. Nevertheless, we mustn't be too dogmatic regarding a phylogenetic sequence for major plant groups. The fossil record is incomplete, and there are numerous gaps in the evolutionary sequence. For example, the fungi leave poor fossils except for their spores, and so they have been omitted from our discussion, as have the monerans and protists. Perhaps you are also concerned about the bryophytes, another group that does not appear in our simplified evolutionary scheme. This is because the bryophytes appear to be a dead end in evolution—a group that does not appear to have contributed to the main evolutionary sequence leading to angiosperms (see Figure 9–4). Could we consider the fungi to be a similar dead end in evolution?

Another point that should be made is that a phylogenetic tree is traced *backward* in time. Beginning with the angiosperms, for example, we find a stepwise, almost linear sequence of simplification. Loss of flowers (gymnosperms) is followed by loss of seeds (ferns), then loss of vascular tissue, and so on. However, tracing the phylogenetic tree *forward* in time isn't nearly as simple. There are two major groups within the angiosperms—the *monocots* and *dicots*. Within the gymnosperms are many groups, including pines, firs, cedars, ginkgoes, and cycads. The question naturally arises as to which kind of plant came first and gave rise to the rest? Did the monocots give rise to the dicots or the other way around? Which of the gymnosperms are the most ancient and which are derived? Are the fern allies more primitive and the true ferns derived from them, and if so, which of the fern allies is the granddaddy of them all? For many of these questions, we can only surmise an answer or partial answer due to the paucity of evidence. In any case it would appear that the picture is one of a branching, occasionally dead-end-producing phylogenetic tree, rather than a simple linear sequence.

Finally, we indicated earlier that organisms are grouped into kingdoms because all organisms within one kingdom are more closely related to one another than they are to organisms in other kingdoms. Let us reconsider that point now in relation to a phylogeny of life forms. Remember, all life arises from pre-existing life. So, if we can trace the phylogeny within the plant kingdom to the algae as the most primitive group, where did the algae come from? The answer would appear to be from preexisting life, probably some single-celled organisms which today we would call a protistan. From the myriad of primitive single-celled forms which must have evolved before multicellular organisms, a major branch developed which eventually gave rise to increasing complexities and advanced forms as we see them today. But even today some of those primitive forms are still around, including many simple algae. Does it not follow then that some simple algae may in fact be more closely related to *Euglena*, for example, than they are to angiosperms? This would appear to be the case. Near the fork in the tree, the two branches are closer together than either one is to the tip of its branch, but we still recognize two distinct branches. The reason that we draw the lines separating kingdoms the way we do is because the *differences* between algae and protists are greater than the differences among the brown, green, and red algae, and between algae and bryophytes, and between bryophytes and ferns, and so on. Again, the differences between the fungi and other groups are greater than the differences among the various groups of fungi. Thus, the combination of divergently branching phylogenetic relationships among organisms, together with a classification based as much on differences as on similarities, makes life complex.

Selected Readings

Cohen, S. S. 1970. Are/were mitochondria and chloroplasts microorganisms? *American Scientist* 58:281–289. Interesting speculation concerning the evolution of microorganisms.

Darwin, C. 1962. *The Voyage of the Beagle.* Garden City, N.Y.: Doubleday, Anchor Books. More entertaining than you might imagine.

Echlin, P. June 1966. The blue-green algae. *Scientific American* 214-(6):75–81. More about these primitive organisms.

Emerson, R. January 1952. Molds and men. *Scientific American* 186 (1):28–32. (Offprint no. 115.) Nice discussion of fungi in human history.

Fenton, C. L. 1966. *Tales Told by Fossils.* Garden City, N.Y.: Doubleday. Title is good description of subject matter.

Gibor, A. November 1966. Acetabularia: A useful giant cell. *Scientific American* 215(5):118–124. Interesting facts about a most unusual algal cell.

Kavaler, L. 1965. *Mushrooms, Molds, and Miracles.* New York: John Day. Popular book on fungi.

Whittaker, R. H. 1969. New concepts of kingdoms of organisms. *Science* 163:150–160. Evolutionary relations are better represented by new classifications than by the traditional two kingdoms.

Whittier, D. P. 1971. The value of ferns in an understanding of alternation of generations. *BioScience* 21:225–227. Check your understanding of the importance of an alternation of generations.

Williams, G. 1959. *Virus Hunters*. New York: Knopf. Entertaining stories of many researchers in the field of viruses.

10 Biomes

The land where you live now was once barren, the rocks lying exposed to the sun and weather. There was no soil—only sand, gravel, and rocks. The entire continent was devoid of vegetation. This was the landscape of North America throughout most of its history, up to about 500 million years ago. During the period of time called the Ordovician (Table 10–1), primitive nonvascular plants probably began to establish themselves in suitable habitats on land. They were probably confined to the more or less permanently moist and protected places near the seashore where their algal ancestors were already abundant. These earliest invaders of the land passed countless times through their life cycles and gradually changed the land by adding humus, nitrates, and other components. Thus, once plants were established on land, the stage was set for more substantial changes. These changes occurred during the 50 million years of the Silurian period and must have been dramatic, for we find that during the next geologic period, the Devonian, a variety of land plants abound. These plants included small herbaceous types and also trees more than 40 feet tall with trunks over 3 feet in diameter. The small types were primarily ferns and fernlike plants growing in swamps, but seed-bearing types were slowly evolving and advancing over the landscape.

The history of terrestrial vegetation covers about 500 million years, scarcely more than one-tenth the age of our earth and perhaps a sixth of the time since the appearance of the most primitive life in the seas. It is a long, complex, and incompletely known story. In this chapter we will concentrate our efforts on the more recent history of our vegetation and discuss those factors responsible for molding our present landscape. An appropriate place to begin is with a short account of the kind of evidence used to date fossil remains.

The Evidence

The ages of rocks and fossils can be estimated fairly accurately today by a technique called radioactive dating. This technique makes use of the fact that some of the elements constituting the fossils change with time. We can illustrate this with the element carbon (C). The most abundant and stable form of carbon has a molecular weight of 12 and, to abbreviate, this information is written ^{12}C. There is an abundance of ^{12}C all around us in many forms which look very different from one another. However, we are initially concerned with the carbon in our air which is in the form of carbon dioxide (CO_2), a colorless gas. The vast majority of the CO_2 in the air contains carbon of the ^{12}C variety; however, a certain fraction of this CO_2 contains ^{14}C. Carbon-14 is unstable and over a period of time changes to the stable element ^{14}N by emitting a particle from its nucleus. Fortunately the emission of such particles can be accurately measured by suitable instrumentation. Furthermore, carbon is especially well suited for such studies since it is readily incorporated into plant material and the unstable form of carbon, ^{14}C, decays very slowly. But we're getting a little ahead of ourselves, so let's back up and take it step by step.

In our upper atmosphere a small amount of ^{14}C is continually forming. In the atmosphere as well as everywhere else ^{14}C is slowly decaying to ^{14}N. As a result, the relative amounts of ^{12}C and ^{14}C in the atmosphere remain relatively constant. As plants photosynthesize, they incorporate $^{12}CO_2$ and $^{14}CO_2$ in their relative proportions from the air. Therefore plants have organic constituents composed of ^{12}C and ^{14}C in the same proportion as that found in the air. If the plant happens to die and become fossilized, the ^{14}C slowly decays to ^{14}N as noted above, but as a fossil it absorbs no more carbon. Since this decay occurs very slowly (the half-life of ^{14}C is about 5,500 years), we can measure how much ^{14}C and ^{12}C there are in the fossil. From such measurements we can estimate how long it has been since the plant originally incorporated those materials into its organic molecules, and that can tell us how old the fossil plant is. Carbon is not the only element that can be utilized for radioactive dating; a number of other elements have also proved useful. The principle is the same in all cases, however.

Turning the Clocks Back

We know that the oldest rocks on the earth are about 4.5 billion years old, and the earth itself must be at least that old. The first "living" thing probably first appeared about 3.0–3.5 billion years ago. It may have been a primitive sort of microbe, perhaps not even with a cellular structure. By "first living thing" we mean the first object capable of reproduction. Fossils of multicellular organisms are first abundant in rocks about 500 million years old (see Table 10–1). Multicellular organisms must have arisen some time before that, perhaps as early as 750 million to 1 billion years ago. These organisms were algae, and the fossil record includes no land plants from that period. During the next several

ERA	PERIOD	EPOCH	DURATION IN MILLIONS OF YEARS	TIME FROM BEGINNING OF PERIOD TO PRESENT (MILLIONS OF YEARS)	GEOLOGIC CONDITIONS
Cenozoic (Age of Mammals)	Quaternary	Recent	0.025	0.025	End of last ice age; climate warmer
		Pleistocene	1	1	Repeated glaciation; 4 ice ages
	Tertiary	Pliocene	19	20	Continued rise of mountains of western North America; volcanic activity
		Miocene	15	35	Sierra and Cascade mountains formed; volcanic activity in northwest U.S.; climate cooler
		Oligocene	10	45	Lands lower; climate warmer
		Eocene	20	65	Climate warmer
		Paleocene	10	75	

Rocky Mountain Revolution (Little Destruction of Fossils)

ERA	PERIOD	EPOCH	DURATION IN MILLIONS OF YEARS	TIME FROM BEGINNING OF PERIOD TO PRESENT (MILLIONS OF YEARS)	GEOLOGIC CONDITIONS
Mesozoic (Age of Reptiles)	Cretaceous		60	135	Rockies formed late; earler, inland seas and swamps
	Jurassic		30	165	Continents fairly high; shallow seas over some of Europe and western U.S.
	Triassic		60	225	Continents exposed, widespread desert conditions

Appalachian Revolution (Some Loss of Fossils)

ERA	PERIOD	EPOCH	DURATION IN MILLIONS OF YEARS	TIME FROM BEGINNING OF PERIOD TO PRESENT (MILLIONS OF YEARS)	GEOLOGIC CONDITIONS
Paleozoic (Age of Ancient Life)	Permian		15	240	Continents rose; Appalachians formed; increasing glaciation and aridity
	Pennsylvanian		35	275	Lands at first low; great coal swamps
	Mississippian		50	325	Climate warm and humid at first, cooler later as land rose
	Devonian		50	375	Smaller inland seas; land higher, more arid; glaciation
	Silurian		50	425	Extensive continental seas; lowlands increasingly arid as land rose
	Ordovician		80	505	Great submergence of land; warm climates even in Arctic
	Cambrian		80	585	Lands low, climate mild; earliest rocks with abundant fossils

Second Great Revolution (Considerable Loss of Fossils)

ERA	PERIOD	EPOCH	DURATION IN MILLIONS OF YEARS	TIME FROM BEGINNING OF PERIOD TO PRESENT (MILLIONS OF YEARS)	GEOLOGIC CONDITIONS
Proterozoic			1000	1500	Great sedimentation; volcanic activity later; extensive erosion, repeated glaciations

First Great Revolution (Considerable Loss of Fossils)

ERA	PERIOD	EPOCH	DURATION IN MILLIONS OF YEARS	TIME FROM BEGINNING OF PERIOD TO PRESENT (MILLIONS OF YEARS)	GEOLOGIC CONDITIONS
Archeozoic			2000	3500	Great volcanic activity; some sedimentary deposition; extensive erosion

TABLE 10–1. GEOLOGIC TIME TABLE

Source: Modified from Claude A. Villee. 1972. *Biology*, 6th ed., Table 33–1. By permission of the author and W. B. Saunders Company.

PLANT LIFE	ANIMAL LIFE
Present day biomes differentiate	Age of man
Great migrations of biomes	Extinction of great mammals; first human social life
Decline of forests; spread of grasslands	Man evolved from manlike apes
Tropical forests retreat; dry-adapted types expand	First manlike apes
Maximum spread of forests; rise of monocotyledons	Rise of anthropoids; forerunners of most living genera of mammals
Tropical forests widespread	Hoofed mammals and carnivores established
	Spread of archaic mammals

First monocotyledons; first oak and maple forests; gymnosperms declined	Dinosaurs reached peak, became extinct; first modern birds; archaic mammals common
Increase of dicotyledons; cycads and conifers common	First toothed birds; dinosaurs larger and specialized
Gymnosperms dominant	First dinosaurs and egg-laying mammals

Decline of fern allies; gymnosperms expand	Many ancient animals died out; mammal-like reptiles, modern insects arose
Great forests of seed ferns and gymnosperms	First reptiles; insects common
Fern allies dominant; gymnosperms increasingly widespread	Spread of ancient sharks
First forests; first gymnosperms	First amphibians
First definite evidence of land plants; algae dominant	First (wingless) insects; rise of fishes
Land plants probably first appeared	First fishes
Marine algae	Most modern phyla established

Primitive aquatic plants—algae, fungi	Various marine protozoa; toward end, various marine invertebrates

No recognizable fossils; indirect evidence of living things from deposits of organic material in rock

100-million years, plants invaded the land and became increasingly advanced and complex. Ferns and related plants that lack seeds were the dominant form about 200–400 million years ago; gymnosperms were the dominant vegetation around 150–200 million years ago (Figure 10–1), and the flowering plants came into their own something over 100 million years ago. Thus, all of the major groups of plants have been here for well over 100 million years. This does not mean, however, that plants have remained evolutionarily stagnant for the last 100-million years. Far from it. Many new species have arisen, while others have died out, and of course plant distributions have changed in response to the climate and topography.

Continental Drift—A Slow Freight to Anywhere

An interesting problem which we can only refer to in passing is the role of continental drift in distributions of vegetational types. From time to time in geologic history, shallow seas have invaded large areas of the continental shelf, and oceanic islands have developed through volcanic activity near the outer margins of the shelf. In spite of these changes there is little doubt that the continental blocks themselves are permanent features. We do *not* find that old continents erode away and wash out to sea to be replaced by new continents arising out of the ocean. The existing continents actually move around—they drift. Thus, it is believed that South America was once adjacent to Africa, and the mid-

FIGURE 10–1.

The fossil on the left is of a *gymnosperm*—probably pine—from the Petrified Forest of Arizona; it is of the Triassic period and over 150 million years old. The log on the right with which it is compared was freshly cut at the time the photograph was taken.

Atlantic ridge is the line where the sea floor is spreading out east and west, causing those two continents to move farther away from one another. This movement continues today at about 6–10 cm per year. Other continents drift, too (Figure 10–2). This being the case, you might inquire into the fate of the leading edge of these drifting continents. As one example, the west coast of North America is being pushed against the Pacific oceanic plate which is made of heavier material. As the lighter North American continental block collides with the heavier oceanic basin, it slides over the top, forcing the heavier material down into the earth's interior. In another example, two or more continents may collide and together form a larger continent, as seems to be the case with India and Eurasia.

Can you see how continental drift relates to the problem of plant distributions? Continental drift may help *explain* the distributions of some plants. For example, southern Africa, South America, Australia, and Antarctica were all contiguous in late Mesozoic. The climate was temperate, and no doubt some kinds of plants grew throughout that entire land mass (called Gondwanaland). As the continents separated and drifted apart, plants were carried along. With time and changing conditions, new species arose through evolution, and some of the original species became extinct. In such a way each continent would be expected to have somewhat unique species. In spite of these changes, however, a certain species affinity among those once contiguous areas should persist. And so today, for example, we see that at comparable latitudes the flora of South America is more similar to that of Australia than to North America.

To Make a Long Story Short—
The History of North American Vegetation

At the beginning of the Tertiary (75 million years ago), the climate over North America was humid and warm all year. We know this because fossil plants from that time indicate that a tropical forest existed as far north as southern Canada, even into Greenland. Farther north the tropical forest was replaced (in the present Arctic region) by vegetation we generally associate with places that today are cool at least part of the year, such as at higher elevations in the mountains, the northern coastal regions, or farther north into Canada. The kinds of plants in northern latitudes would include pines, firs, redwoods, spruces, hemlocks, poplars, and alders. At that time there were no deserts and no dry scrub, presumably because the climate was too moist.

Later in the Tertiary period the climate, as reflected by vegetation changes seen in the fossil record, became increasingly drier and cooler. The tropical vegetation was gradually restricted southward, until today such vegetation is seen only in the tropics and as far north as Florida and north-central Mexico. The cooler temperatures allowed the Arctic vegetation to expand its distribution somewhat southward during that time, particularly into the cool coastal areas of western North America and also into most of the eastern half of North America. At this time, around 50–60 million years ago, a new vegetation type appeared in southwestern North America which was adapted to the drier conditions. These plants were primarily small-leaved, drought-resistant types, including chaparral, thorn scrub, desert sage, and dry oak. Today these types of plants cover most of southwestern North America.

During the latter part of the Tertiary, between 1–15 million years ago, two major changes occurred which set the stage for

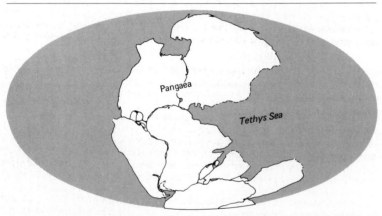

a. The universal landmass of Pangaea at the end of the Permian (225 million years ago).

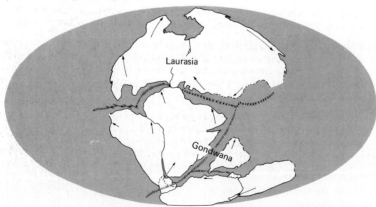

b. A rift forms across Pangaea at the end of the Triassic (180 million years ago), breaking it into the two large continents of Laurasia and Gondwana.

FIGURE 10–2.

The shifts in relative positions of the continents since the end of the Paleozoic era.

From "Reconstruction of Pangaea: Breakup and dispersal of continents, Permian to present," by R. S. Dietz and J. C. Holden, in *Journal of Geophysical Research* 75(26):4943–4951, Figs. 2–6, (September 10, 1970). 1970, copyright by American Geophysical Union.

⁓⁓⁓	Sea-floor spreading
××××××××	Continents colliding
⟶	Direction of continental drift
------	Boundaries that have greatly changed

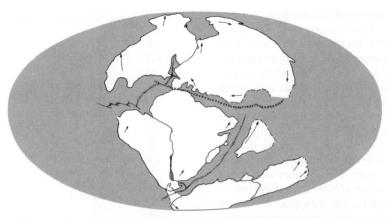

c. The continental drift dispersion of the continents during the late Jurassic (135 million years ago).

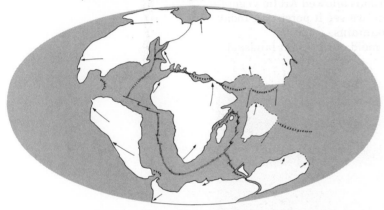

d. The continental drift dispersion of the continents at the end of the Cretaceous (65 million years ago).

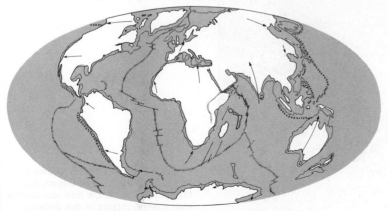

e. The position of the continents today, showing sea-floor spreading and collision of the continents during the Cenozoic.

our present-day landscape. One was mountain building. During most of the Tertiary the North American topography was fairly flat; the Sierra Nevada, for example, was only a low range of 3000-foot elevation or so. Beginning around 15 million years ago, however, the Sierra Nevada and other ranges began to be uplifted, and this trend continues today. This uplift has had some important effects on the environment and therefore on vegetation. For example, the prevailing storms arising near the coasts tend to lose most of their moisture as they pass over the high mountain ranges; as a consequence a dry area occurs on the lee side (Figure 10–3). This rain shadow phenomenon caused a striking differentiation of climates over fairly short distances and allowed for a substantial expansion of desert vegetation.

A second effect of the mountain building was that an arctic type of climate was found much farther south than would otherwise be possible. That is, the annual temperature extremes resulting from high mountain elevations allowed Arctic vegetation to migrate far southward where we see it persisting today in the Sierra Nevada and Rocky Mountains.

Another major event that helped mold our present landscape

FIGURE 10–3.

The rain shadow effect is shown in this view of the Mojave Desert, California. Note the snow and rain clouds over the mountains in the distance, while the sky is clear above the desert.

was the Ice Age. During the last million years we have had at least four separate and distinct cold snaps lasting on the order of 10,000–15,000 years each. Great ice sheets covered most of North America, and glaciers formed in the mountain valleys (Figures 10–4 and 10–5). Yosemite Valley, for example, was carved and molded into its present shape during the Ice Age (Pleistocene). Ice sheets and glaciers formed because more snow fell than melted each year, and it is concluded that temperatures cooler than what we know today prevailed.

Plants respond dramatically to changes in temperature and precipitation. During the periods when the glaciers were advancing, the kinds of plants that require warmer temperatures could only survive toward lower elevations and farther south than we find them today. In fact, a southward migration was generally true for all species in North America. When the temperatures warmed up between each of the colder periods, the various plant communities would migrate northward and to higher elevations. During the million years of the Pleistocene epoch, plant communities shifted their distribution drastically on four separate occasions in response to a cool, moist climate

FIGURE 10–4.

Pleistocene glaciers carved this bowl-shaped cirque at 12,000 feet elevation in the Sierra Nevada.

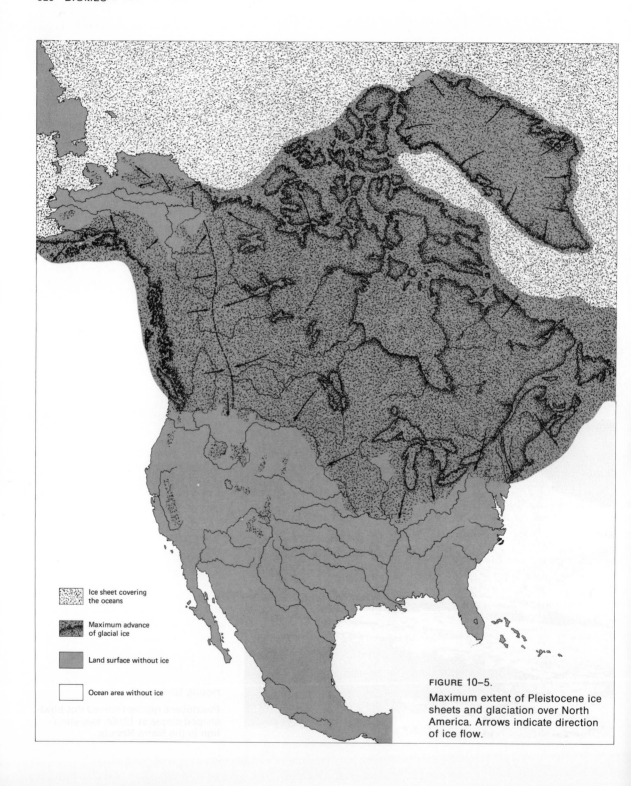

FIGURE 10–5.

Maximum extent of Pleistocene ice sheets and glaciation over North America. Arrows indicate direction of ice flow.

Ice sheet covering the oceans

Maximum advance of glacial ice

Land surface without ice

Ocean area without ice

alternating with warmer, drier conditions. It has been around 10,000–12,000 years since the end of the last glacial period. Some geologists tell us that we may not have seen the last of the Ice Age, that the climate may again become cool and the glaciers may advance once more.

Biomes: Plants and Where to Find Them

After the events of the last million years, plants have sorted themselves out into groups of different kinds corresponding to different ecological conditions. The largest of these groups are termed *biomes*, and the subunits within biomes are called *communities*. For our purposes only the major biomes found in North America will be discussed, leaving to the individual student the task of discovering which of the many plant communities are found in his or her particular area. Topics will also include those vegetation types maintained by man. There are two reasons for this. First, it is felt that these are permanent features of the landscape because man is probably here to stay and will undoubtedly continue to require food and shelter. Second, man is conceived to be a natural, not supernatural, being. So man's alteration of the environment is interpreted in the same light as other animals' doing the same thing. Admittedly, many of man's effects are more profound than those of other animals and therefore qualitatively different, but this is irrelevant to the question of whether he is natural or supernatural.

For each vegetation type a general description of the plants, a rough distribution, and something about the factors limiting their distribution will be given.

Agricultural. The distinguishing feature of this biome is monoculture. Large tracts varying in size from only a few to thousands of acres are maintained in a single kind of plant (Figure 10–6). The particular kind of plant normally differs from tract to tract, but in all cases they are called crops. Crops are predominantly utilitarian rather than ornamental. Most, perhaps all, crops could not maintain themselves without man's constant input of energy and materials. Thus, crops must be planted and replanted every generation because they cannot reproduce their own kind without man's constant attention. They often require extensive fertilization and irrigation, and they must be protected from diseases, pests, and competitors. Man has developed elaborate techniques and machinery to accomplish these ends. Perhaps you can see that there is a mutual, interdependent relationship between man and his crops and that neither could persist in the present form without the other. The kinds of plants that make up this vegetation type often are not known to grow in the wild, and in many cases their wild ancestors are not even known. Clearly, these plants have been in cultivation a long time, probably from 7,000–10,000 years.

Urban. This biome is characterized by many different kinds of plants clustered about places of human habitation (Figure 10–7). As with agricultural plants, urban vegetation requires man's constant attention, or much of it would die out in a single generation. Unlike agricultural plants, urban plants are primarily ornamental, although a few have edible parts or perform useful functions. The kinds of plants that form the urban type of vegetation are normally special strains of plants which grow wild in parts of the world other than the region where they are cultivated. This is the most recent of all vegetation types, probably about 5,000–8,000 years old. Why are such types only of recent origin? We don't really know, but it seems plausible that the sedentary habit was required before any cultivation could be practiced and that man's first needs were for food rather than ornament.

Strand. Plants found on sandy beaches and dunes belong to this biome (Figure 10–8). Their distribution is determined largely by strictly local conditions including soil, weather, and salt spray. Adjacent to large bodies of water, strand plants are subjected to more moderate extremes of temperature than those that occur only a few miles inland. The plants are usually low,

FIGURE 10–6.

Monoculture characterizes the agricultural vegetation type.

even prostrate, and often succulent. The kinds of plants differ from region to region, their distributions presumably determined largely by precipitation and temperature.

Marsh. Plants growing in quiet or slow-moving water form the marsh biome (Figure 10–9). The plants face special problems in relation to the soil which is submerged in water. Oxygen, for ex-

FIGURE 10–7.

The urban vegetation type is represented by many kinds of plants clustered about places of human habitation and primarily ornamental in function.

FIGURE 10–8.

Strand vegetation is low, often succulent, and adapted to the constant breezes and usually sandy soil conditions found adjacent to large bodies of water.

ample, may limit root growth in some cases. Salinity is also very high in salt marshes and excludes most kinds of plants except for a few specially adapted species.

Tropical Forest. Tall trees with abundant vegetation at several levels underneath, many different kinds of plants intermixed with no single species dominating an area, and the evergreen habit characterize tropical forests (Figure 10–10). Although during the early Tertiary, tropical forests covered most of what is now the United States, today they are found only in southern Florida and Texas. They require warm, frost-free temperatures all year and abundant rainfall.

Evergreen Coniferous Forests. Pines, firs, redwoods, and other plants that reproduce by seeds but have no flowers are characteristic of coniferous forests (Figure 10–11). In contrast to tropical forests, they often occur as dense stands with one or only a few species dominating. Always abundant in northern latitudes, coniferous forests are found over vast areas of Canada and in more recent geologic times have extended their distribution southward into the higher mountains of western North America, through the Sierra Nevada and the Rocky Mountains, and even into northern Mexico. They are found in areas with cold winters and short growing seasons lasting 4–5 months or less. Probably the severe winters prevent other kinds of vegetation from encroaching upon and replacing these hardy forests.

In the southeastern United States large areas are covered by

FIGURE 10–9.

Marsh vegetation grows around, on, and sometimes in water.

evergreen pine forests which are really different in character from the type of vegetation discussed in the preceding paragraph. The winters are milder than those in Canada and the western mountains. When undisturbed the southeastern pine forests are gradually being replaced by oak, hickory, and magnolia. These forests continue to persist because they have been burned repeatedly, and the pines can reseed or sprout new shoots from their roots after fires whereas the oaks, hickories, and magnolias cannot.

Deciduous Forest. Unlike evergreens, which drop their leaves gradually, deciduous trees and shrubs sprout a new set of leaves each growing season and then lose them all and become leafless during seasons unfavorable for growth. A number of different kinds of deciduous trees intermix and form the dominant vegetation over much of the eastern half of the United States (Figure 10–12). These forests enjoy cool but not too severe winters and warm summers with adequate rainfall. Most of the deciduous forests have been replaced by agricultural and urban vegetation, although they may still be seen in some areas that have been protected, especially in the Appalachian Mountains.

Mixed Hemlock-Deciduous Forest. North of the deciduous forest, in the area surrounding the Great Lakes and eastward through New England to the east coast, the character of the deciduous forest changes. Only a few dominant species from the deciduous forest, notably beech and maple, extend their distribu-

FIGURE 10–10.

Tropical vegetation is abundant and diverse.
Photograph by George W. Cox, San Diego State University, California.

FIGURE 10–11.

a. Evergreen coniferous forest is typified by this stand of yellow pine in southern Oregon. b. Pine forests in the southeastern United States can tolerate the frequent ground fires which eliminate the deciduous hardwood trees and thereby maintain the coniferous species.
(a) Photograph courtesy of Weyerhaeuser Company, Tacoma, Washington. (b) U.S. Forest Service photo.

a.

b.

tions northward into this region. Also, the evergreen tree hemlock (*Tsuga canadensis*) is a common part of the vegetation throughout the area (Figure 10–13). (The poison hemlock that killed Socrates and countless other Greeks is an herb, not a tree, and is unrelated to our native species of the mixed hemlock-deciduous forest.)

FIGURE 10–12.

The deciduous forests of much of the eastern half of the United States are typified by this stand of white oak in Indiana.
U.S. Forest Service photo.

FIGURE 10–13.

Mixed hemlock-deciduous forest is the native vegetation of much of the northeastern states, as shown here in Pennsylvania.
U.S. Forest Service photo.

Scrub-Desert. A number of distinctive plant communities of
western North America may be grouped together as desert and
scrub biomes (Figure 10–14a). They are scattered, low to me-
dium-sized shrubs. Often a brilliant display of annual wildflowers appears following a particularly wet season (Figure 10–14b).
This desert-scrub vegetation type is common throughout the
Great Basin (between the Sierra Nevada and the Rockies) and
much of southwestern North America. These plants can tolerate
long periods of extreme drought too severe for other vegetation
types which might otherwise invade this region.

a.

b.

FIGURE 10–14.

a. Desert plants, such as the scraggly
ocotillo shown here, must withstand
long periods of extreme drought.
b. Following the infrequent rains, annual wildflowers often make a fine
display in the openings among the
shrubs. Those shown are desert dan-
delions (*Malacothrix glabrata*) on the
Mojave Desert.

Chaparral.   In chaparral the shrubs grow so densely that they can scarcely be penetrated (Figure 10–15). Indeed, to penetrate a mature stand you might be obliged to crawl over the tops of the shrubs or wiggle through them in the manner of a snake. Occasionally there are open, grassy meadows or patches of oaks in the canyons. For the most part, however, this vegetation type is an exceedingly dense growth of small-leaved shrubs on hills and mountain slopes of moderately dry areas. In North America

a.

b.

FIGURE 10–15.

Chaparral is a very dense, drought-resistant, shrubby vegetation type. a. At this roadside in southern California, leafless dead branches left after a fire about five years earlier can be seen extending beyond the vigorous new growth. b. An older stand of chaparral shows no sign of fire, even though it was burned about ten years earlier.

chaparral occurs in southern California where cool, moist winters are combined with warm, dry summers. Plants growing in these areas must be adapted to fires which frequently burn the vegetation to the ground. Toward that end, some plants sprout new shoots from the underground roots, while others produce seeds which germinate readily after a fire but rarely or not at all in the intervals between fires.

Grassland-Savanna.   The difference between grassland and savanna is that a savanna has scattered trees, while grasslands are treeless (Figure 10–16a). In the vast, open prairies of central North America, in California, Oregon, Washington, and Idaho, the typical grassland vegetation type dominates (Figure 10–16b). Apparently tolerant to a wide range of temperatures, these plants generally occur in areas of only moderate precipitation. Man has succeeded in converting most of what was until recently grassland into agricultural vegetation types, principally grasses such as corn, wheat, and sorghum.

Tundra.   Vast, open, treeless northlands, with herbs and low shrubs covering the permafrost, characterize the tundra (Figure 10–17). The soil remains frozen a few feet below the surface, and above that it thaws for only 2–3 months a year. Small wonder that trees are not found in the tundra. It is the northernmost biome we have; only rocks and ice are found beyond. Southward, where the growing season becomes longer, tundra is gradually replaced by coniferous forest in a broad band extending from the Pacific to the Atlantic oceans. But even in these more temperate latitudes, tundra is found above timberline on the higher mountains of western North America.

FIGURE 10–16.

The difference between grassland and savanna is that savanna has trees scattered about. a. Here we see grassland surrounded by savanna. b. The grassland vegetation type is best seen in the vast, open prairies of central North America, as in this scene near Boulder, Colorado.

a.

b.

Man's Utilization of the Forest

The vegetation types listed have persisted for thousands of years. Man has introduced two new types, the agricultural and urban, but until now he has not eliminated any. Some types, however, particularly the grassland and deciduous forest, have been greatly modified.

Let's turn our attention now to man's utilization of one vegetation type—the coniferous forest. It is one of the three vegetation types of greatest direct economic importance; the others are the agricultural and urban types. We have largely ignored the economic role of coniferous forests up to now, whereas examples from agricultural and urban plants are found throughout this book.

Forests and forest management, or at least certain aspects thereof, are dear to the hearts and pocketbooks of us all. Paper, lumber, and recreation are probably the three most apparent assets of forests, but others less obvious are just as important to our way of life.

Of the 753.5 million acres of forestland in America, one-third its total area, 233.9 million acres, is unproductive, and 19.9 million acres are withdrawn from commercial classification by legislative or administrative action and are included in parks and wilderness areas. America's commercial forests therefore include nearly 500 million acres. Of this, 296.2 million acres (59%) is owned by private individuals, mostly small farms and plots of 50 acres or less. The forest industry owns 67.3 million acres (13%), the federal government (including National Forests) 107.1 million acres (22%), and other public lands such as owned by state governments amount to 29.1 million acres (6%) (Figure 10–18).

Let us begin our discussion of forest practices with a review of the forest ecosystem in the absence of man's activities, and then we will be in a better position to assess current forest management practices and their impact on the ecosystem. To do this you should take a real or imagined trip to some local wooded area.

This is the Forest Primeval

Step into the woods. Go far enough to be away if possible from man's direct and immediate impact on the environment. In that way you can see how things are in man's absence. Your first thought is probably a question: Where can I go? What can I do? This is a lesson in itself. Man's direct and immediate impact on our environment is so widespread as to be almost universally

FIGURE 10–17.

There are no trees, and even the shrubs remain low and prostrate in the tundra, as shown near Point Barrow in northern Alaska. Photograph by Albert W. Johnson, San Diego State University, California.

present, it seems. You may wish to dwell on that problem, but in the meantime let's get on with our walk. The lessons to be learned are so obvious that they cannot escape even the most casual observer.

The picture we wish to paint and the point we wish to make concern the dynamic stability of natural ecosystems, the forest in particular. You can see that the whole system is in dynamic equilibrium everywhere you look. There is a fine, old tree which has died; its bark is peeling and falling to the ground to lie among the dead branches which were pruned by the winter storms. The bark may be charred, or the uppermost branches stunted by lightning, suggesting the natural misfortunes that this tree was able to endure. The base of the tree may have been sawed, suggesting a different sort of misfortune to which this species was not well adapted. Nearby, as distances are judged in wild places, another tree stands alive and well; it has replaced the dead one as the reigning neighborhood giant. Somewhere in these woods there is probably another one even larger. All around is evidence of reproduction and replacement.

Perhaps you may be wondering about the number of seeds produced by the trees around you. Precise numbers are hard to come by, but 1,000–10,000 viable seeds is a fair estimate of the average annual reproductive potential of a gymnosperm. Probably 100,000 seeds is a minimum lifetime figure for most of our forest dominants, and 1 million or more may be closer to the truth. Of all those seeds, how many will realize their potential to germinate and grow into mature adults? The answer is very few. Perhaps none will reach adulthood because a mature forest is a balanced system which responds to many different environmental conditions (Figure 10–19).

Elsewhere on the forest floor are shrubs, and during late spring through early fall herbs abound. If this particular woodland has been protected from fire for a number of years, the undergrowth and litter may be building up to dangerous levels, setting the stage for a real holocaust. If a costly fire occurs, politicians will then make points by obtaining funds for the purpose of preventing fires, thereby perpetuating the cycle. On the other hand, if fires have been allowed to run their course every

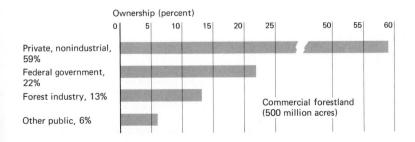

FIGURE 10–18.

Ownership of commercial forestland in the United States.

few years, you'll see little fuel accumulating, and the forest floor will be open and parklike (Figure 10–20).

Some forests, such as the Douglas fir forest in Oregon, seem adapted to natural disasters such as fire and insect attack. Douglas fir seedlings establish themselves quickly in these areas and grow rapidly when free from shade. Presumably, natural catastrophes periodically decimate local populations of Douglas fir.

FIGURE 10–19.

Douglas fir seedling in an open area in west-central Oregon.

FIGURE 10–20.

A periodic fire helps to maintain an open, parklike aspect in this yellow pine forest in the mountains of southern California and also helps to return mineral nutrients to the soil. Note the charred trunks on these healthy trees.

When this occurs, the trees quickly reseed, and the resulting forest trees are about the same age.

The forest soil is rich and productive. Mineral nutrients are removed from the soil by the growing plants and incorporated into their tissues. These tissues may serve as forage for browsing animals, who return many of these nutrients back to the soil. Animals and plants eventually die, of course, and so in time all nutrients return to the soil. Fire is simply one way in which the cycle can be completed.

Having read the foregoing chapters in this book, your mind may dwell for a moment on energy. You see the light filtering among the trees' upper branches and down onto the objects below (Figure 10–21). You notice the trees competing for all the sunshine they can get in the canopy, while plants in the understory must be content with the remainder. Think back to the role of auxins in controlling plant growth and also to that phrase "compensation point." You realize that most of the light is reflected or simply lost as heat without ever entering a biological energy cycle. But now you see the vital role played by the small part of light that drives photosynthesis. The trees, shrubs and other plants, the herbivorous animals which eat them, the animals at higher levels of the food pyramid, the humus in the soil, and the bacteria and fungi (Figure 10–22) in the soil all depend on that one step you learned as "the light reactions of photosynthesis." Finally, you look about one last time and think about materials moving in cycles everywhere you look—energy moving one way. You realize the forest is one whole complex system. No longer are water, soil, plants, animals, air, and sunshine discrete segments more or less independent from the others. Rather, they are seen as integrated parts of a whole system, an ecosystem. The only dispensable part of that ecosystem is the animals. If any other part were to be removed or cease its normal function, the whole system would collapse. Animals, including man, depend on the rest of the ecosystem for all their necessities of life —food, shelter, water, even air. That dependence is not likely to change in the near future.

This is Another Forest

Man, as any other animal, derives the materials for his livelihood from his environment. A great many of these materials are taken from forests such as the one described above. Therefore, let us compare this picture with that of a logged forest.

The purpose of logging is to remove the wood from the forest and deliver it to the sawmill. The forester must first decide the method to use, whether it would be best to clear-cut or selectively cut. Clear-cutting means that every tree in the area is

FIGURE 10–21.

Sunlight filters down through the forest's upper canopy to the herbs and shrubs below and represents the beginning of energy flow through the forest ecosystem and the building of complex organic material from carbon dioxide and water.

FIGURE 10–22.

Fungi on the forest floor represent the end of energy flow through the forest ecosystem and the return of carbon, hydrogen, and oxygen to their inorganic forms.

removed, and nothing is left standing (Figure 10–23). This is the normal mode of operation in many areas, such as the Douglas fir forests in central Oregon and northward. After being clearcut an area may be replanted or allowed to reseed naturally. In either case a number of other shrubs and small trees grow fairly quickly and dominate for perhaps 10 years while the Douglas fir seedlings are getting started. In the end the Douglas fir

a.

FIGURE 10–23.

a. The sizes of most clearcuts today are small (30–50 acres). b. Some, however, are several square miles in area. c. Closeup of a clearcut is shown.

b.

c.

usually comes to dominate, although not always (Figure 10–24). All of these second-growth trees are of the same age, and so clear-cutting is also called even-age management. After about 15–20 years the smaller trees are removed so that the faster-growing ones can grow even more rapidly. The competition for water, minerals, and sunlight is reduced. The removed saplings are simply burned because they have no commercial value. The whole process is called precommercial thinning. After another 15–20 years the area may be thinned again, but by this time the trees are large enough to be sold, usually to paper mills for pulp. Finally, some years later, another clearcut shaves the area clean, and the cycle begins anew.

Selective logging, in contrast to clear-cutting, removes only certain trees and leaves others standing (Figure 10–25). This is practiced in some areas of mixed forests, such as where ponderosa pine, Douglas fir, incense cedar, and white fir grow together in south-central Oregon. With selective cutting the trees left standing are of various ages, and so there is no need for precommercial thinning; every tree is sold. At least some of the forests now managed by the large timber companies on a selective cutting basis are to be put on a system of clear-cutting at some future time. It is more economical. The practice of selective logging would then appear to be the first phase of a long-range plan.

After a tree is felled, all limbs are removed and the main trunk is cut into about 30-foot lengths and taken to the mill by logging trucks. When cool, damp weather reduces the risk of forest fires,

FIGURE 10–24.

An example of a "mistake" made by a large timber-growing company. This area was harvested about 30 years earlier, and the original vegetation type has not yet recovered.

the limbs and other debris are burned on the logging site. Meanwhile, the logs may be processed at a lumber mill or paper mill (Figure 10–26). The lumber mills first remove the bark and slice the logs longitudinally into large planks and then into smaller sizes according to the needs of their customers. Paper mills also remove the bark and then run the logs into a machine which re-

FIGURE 10–25.

Selective logging means that only the larger trees are cut, leaving the smaller ones to grow to commercial size.

FIGURE 10–26.

Redwood logs are delivered to the mill by truck and stored temporarily in piles such as these. The older logs here are probably 400 to 600 years old.

duces an average tree trunk 1–2½ feet in diameter and 20–30 feet in length into chips smaller than your little finger in 10–20 seconds. The chips are then made into paper in the manner described in Chapter 7.

Timber companies own a great deal of land, and much of it is open to the public for recreational purposes except when logging operations make it unsafe. Nevertheless, our present market for wood could not be satisfied without the trees from public lands which are sold to lumber companies under the federal jurisdiction of the Forest Service and Bureau of Land Management.

Forest Service timber is managed as follows. Each national forest is administered by a forest supervisor who is responsible for all aspects of managing that particular national forest. A long-term timber management plan is prepared by the forest supervisor, including the potential yield and harvest levels based on the standing crop, growth rates, types of terrain, and so on. With this as a guide, five-year plans are prepared by smaller administrative units called Ranger Districts which list planned individual sales by volume, location, and year. The next step is to prepare a listing of individual sales of timber to be offered that fiscal year. Finally, bids are submitted by the local timber-harvesting companies, and a contract for removal of the timber is completed with the high bidder. The contracts include specifications for road building, times and methods of logging, and price for the timber.

One of the more serious problems facing the public as well as the timber industry is the concept of multiple use under which timber is harvested from national forest land. In theory multiple use implies that the national forests can be shared and used jointly for several purposes including recreation, timber production, grazing land for stock animals, watershed, and hydroelectric power. In practice, however, priorities must be set among these various interests. For example, logging operations are too dangerous to allow others in the same area, and after the trees are removed, the area is unsuitable for recreation. Grazing operations must also be curtailed after logging to allow the young trees to be established. Realizing that logging operations preclude recreation activities in the same area, conservation groups have succeeded in getting additional protected acreage in the form of wilderness areas and state and national parks. In general, industry representatives seem to resent this and have questioned whether more land is necessary for recreational purposes. It is likely that both sides are biased for the following reasons. First, it is obvious that the industry's economic growth is restricted by reducing the acreage available to them. Second, foresters tend to work in forests, that is, away from congested cities. Thus, their needs for the benefits of wild places away from population centers are well satisfied. Conservationists on the other hand are often from distant urban areas where they see the needs for wilderness areas growing more desperate daily. Incidentally, when the question is turned around and a timber man is asked how much land he needs, he also has no answer but points instead to the ever-increasing need for wood and wood products. It might be useful for us here to examine briefly this need so that we may better understand the problems of the industry and its relation to society as a whole.

The two biggest consumers of wood are probably the construction and paper industries. In 1968 the U.S. Congress set a goal of 26 million new housing units by the end of the decade. If the average housing unit holds four persons, then 104 million people or nearly half of the present population would live in a new dwelling. What will become of all the old houses and the wood in them at the rate we are to construct new ones!

This perspective suggests that a reconsideration of housing goals in relation to recreational needs is in order.

In regard to the need for paper, the United States uses over 500 pounds of paper per person each year. The next highest consumer is Canada with less than half that rate, and other countries go down from there. What do we do with all that paper? Well, as any business executive can explain, a new product creates a need for itself. Thus, until there were such things as paper cups, for example, we got along fine without them. It would seem, then, that this need for more and more wood and wood products is not so much a real need as it is a market. So long as there is someone willing to pay for it, someone else will be willing to produce it. On the surface it would appear that economics and the profit motive, rather than an altruistic drive to provide for the needs of others, are the motivating forces behind much of the timber industry's policies and decisions. But can the industry be entirely to blame? A market would not exist without buyers, and this obviously refers to society as a whole, including governmental decisions determining management policy of our public lands.

Another problem we should all be concerned with is whether or not our forest lands are capable of sustaining constant logging generation after generation. We don't have the answer, but it would appear that the most important consideration has to do with minerals in the soil.

Let us first consider the fate of mineral elements that are incorporated into trees and then consider a separate but related problem involving erosion and leaching. With each harvest of timber, some minerals are lost because part of the tree is removed. Does this represent a major loss or is it trivial? The answer is complex and appears to vary from species to species, location to location, and on the particular mineral in question

Part of Tree	Nitro-gen	Phos-phorus	Potas-sium	Cal-cium	Magne-sium	So-dium
Harvested and removed	19%	19%	26%	35%	34%	9%
Crowns and roots of harvested trees left at the site	81%	84%	74%	65%	66%	91%

TABLE 10–2. DISTRIBUTION OF MINERALS IN *Pinus sylvestris*

(Table 10–2). As you can see in the table, a relatively large proportion of the minerals originally contained within the plant remain at the site of harvest, but this does not necessarily mean they will be efficiently recycled. Those minerals in the underground parts of a tree slowly return to the soil through the normal process of decay. The branches and crowns of the trees on the other hand are trimmed from the trunk where it was felled (Figure 10–27), and all of this debris is raked into a pile and burned at a time when the danger of forest fire is minimal. Burning returns the organic matter to carbon dioxide and water in the air, while most of the minerals are left behind in the ash. Thus, while some major fraction of the total mineral content remains at the logging site, its distribution is now very low over most of the area and excessively high at the spot where the slash was burned. Furthermore it is now lying on the surface and is exposed to wind and precipitation.

FIGURE 10–27.

Branches of felled trees are removed on the site before the trunk is hauled to the mill.
Photograph courtesy of Weyerhaeuser Company, Tacoma, Washington.

One consequence of this is that the loss of minerals through surface runoff is very high during the first year or two following logging, depleting the soil and increasing eutrophication of the waterways. Indeed, perhaps the major long-range problem in the area of applied forest ecology concerns the action of water on soil. Foresters explain that it is not the logging operation itself that causes erosion but the roads. Realistically speaking, however, can an area be logged without roads? Helicopters and balloons can be used to remove cut logs from steep terrain where roads might prove dangerous or destructive (Figure 10–28). These techniques are rather expensive and beyond the means of most logging companies, and consequently their use is limited at the present time to experimental work. The role of water in leaching minerals from the soil is a related problem. As water percolates down through the soil, it carries in dissolved form some of the minerals originally on the surface or in the topsoil. This water either flows fairly rapidly through an underground channel to become a spring at a lower elevation or else becomes part of the more permanent ground water. We need more quantitative and qualitative information on the effects of logging on this water and on the soil it passes through. How much water moves in these ways, how fast, and how far? Is it leaching minerals from the upper soil layers faster before or after logging? Do the leached minerals remain in solution until they

FIGURE 10–28.

Experimental techniques for removing logs from steep slopes. a. The balloon is controlled by a cable which runs by means of a pulley to the top of the hill. b. The helicopter's cable is seen hanging from its midsection.

a.

b.

resurface in a spring or artesian well, or are they removed at a lower soil level? This information is important because in an attempt to replace mineral elements that are depleted, fertilizers are being tried in some areas to increase timber growth. If this practice becomes widespread, an understanding of soil water becomes critical. What is to become of our remaining clear mountain lakes and streams if, for example, 50% of the nutrients in fertilizers used on forests does not go into the trees but moves into our waterways as run-off or soil water? The answer, eutrophication, is discussed elsewhere in this book.

It is interesting to analyze what foresters have to say about these kinds of criticism. Most tell us to relax; they are the last ones to want to destroy our forests. Managing forests is their profession, and therefore they know more about it than anyone else. This may indeed be the case, but we must also realize that most of their information concerning future yields is based on practices elsewhere rather than based on a particular logging site. Foresters tell us that in Germany, for example, the forests have been logged for 500 years and their forests are still producing, and therefore we can do the same here. The foresters do not tell us whether the trees in Germany are growing faster or slower, larger or smaller, than they used to, and neither do they mention that the rivers in Germany are among the most polluted in the world. They do tell us that good forestry is practiced almost acre by acre and that timber management practices in one area may not be appropriate for another area even within the same state. How then, one might ask, can the practices in Europe be used to justify our own here in America? And so the question persists, what might happen? One possible answer is seen in the fate of *Cedrus lebani*.

Whatever Happened to the Cedars of Lebanon?

Cedrus lebani, the cedar of Lebanon, has long been recognized in art and literature as a symbol of power and longevity. "The voice of the Lord breaketh the cedars; yea, the Lord breaketh the cedars of Lebanon" (Psalm 29:5). Legend holds that Seth planted a cedar of Lebanon on the grave of his father Adam, and 5000 years later wood was cut from that same tree to construct the Holy Cross. The oldest specimens today are about 2500 years old, but the plant is thought to have attained at least twice that age in earlier days. The cedar of Lebanon first appears in a major historical role as the base for the great trade network of the early Phoenicians. The Egyptians are known to have traded cedar at least as early as 2500 B.C., the wood highly prized for shipbuilding because of its durability. During these ancient times most of Lebanon was covered with timber, and only a small area in the northeastern part of the country was open grassland. Today in Lebanon you'll find barren slopes at the higher elevations, and the wild lands are covered by scrub. The cedar of Lebanon persists in a few small groves, and it is occasionally found in cultivation. The best-known groves are in a tiny part of Kedisha Valley at an elevation above 6000 feet, and it is more plentiful in the Taurus Mountains. But in the land of its namesake, the cedar of Lebanon is found as the emblem on the Lebanese flag and as a tourist attraction in a grove of about 300 individuals protected by a wall. What did happen to the cedars of Lebanon?

Cedar wood is more than durable; it is fragrant and very attractive as well. Solomon, in his wisdom, recognized this and used the wood to build his holy temple and adjoining palace, together with a number of other structures. The demand for cedar wood for commerce, shipbuilding, and construction became excessive, and the means for satisfying this demand are scarcely comprehensible. Solomon, for example, used 100,000 slaves steadily for 20 years felling timber for his temple. At one point during the tenth century B.C., the slave army of the Phoenician King Hiram was supplemented by 30,000 Israelites and 150,000 slaves supplied by Solomon, all cutting timber in Lebanon. In modern parlance this would be termed clear-cutting or even-age management, and the public relations office of the local timber company would explain that we should simply relax now and wait for the next crop of trees to mature. The cedars of Lebanon were clear-cut nearly 3000 years ago, and we are still waiting.

And so on the one hand you have the scientist asking a number of questions. What is the evidence that today's so-called needs for timber are worth sacrificing our recreational sites—possibly forever? How can we monitor the soil and soil water so as to be assured that the forestland will remain productive? Industry on the other hand recognizes a few mistakes but otherwise expresses little concern for the long-lasting damage to the forest ecosystem. Since the critical facts are not available, the evidence is necessarily lacking, and so our forests are being used as one means to maintain our economy for a while longer. One wonders for how long.

Selected Readings

Deevey, E. S. October 1958. Bogs. *Scientific American* 199(4):114–122.
Kendeigh, S. C. 1961. *Animal Ecology*. Englewood Cliffs, N.J.: Prentice-Hall, Chapters 21–27. A good review of the major biomes and their history during the Tertiary period.

Piel, G. (ed.). 1968. Gondwanaland revisited: New evidence for continental drift. Proc. Am. Phil. Soc. 112:335–343. Details the evidence for continental drift.

Report of the President's Advisory Panel on Timber and the Environment. 1973. Washington, D.C.: U.S. Government Printing Office. Contains a summary of major recommendations together with extensive data on supply and demand of timber.

Wood, N. 1971. *Clearcut.* San Francisco and New York: Sierra Club. A critical review of forestry practices in the United States.

Appendix One: Go, Metric!

The system of measurement commonly used in the United States is antiquated and quite arbitrary. That is, a person has to remember certain units and conversion factors. For example, there are 12 inches in one foot, 3 feet in a yard, and 5,280 feet in a mile. This makes it difficult to change a measurement from one unit to another. It would be much easier to use a system where conversions went by multiples of ten. Changing units then simply involves moving a decimal point the proper number of places. Such a system—the *metric system*—is used by many countries other than the United States and is the accepted system of measurement for all scientific work in all countries. The three main units—meter, liter, and gram—are modified with prefixes to express the units as larger (kilo:1000) or smaller (milli:0.001) quantities.

The United States is changing over to the metric system. The ramifications of this change in measurement systems are numerous. Consider the problems involved in purchasing replacement parts for maintenance of the automobile. People who own foreign cars are familiar with millimeters and liters and will certainly have an advantage. In the supermarket, food cans will be marked in grams rather than ounces and milk will sell by the liter. Road signs will be in kilometers and the famed 50 yard line will no longer exist. Nevertheless, the changeover will take place, and in the long run it will be a useful change; so, go metric now and beat the rush.

Metric Units of Length

1 kilometer (km)	= 1000 m		
1 meter (m)	= 100 cm	= 1000 mm	
1 decimeter (dm)	= 10 cm	= 0.1 m	
1 centimeter (cm)	= 10 mm	= 0.01 m	
1 millimeter (mm)	= 0.1 cm	= 0.001 m	
1 micron (μ)	= 10^{-4} cm		
1 millimicron (mμ)	= 10^{-7} cm		
1 angstrom (Å)	= 10^{-8} cm		

Conversion Factors for Length

1 inch (in.)	= 2.540 cm	
1 foot (ft)	= 30.48 cm	
1 mile (mi)	= 1.609 km	
1 centimeter	= 0.3937 in.	
1 meter	= 3.281 ft	= 39.37 in.
1 kilometer	= 0.6214 mi	

Metric Units of Volume

1 kiloliter (kl) = 1000 liters
1 liter (l)   = 1000 milliliters (ml) = 1000.027 cc (cm³)
1 milliliter =   1.000027 cc   =   0.001 l

Conversion Factors for Volume

1 liter   =   1.057 quarts (liquid)
1 quart   =   0.9463 liter
1 gallon   =   3.785 liters
1 cubic foot = 28.32 l = 7.84 gallons
1 cubic inch = 16.39 cc or ml
1 fluid ounce = 29.57 cc or ml

Metric Units of Mass (Weight)

1 kilogram (kg) = 1000 grams
1 gram (g)   = 1000 milligrams (mg)

Conversion Factors for Mass

1 pound   = 453.6 g
1 ounce   =   28.35 g
1 kilogram =   2.205 lb

Useful Conversion Factors for Area

1 square centimeter (cm²) = 0.1550 square inch (in.²)
1 square inch   = 6.452 square centimeters
1 square yard (yd²)   = 0.8361 square meter (m²)
1 square mile (mi²)   = 2.59 square kilometers (km²)

Temperature Conversion

On the centigrade scale (°C) the boiling point of water is 100 degrees; the freezing point is zero degrees. On the Fahrenheit scale (°F) the boiling point of water is 212 degrees; the freezing point is 32 degrees.

To convert from the Fahrenheit to the centigrade scale:
$$(°F - 32) \times 5/9 = °C$$
To convert from the centigrade to the Fahrenheit scale:
$$(°C \times 9/5) + 32 = °F$$

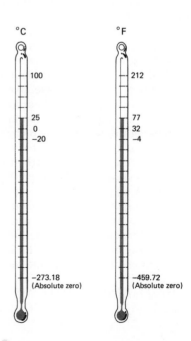

Appendix Two: Respiration

Respiration is the process by which foodstuffs (organic molecules) react with oxygen and are broken down to carbon dioxide and water with a net gain of captured energy (in the form of ATP). The most common form of respiration (aerobic) proceeds through three stages: glycolysis, citric acid cycle, and oxidative phosphorylation. Plants utilize the processes of respiration to provide energy for cellular maintenance and for the production of starting materials for the synthesis of needed compounds.

The cellular process used for the initial breakdown of 6-carbon sugars is called *glycolysis*. The first stage of glycolysis is the breakdown of one glucose molecule (six carbons) into two molecules of the 3-carbon sugar glyceraldehyde phosphate. This reaction sequence requires the input of energy in the form of two ATP molecules. In the second stage of glycolysis, the glyceraldehyde phosphate is converted to another 3-carbon compound, pyruvate. This process is coupled to the formation of four molecules of ATP per molecule of glucose. The overall equation for glycolysis can be written

$$C_6H_{12}O_6 + 2ATP + 2ADP + 2P_i + 2NAD^+ \rightarrow$$
$$2(C_3H_4O_3) + 4ATP + 2NADH + 2H^+$$

showing a net gain of two molecules of ATP and two pairs of hydrogen atoms. A more detailed view of glycolysis is shown in Figure A–1.

After its formation in the cytoplasm, pyruvate is transported into the mitochondria where its further metabolism to carbon dioxide and water takes place. This final breakdown is accomplished by a series of reactions known as the *citric acid cycle*. Before beginning the cycle, pyruvate is converted to a 2-carbon fragment with the loss of carbon dioxide and 2 hydrogen atoms. This 2-carbon fragment combines with a 4-carbon compound (oxaloacetate) to form the 6-carbon compound, citrate. Through a series of reactions citrate is converted back to oxaloacetate with the release of two carbon dioxide molecules, four pairs of hydrogen atoms, and the production of one molecule of ATP (Figure A–2).

In both glycolysis and the citric acid cycle, released hydrogen atom pairs are captured by a molecule of NAD^+ to form NADH and H^+ (except for one reaction in the citric acid cycle in which the hydrogen acceptor is FAD). The overall citric acid cycle reaction beginning with one molecule of pyruvate is

$$C_3H_4O_3 + 4NAD^+ + FAD + ADP + 3H_2O \rightarrow$$
$$3CO_2 + 4NADH + 4H^+ + FADH_2 + ATP$$

The coupling of oxygen utilization to ATP formation is a part of respiration known as *oxidative phosphorylation*. This process utilizes the reduced hydrogen carriers produced by glycolysis and the citric acid cycle for the production of water and the coupled formation of three molecules of ATP (Figure A-3). The oxidative phosphorylation process occurs in the mitochondria. NADH and $FADH_2$ produced in the citric acid cycle are used directly, but glycolytically generated NADH must first transport its hydrogens into the mitochondria. This transport process is accomplished by extramitochondrial NADH transferring its hydrogens to an internal mitochondrial FAD.

We can now compute the total number of ATP molecules formed from each glucose molecule consumed. During glycolysis we generate two ATP molecules and two extramitochondrial NADH. During *two* turns of the citric acid cycle we generate two ATP molecules, eight mitochondrial NADH, and two $FADH_2$. From oxidative phosphorylation each mitochondrial NADH yields three ATP molecules and each extramitochondrial NADH or mitochondrial $FADH_2$ yields two molecules of ATP. From oxidative phosphorylation we can produce 32 ATP molecules, for a grand total of 36 molecules of ATP manufactured from each glucose that is oxidized to carbon dioxide and water.

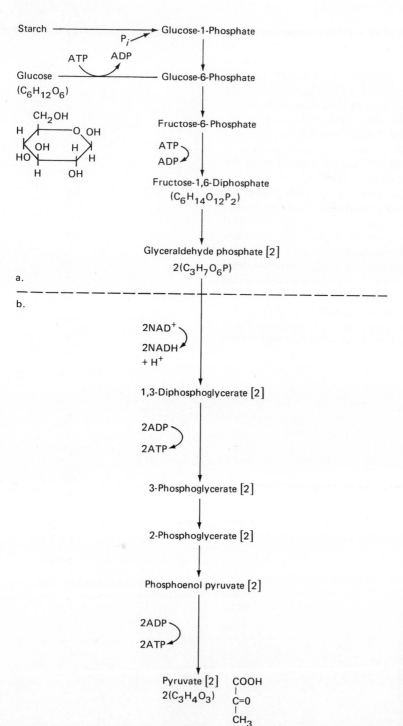

FIGURE A–1.

The reactions of glycolysis to form pyruvate. Each step in the sequence is catalyzed by the action of a separate enzyme in the cytoplasm. a. The first stage of glycolysis: the production of two 3-carbon sugars from glucose. b. The second stage of glycolysis: the formation of NADH and ATP.

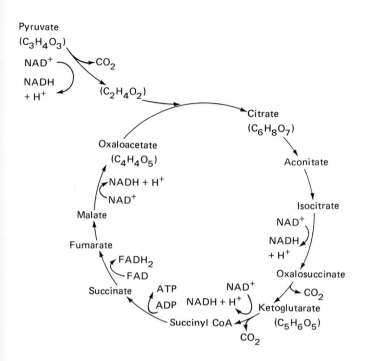

FIGURE A–2.

The enzymatic steps for the breakdown of pyruvate by the citric acid cycle.

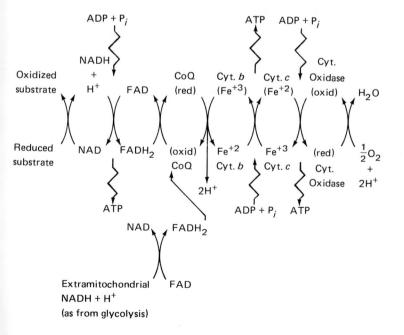

FIGURE A–3.

Tentative model of the mitochondrial electron transport or cytochrome system for oxidative phosphorylation. Sites of ATP production are not well established, especially the second one. Except for CoQ (the function of which is tentative) and perhaps NAD (and NADH), the participants are tightly bound to proteins associated with the mitochondrial membranes. Electrons can be driven in the reverse direction if ATP is added.

From *Plant Physiology* by Frank B. Salisbury and Cleon Ross. © 1969 by Wadsworth Publishing Company, Inc., Belmont, California 94002. Reprinted by permission of the publisher. Fig. 15–5, p. 310.

Appendix Three: Amino Acids and Codons

A specific sequence of three bases in an mRNA chain is termed a codon, and the complementary sequence on DNA or tRNA is termed an anticodon (see Chapter 5). This "triplet" code has four major features.

1. The code is degenerate. Degeneracy means that more than one 3-base sequence will code for the same amino acid. Although the code is degenerate it is not ambiguous since under normal conditions no codon specifies more than one amino acid.

2. The only punctuation in the code is for initiation and termination. The codons AUG and GUG code for the initiation of protein synthesis when they occur at the beginning of a mRNA strand or after a stop signal. A special termination signal ("stop") is always required before a completed protein can be released in free form. Three codons (UAG, UAA, and UGA) terminate protein transcription. These codons do not code for amino acids and were, therefore, originally termed "nonsense codons."

3. The code appears to be universal in nature. A given mRNA will be "read" in the same way in even the most distantly related species.

4. The code is directional. Due to the chemical structure of the bases, transcription only takes place in one direction on the mRNA. UUC which codes for phenylalanine will not be confused with CUU which specifies leucine. Listed below are the most common amino acids and their codons and anticodons.

AMINO ACID	SYMBOL	CODON (mRNA)	ANTICODON (tRNA)
Alanine	Ala	GCU	CGA
		GCC	CGG
		GCA	CGU
		GCG	CGC
Arginine	Arg	AGA	UCU
		AGG	UCC
		CGU	GCA
		CGC	GCG
		CGA	GCU
		CGG	GCC
Asparagine	Asn	AAU	UUA
		AAC	UUG
Aspartic acid	Asp	GAU	CUA
		GAC	CUG
Cysteine	Cys	UGU	ACA
		UGC	ACG
Glutamic acid	Glu	GAA	CUU
		GAG	CUC

AMINO ACID	SYMBOL	CODON (mRNA)	ANTICODON (tRNA)
Glutamine	Gln	CAA	GUU
		CAG	GUC
Glycine	Gly	GGU	CCA
		GGC	CCG
		GGA	CCU
		GGG	CCC
Histidine	His	CAU	GUA
		CAC	GUG
Isoleucine	Ile	AUU	UAA
		AUC	UAG
		AUA	UAU
Leucine	Leu	UUA	AAU
		UUG	AAC
		CUU	GAA
		CUC	GAG
Lysine	Lys	AAA	UUU
		AAG	UUC
Methionine	Met	AUG	UAC
Phenylalanine	Phe	UUU	AAA
		UUC	AAG
Proline	Pro	CCU	GGA
		CCC	GGG
		CCA	GGU
		CCG	GGC
Serine	Ser	UCU	AGA
		UCC	AGG
		UCA	AGU
		UCG	AGC
		AGU	UCA
		AGC	UCG
Theronine	Thr	ACU	UGA
		ACC	UGG
		ACA	UGU
		ACG	UGC
Tryptophane	Try	UGG	ACC
Tyrosine	Tyr	UAU	AUA
		UAC	AUG
Valine	Val	GUU	CAA
		GUC	CAG
		GUA	CAU
		GUG	CAC

Glossary

ABORT To cease development before maturity.

ABSCISSION The dropping of various plant parts (e.g., leaves, flowers, and fruits) after a separation layer has been formed.

ABSCISSION ZONE A special layer of cells, the breaking of which leads to abscission.

ABSORPTION The taking up of water and other materials into the plant cell.

ABSORPTION SPECTRUM Measurement of the kinds of light waves absorbed.

ACTION SPECTRUM Elucidation of the light waves which initiate a particular response.

ACTIVE TRANSPORT The movement of materials across a cellular membrane with the expenditure of energy from respiration, frequently against a diffusion gradient.

ADAPTATION A change in an organism or group of organisms that results in increased ability to survive and reproduce in a particular environment.

ADENINE A purine base in DNA, RNA, and some compounds associated with energy (i.e., ATP).

ADENOSINE TRIPHOSPHATE (ATP) An energy-storing nucleotide containing three phosphate groups. Energy for metabolism is released when a phosphate group is removed forming ADP (adenosine diphosphate).

ADHESION The adhesion of molecules in the solid, liquid, or gaseous state onto the surface of a solid body, forming a thin layer.

ADVENTITIOUS Plant structures which arise from an organ where they are not normally found, such as roots growing out of leaves.

AEROBIC Requiring the presence of oxygen.

AGAR An extract from red algae of gelatinous consistency used to solidify nutrient media.

ALLELE Alternate forms of a gene which may occur at a specific position or locus on a chromosome.

ALTERNATE The distinctive arrangement of leaves on a stem due to the occurrence of only one leaf at each node.

ALTERNATION OF GENERATIONS In the life cycle of a plant, the alternation of the sporophytic, or diploid, phase and the gametophytic, or haploid, phase.

AMINO ACIDS The "building blocks" of protein. They are organic molecules containing an amino group (NH_2).

AMMONIFICATION The decomposition of proteinaceous compounds by bacteria and other organisms with the release of ammonia or ammonium.

AMYLASE An enzyme which breaks down starch through the process of hydrolysis.

ANAEROBIC Occurring in the absence of free oxygen.

ANAEROBIC RESPIRATION A special type of respiration which occurs in the absence of free oxygen.

ANAPHASE The stage of mitosis during which the chromatids separate from one another and move to opposite poles in the cell.

ANGIOSPERM A flowering plant whose seeds are enclosed in a fruit (mature ovary).

ANGSTROM A unit of linear measurement equal to one ten-thousandth of a micron.

ANNUAL A type of plant which completes its entire life cycle within a year.

ANNUAL RING The ring of vascular tissue formed during one growing season, seen in a cross-section.

ANTHER The uppermost part of a stamen which contains the pollen.

ANTHERIDIUM The male sex organ of plants which contains the sperm.

ANTHOCYANINS A group of natural, water-soluble pigments found in the cell sap, ranging in color from red through blue.

APICAL DOMINANCE The growth-suppressing influence of the terminal bud upon the lateral buds mediated by plant hormones.

APICAL MERISTEM Tissue at the tip of a shoot or root which is composed of meristematic (dividing) cells.

ARCHEGONIUM The female sex organ of plants which contains the egg.

ASEXUAL Reproduction which occurs without the union of gametes.

ATOM The smallest unit of a chemical element that still retains the properties of that element; composed of protons, neutrons, and electrons.

ATP See adenosine triphosphate.

AUTOTROPHIC Referring to organisms which are able to synthesize their own food from inorganic materials (e.g., green plants and some bacteria).

AUXIN A natural or synthetic plant-growth regulator responsible for cell elongation and other physiological effects.

AXIL The angle between a leaf and the stem from which it arises.

AXILLARY   Located in the axil of a leaf; referring to structures, such as buds, found in this position.

BACK-CROSSING   The process of crossing an F_1 hybrid with one of its parents or with an organism of the same genetic constitution.

BACTERIOPHAGE   Any virus which attacks bacterial cells.

BARK   Those tissues of a woody stem which occur exterior to the vascular cambium.

BASIDIOCARP   The reproductive structure of mushrooms, puffballs, and shelf fungi; the structure in which basidiospores are formed.

BASIDIOSPORE   A spore of the basidiomycetes; produced after fusion of two haploid nuclei and a subsequent meiotic division.

BIENNIAL   A type of plant which completes its life cycle during two growing seasons—growing vegetatively the first growing season and producing flowers and fruit the second season.

BILATERALLY SYMMETRICAL   Descriptive of the structural arrangement of a flower in which one side of a single longitudinal plane is a mirror image of the other side (an irregular flower). This is opposed to radially symmetrical.

BINOMIAL SYSTEM   The system of nomenclature in which a plant is given two names, the first indicating genus and the second indicating species.

BIOSPHERE   The part of the globe, including land, water, and air, in which living organisms can exist.

BISEXUAL   Referring to an organism which produces both male and female gametes.

BLADE   The part of the leaf which is broad and flat.

BORDERED PIT   A cavity in the cell wall where the secondary wall arches over the pit cavity.

BRYOPHYTE   One of a group of nonflowering plants, including liverworts and mosses.

BUD   (1) The embryonic, underdeveloped shoot of a plant.
(2) The outgrowth of a yeast cell formed during vegetative reproduction.

BUD MUTATION   A mutation in a single bud resulting in the branch, flower, or fruit formed from that bud being different from others on the plant.

BUD SCALES   The protective sheath covering the embryonic shoot apex and leaves in a bud.

BULB   A short, modified stem with enlarged fleshy leaf bases found underground and serving as a storage organ.

BUNDLE SHEATH   The parenchyma or sclerenchyma cells surrounding the vascular bundle in a leaf.

BUTTON   A mushroom which is not yet mature, lacking the expanded cap.

CALLUS (1) A tissue of thin-walled, parenchymous cells formed at an injured surface of a plant.
(2) Undifferentiated tissue in a tissue culture.

CALORIE The amount of heat energy required to raise the temperature of one gram of water 1°C.

CALYPTRA The hoodlike structure formed from the archegonial wall which covers the capsule in a moss plant.

CALYX The outermost part of the flower, consisting of the sepals. It is usually green and encloses the flower when in the bud stage.

CAMBIAL ZONE The zone of tissue between the xylem and phloem which is composed of thin-walled cells, including cambium and its undifferentiated daughter cells.

CAMBIUM INITIALS The cells of the vascular cambium and cork cambium (phellogen); of two types: fusiform initials and ray initials.

CAP In fleshy types of fungi, the top structure which is supported by the stalk.

CAPILLARITY The movement of a liquid as a result of cohesive forces between molecules of the liquid together with adhesive forces between the liquid molecules and a solid surface, such as occurs in a capillary tube.

CAPILLARY SOIL WATER Water which is held in soil spaces and on soil particles against the force of gravity and is thus available to the plant.

CAPSULE (1) A simple, dehiscent, dry fruit which develops from two or more carpels and may open in a variety of ways.
(2) The spore case found in mosses and liverworts.
(3) A slimy layer surrounding certain bacterial cells.

CARBOHYDRATES A group of organic compounds (e.g., sugar, starch, and cellulose) composed of a carbon chain with hydrogen and oxygen atoms attached in a 2:1 ratio.

CARNIVOROUS That which feeds upon animals, as opposed to herbivorous. It includes plants that trap animals such as insects, utilizing their proteins.

CAROTENES Pigments of plant cells which are yellow or orange. An example is pro-vitamin A.

CAROTENOIDS The yellow, orange, or sometimes red pigments found in plant plastids; includes the carotenes.

CARPEL The floral organ, leaflike in structure, which bears the ovules. The pistil is composed of one or more carpels.

CATALYST A substance which influences the rate of a reaction but is not changed or used up by the reaction (e.g., an enzyme).

CELL The structural unit of plants and animals consisting of cytoplasm with one or more nuclei surrounded by a cell membrane. In plants it is usually surrounded by a cell wall.

CELL DIVISION The process whereby the cytoplasm is divided into two equal parts, frequently occurring by the formation of a cell plate.

CELL PLATE A membranelike structure which forms during early telophase at the equator of the spindle; the predecessor of the middle lamella.

CELL SAP The liquid contents of plant cell vacuoles.

CELLULOSE A complex carbohydrate that is made up of glucose molecules; the major component of the cell wall in most plants.

CHLOROPHYLLS The green pigments which are found in chloroplasts and are necessary for photosynthesis.

CHLOROPLAST A specialized, membrane-bound organelle in the cytoplasm of green plants which contains chlorophyll; the site of photosynthesis.

CHROMATID One of the two strands of a duplicated chromosome joined by the centromere.

CHROMATIN Primarily nucleoprotein material of the cell which stains intensely with basic dyes.

CLASS A taxonomic category ranking between an order and a division (phylum); consisting of closely related orders of plants.

CLIMAX The terminal, stable community of an ecological succession which is in equilibrium with the existing environment for the most part. It will not change unless the environment changes.

COENOCYTIC Referring to a cell with more than one nuclei (multinucleate) which are not separated by cell walls.

COHESION THEORY The theory explaining the rise of water in a plant based on the cohesion and adhesion of water molecules and the effects of transpiration.

COLEOPTILE The protective sheathlike structure covering the young shoot in grass seedlings. It is considered a part of the cotyledon morphologically.

COLLENCHYMA A flexible, supporting tissue or the cells of which it is composed; living, elongated cells with irregularly thickened primary walls.

COMMUNITY All the organisms living together and interacting with one another in a common environment or natural habitat.

COMPANION CELL A small, specialized parenchyma cell closely associated with the sieve-tube elements of the phloem in flowering plants.

COMPENSATION POINT The light intensity where the rates of photosynthesis and respiration are equal.

COMPETITION The struggle between two or more organisms for life's necessities (e.g., food, water, light, and minerals) when they are in limited supply.

COMPLETE FLOWER   A flower having four types of floral organs: sepals, petals, stamens, and pistil.

COMPOUND LEAF   A leaf which consists of two or more distinct leaflets.

CORK   Protective, secondary tissue formed by the cork cambium which prevents the passage of gases and water vapor. The cells are polygonal and nonliving at maturity, with suberized (waxy) walls.

CORK CAMBIUM   Also known as phellogen, the lateral meristem of woody and some herbaceous plants which produces cork toward the outside of the stem (and phelloderm toward the inside).

CORTEX   The primary tissue of stems and roots, consisting primarily of parenchyma cells and located between the epidermis and primary phloem (or endodermis in roots).

COTYLEDON   The seed leaves of an embryo which function to store or absorb food.

COVALENT BOND   A physical force which holds electrons together when they are shared more or less equally between two atoms.

CROP ROTATION   Growing different crops in regular rotation to aid in the control of insects and diseases and to increase soil fertility, particularly fixed nitrogen.

CROSSING OVER   The exchange of corresponding chromatid segments of a homologous chromosome pair during meiosis.

CROSS-POLLINATION   The process in which pollen is transferred from the anther of a flower in one plant to the stigma of a flower in another plant.

CUTICLE   The waxy layer covering the outer wall of epidermal cells.

CUTTING   A portion of a plant, such as stem segment or leaf, used for vegetative propagation.

CYTOLOGY   That branch of biology dealing with the structure and functions of the cell and its components.

CYTOPLASM   All the photoplasm of the cell with the exception of the nucleus.

DECIDUOUS   (1) Referring to plant parts (e.g., leaves, fruits, or flower organs) which fall off at the end of the growing period.
(2) Broad-leafed plants which lose all of their leaves regularly each year at the end of the growing season, as opposed to evergreens which always retain some leaves.

DECOMPOSERS   Organisms which recycle organic matter by breaking it down into smaller units.

DENITRIFICATION   The reduction of nitrates in the soil to gaseous nitrogen, carried out by certain bacteria.

DEOXYRIBONUCLEIC ACID (DNA)   The complex nucleic acid molecules which carry the genetic information in the cell. It is composed of phosphate, sugar (deoxyribose) and the purine and pyrimidine bases.

DEOXYRIBOSE   The 5-carbon sugar found in DNA which has one less oxygen than ribose.

DICHOTOMY   A system of branching in which the branches fork into two equal branches.

DICOTYLEDON   A subclass of flowering plants (angiosperms) which have two cotyledons or seed leaves; abbreviated, dicot.

DIFFERENTIALLY PERMEABLE   Referring to membranes which allow the passage of some materials but which slow or prevent the passage of others.

DIFFERENTIATION   The modifications of a cell, tissue, or organ in its physiology or morphology that occur during development, resulting in specialization for particular functions.

DIFFUSE-POROUS   A type of hardwood in which the vessels are approximately uniform in size and distribution throughout the growth ring.

DIFFUSION   The net movement of a substance from a region of high concentration to one of low concentration due to the random motion of the molecules. The end result is the uniform distribution of the molecules or particles throughout the available area.

DIFFUSION PRESSURE   The tendency of a particular kind of molecule to escape or diffuse due to the effects of concentration, temperature, and pressure.

DIGESTION   The process of transforming complex, usually insoluble, foods into simple, usually soluble, substances by the action of enzymes.

DIHYBRID   Referring to a cross between two organisms that differ in two pairs of genes.

DIOECIOUS   Bearing the male and female organs (staminate and pistillate flowers or pollen and seed cones) on separate individuals of the same species.

DIPLOID   Referring to the presence of two sets of chromosomes usually associated with the sporophyte generation; the $2n$ number of chromosomes.

DNA   *See* deoxyribonucleic acid.

DOMINANT   (1) A gene that expresses itself phenotypically and masks the effect of the partner allele; also referring to the trait controlled by such a gene.
(2) Referring to the one or more species which are most prevalent in an ecological community and which influence other plants in the community through their controlling effects on natural conditions.

DORMANCY A period of lowered physiological activity in various plant organs (e.g., seeds, bulbs, and buds).

DOUBLE CROSS A cross between hybrids originating from four inbred strains; referring especially to corn.

DOUBLE FERTILIZATION In flowering plants, the simultaneous fusion of the egg with the sperm (forming a $2n$ fertilized egg) and the other male gamete with the polar nuclei (forming a $3n$ primary endosperm nucleus).

DRY WEIGHT The weight of material after it has been dried at high temperatures for a sufficient time period and is thus devoid of moisture.

EARLY WOOD The part of the annual ring which is formed during the early part of the growing season, usually consisting of larger and less dense cells than are found in the late wood.

ECOLOGY The branch of science dealing with the relationship between organisms and their environment.

ECOSYSTEM An interacting system consisting of living organisms together with their nonliving environment.

EMBRYO The young sporophyte plant in a seed or archegonium before the beginning of the rapid growth period.

ENDODERMIS A single layer of specialized cells found between the cortex and pericycle in most roots and some stems.

ENDOPLASMIC RETICULUM An extensive network of double membranes forming channels and compartments in the cytoplasm of a cell.

ENDOSPERM The triploid, food storage tissue formed in the ovule of flowering plants by the union of a male gamete with the two polar nuclei of the embryo sac. Persisting in some seeds, it is digested and used by the growing sporophyte.

ENERGY The ability or capacity to do work, including such forms of energy as radiant, heat, electrical, chemical, and kinetic.

ENZYME A complex protein, synthesized in a living cell, that functions as an organic catalyst, speeding the rate of a chemical reaction.

EPIDERMIS The outermost layer of cells of various plant parts (e.g., leaves, young stems, and roots).

EROSION The removal of soil by the action of water, wind, and sometimes other natural agents.

ESSENTIAL ELEMENTS Chemical elements from the soil and air necessary for normal plant growth and development.

EUCARYOTIC Referring to cells which possess membrane-bound nuclei and other organelles, as opposed to procaryotic, and which undergo mitosis.

EVOLUTION The history of gradual changes that have occurred in a race, species, or larger group from generation to generation resulting in the differences characterizing various groups.

EYESPOT A small, pigmented structure in certain algae that is possibly light-sensitive.

F_1 The first (filial) generation after a cross. F_2 and F_3 are the second and third generations.

FATS Organic compounds consisting of carbon, hydrogen, and oxygen as in carbohydrates, but with a smaller proportion of oxygen to carbon atoms than in carbohydrates; includes oils which are fats in the liquid state.

FATTY ACIDS Organic acids occurring frequently as part of fat molecules.

FERMENTATION A series of respiratory reactions, usually occurring in the absence of free oxygen, during which the hydrogen released in glycolysis is combined with pyruvic acid to form alcohol, lactic acid, or other products. No additional energy is produced in the process.

FERTILIZATION In sexual reproduction, the fusion of two gametes forming a zygote.

FERTILIZERS Substances which are added to soil in order to improve the nutrient balance in the soil or to supply a nutrient which is lacking but essential for plant growth.

FIBERS Elongated, thick-walled cells with tapering ends which provide strength and support in various plant parts; includes wood and sclerenchyma fibers.

FIBROUS ROOT SYSTEM A root system comprised of secondary roots almost entirely and lacking any one primary root.

FIELD CAPACITY *See* Field percentage.

FIELD PERCENTAGE The greatest amount of available water in the soil after the action of gravity and the movement of free capillary water has ceased.

FILAMENT (1) Part of a stamen, the stalk which supports the anther.
　　　　　(2) In certain algae, referring to the threadlike row of cells.

FISSION Asexual reproduction in which a unicellular organism divides into two daughter cells of equal size.

FLAGELLUM A long, whiplike protrusion found on such cells as sperm, some spores, and certain bacteria which makes locomotion possible. Similar in internal structure to cilia, but longer.

FLOWER The reproductive structure which is characteristic of angiosperms. If complete, it includes the calyx, corolla, stamens, and carpels.

FLOWER BUD A bud that develops into one or more flowers.

FOOD Organic substances which can be broken down to furnish energy or used by an organism to build cellular components.

FOOD CHAIN Organisms in a community which are linked together by their food relationships; primary producers at the bottom of the chain, the largest carnivores at the top.

FOSSIL Plant or animal parts (or impressions of them) preserved in the earth's crust, thus giving information about life in past geological ages.

FRUIT In angiosperms, the mature, ripened ovary or group of ovaries which contain the seeds; also including any adjacent parts which have fused with the ovary.

FUNGICIDE A toxic chemical material which kills fungi or retards their growth.

GAMETE A haploid reproductive cell which may fuse with another gamete to produce a zygote.

GAMETOPHYTE GENERATION The haploid (n) phase of the life cycle which produces gametes.

GENE A fundamental hereditary unit which occurs in a linear arrangement on the chromosomes.

GENE INTERACTION The situation in which the presence of certain genes changes the normal expression of another gene.

GENETIC CODE A group of bases (triplets) which determine the sequence of amino acids in protein synthesis.

GENETICS The branch of science dealing with heredity.

GENOTYPE An organism's genetic makeup, either latent or expressed, as opposed to the phenotype; all of an individual's genes.

GENUS The taxonomy group between species and family which includes closely related species.

GEOTROPISM A growth movement which occurs in response to gravity.

GERMINATION The beginning or resumption of growth in an embryo or spore.

GIBBERELLIC ACID A plant hormone which influences many phases of development including seed germination, dormancy, cell elongation, and flowering; originally isolated from a fungus which caused a disease of rice.

GILL A plate located on the underside of the cap in certain fungi (e.g., basidiomycetes, the gill fungi).

GLUCOSE A simple sugar with the formula $C_6H_{12}O_6$; an important component in cellular respiration and energy conversions.

GLYCOLYSIS The early phase of respiration in which sugar is converted to pyruvic acid with the release of a small amount of useful energy; an anaerobic process.

GRAFTING A method of vegetative propagation in which a scion (a portion of one plant) is inserted on the stock (the stem or root of another plant) resulting in the union of their tissues.

GRAIN The simple, dry, one-seeded fruit, not opening at maturity (indehiscent), which has the pericarp and seed coat completely fused together. Also called a caryopsis, it is the characteristic fruit of the grass family.

GRANA Membrane structures, appearing like small stacked discs in the chloroplast. They contain the chlorophyll and carotenoid pigments and are the site of the light reactions of photosynthesis.

GREEN MANURE Referring to a live crop plowed into the soil for the purpose of increasing the amount of organic matter and nitrogen in the soil.

GROWTH Increase in size (volume or length) due to cell divisions and subsequent enlargement.

GROWTH HORMONE A chemical substance produced in one part of the plant that regulates the growth of another part of that plant. See hormone.

GROWTH INCREMENT The annual growth layer, usually applied when not observed in cross section.

GROWTH RING The growth layer of a woody stem as seen in cross-section; also called annual ring.

GROWTH SUBSTANCE A chemical material, synthetic in nature, that influences plant growth.

GUARD CELLS Cells which occur on either side of stomates (leaf pores) and which regulate the size of the stomates.

GYMNOSPERM A member of a group of vascular plants which have seeds that are not enclosed in an ovary (e.g., the conifers).

HABIT The physical appearance or form that characterizes an organism.

HAPLOID A term which indicates the presence of only one chromosome of each pair. Sometimes associated with gamete formation. A single set of unpaired chromosomes in a cell.

HARDWOOD Wood which is produced by woody dicotyledons, distinctive in that it has vessels; also referring to trees possessing hardwood.

HEARTWOOD The inner layers of wood which are nonliving and in which water conduction has ceased; usually darker in color than the surrounding sapwood.

HEMICELLULOSE Polysaccharides, found especially in cell walls, similar to cellulose though less complex and more soluble than cellulose.

HERB An annual, biennial, or perennial which usually lacks woody tissue and most frequently dies down at the end of the growing season.

HERBACEOUS Referring to any nonwoody plant.

HERBIVORE An animal which eats plant material.

HERBIVOROUS Feeding upon plants, in contrast to carnivorous.

HETEROTROPHIC Referring to organisms which are unable to

manufacture all of their own food and must therefore obtain complex organic material from external sources, as opposed to autotrophs.

HOLDFAST (1) The unicellular or multicellular, basal portion of an algal thallus that serves to attach it to a solid surface.

(2) The tip of a tendril, disklike in appearance, which functions in attachment.

HOMOZYGOUS Having the same allele at corresponding loci on homologous chromosomes.

HORMONE A chemical produced in minute quantities in one part of a plant or animal which has a physiological effect in another part of a multicellular plant or animal; a chemical messenger.

HUMIDITY Dampness. Relative humidity refers to the ratio of actual water-vapor content in the air to the largest amount of water vapor which the air can contain at that same temperature.

HUMUS A dark-colored mixture of organic matter, partially decomposed, in the soil.

HYBRID An individual derived from parents which have opposed genes for one or more characteristics.

HYBRID VIGOR The increased vigor that often occurs in the offspring of a cross between inbred lines or between unrelated forms, varieties, or species.

HYBRIDIZATION The process of producing offspring from a cross between genetically dissimilar parents.

HYDROCARBON An organic compound consisting only of carbon and hydrogen.

HYDROGEN BOND A weak bond between hydrogen in a molecule in which the charge is unequally distributed (polar molecule) and another polar molecule or part of the same molecule; an important factor in determining the structure of water, proteins, and chromosomes.

HYDROLYSIS A decomposition process in which a complex compound is broken down into simpler compounds with the addition of water.

HYDROPONICS The growth of plants in a defined nutrient solution as opposed to growth in soil.

HYPHAE The threadlike filaments of which the fungal plant body is composed.

HYPOCOTYL The part of an embryo or seedling axis that is located between the radicle and the point of attachment of the cotyledons.

IMBIBITION The process in which water molecules are adsorbed onto the internal surfaces of colloidal material resulting in water absorption and swelling by that material.

IMPERFECT FLOWER A flower with the reproductive structures of only one sex.

INBREEDING The breeding of closely related individuals; repeated self-pollination is normal in plants; used for the purpose of establishing the homozygous condition.

INCOMPLETE DOMINANCE The condition produced by two different alleles acting together which is intermediate between the effects of each of these genes when they are homozygous.

INDEPENDENT ASSORTMENT The notion that the inheritance of a trait on one chromosome is not controlled by traits on another chromosome.

INDOLE ACETIC ACID (IAA) A plant-growth hormone which is responsible for initiating cell elongation and indirectly responsible for phototropism and geotropism; also affecting other aspects of plant growth and development.

INTERCALARY MERISTEM A meristem that is located some distance from the apex; may be located between two tissues that are not yet mature.

INTERNODE That portion of the stem situated between any two successive nodes.

IONS Atoms or molecules which carry an electrical charge.

IRRITABILITY The ability to receive and respond to external stimuli, demonstrated by living protoplasm.

ISOGAMY Sexual reproduction by gametes which are alike in size and form, occurring in algae and fungi.

ISOTOPE Atoms of the same element which exhibit an altered atomic weight.

KELP General term for the larger forms of brown algae.

KILOCALORIE Unit of heat measurement equal to 1000 calories (kcal).

KINETIC ENERGY The energy which results from the movement of a body.

KINETIN A compound which initiates cell division; probably does not occur in nature, but related to the cytokinins.

KINGDOM Most general taxonomic division; presently, most scientists recognize five kingdoms.

KREBS CYCLE The series of chemical reactions occurring in the mitochondria involving the oxidation of pyruvic acid to carbon dioxide, hydrogen, and electrons. The hydrogen and electrons, which are bound to a carrier molecule, then undergo oxidative phosphorylation, resulting in the production of ATP and water.

LAMELLA Term meaning a layer of cellular membranes, most often referring to chlorophyll-containing membranes.

LATERAL Situated on one side of an organ rather than in a terminal location.

LATERAL BUD A bud which is located in the leaf axil.

LATERAL MERISTEMS Referring to meristems producing secondary tissue; the vascular cambium and cork cambium.

LAYERING A method of plant propagation in which stems develop roots while they are still attached to the parent.

LEACHING The removal of minerals from the soil due to water flow.

LEAF A lateral plant organ developing superficially from the tissues of the shoot apex; usually composed of a blade and petiole with some variations, and may be simple or compound; major photosynthetic organ in most plants.

LEAF BUD A bud which develops into a leaf-producing shoot, but no flowers.

LEAF GAP An interruption of the vascular cylinder of a stem above the point where the leaf trace bends away from the cylinder into the leaf; composed of parenchyma tissue.

LEAF PRIMORDIUM An embryonic leaf formed from an outgrowth of the apical meristem.

LEAF SCAR The scar left on a twig after the detachment of a leaf.

LEAFLET One of the units comprising a compound leaf.

LICHEN Specialized organism in which there is a symbiotic association of an alga and a fungus.

LIFE CYCLE The sequence of events in the life of a plant or animal; includes zygote formation to mode of reproduction of the mature individual and all pertinent stages in between.

LIGHT REACTIONS Photosynthetic reactions which require light and result in the synthesis of ATP and $NADPH_2$.

LIGNIN An organic substance associated with cellulose in the secondary cell walls of plants; important component of wood.

LINKAGE The tendency for a particular group of genes to be inherited together because they are located on the same chromosome.

LIPASE An enzyme which breaks down fats into fatty acids and glycerol.

LIPID General term for fatlike compounds (e.g., oils, fats, and certain steroids).

LOCUS A particular position on a chromosome.

LYSIS The rupturing of a cell, such as by a bacteriophage, or the disintegration of a compound, such as by an enzyme.

MACROMOLECULES Large organic molecules which usually possess a rather complex structure (e.g., proteins, DNA, and RNA).

MACRONUTRIENTS Those elements required for the growth of plants in relatively large amounts, as contrasted with micronutrients.

MALT Barley, or other cereal, which has been sprouted, dried, and ground, containing amylase and other enzymes.

MALTASE An enzyme responsible for hydrolyzing maltose to glucose.

MATING TYPE Referring to strains of algae or fungi which are unable to reproduce sexually with other members of the same strain, but are able to reproduce sexually with individuals of other strains of the same organism.

MEGASPORANGIUM The plant structure which produces megaspores.

MEGASPORE A type of spore, produced in the megasporangium and generally larger than a microspore, which develops into the female gametophyte.

MEGASPOROPHYLL The leaflike structure which bears one or more megasporangia.

MEIOSIS A sequence of divisions in which a diploid cell ($2n$) produces a haploid cell ($1n$); divisions which halve the chromosome number of a cell.

MEIOSPORE A spore produced in meiosis, having a haploid chromosome number.

MERISTEM Region of undifferentiated tissue in a plant which gives rise to new tissue through frequent cell division continuing throughout the life of the plant.

MESSENGER RNA (mRNA) A macromolecule synthesized using cellular DNA as a template. mRNA can move out of the nucleus to the ribosomes where it participates in protein synthesis.

MESOPHYLL Leaf tissue located between the upper and lower epidermis, composed of parenchyma cells containing chloroplasts.

METABOLISM The sum total of the chemical reactions within a cell.

MICRON 0.001 millimeter or 10^{-6} meters.

MICRONUTRIENTS Those elements required for the growth of plants in relatively small amounts, as contrasted with macronutrients.

MICROPYLE A small opening in the integuments of the ovule where the pollen tube generally enters to reach the female gametophyte.

MICROSPORANGIUM The plant structure producing microspores.

MICROSPORE A type of spore, produced in the microsporangium and generally smaller than a megaspore, which develops into the male gametophyte.

MICROSPOROPHYLL The leaflike structure which bears one or more microsporangia.

MIDDLE LAMELLA The intercellular wall layer, consisting primarily of pectic compounds, which cements together the primary cell walls of adjoining cells.

MILLIMICRON A unit of length equal to one thousandth of a micron ($m\mu$), one millionth of a millimeter, or one twenty-five-millionth of an inch; a nanometer.

MITOCHONDRIA Organelles found in the cytoplasm of cells in which some of the later stages of respiration take place; important in energy conversions and sometimes termed the powerhouse of the cell.

MITOSIS The division of a nucleus which results in chromosome duplication and the subsequent formation of two daughter cells.

MOLECULAR WEIGHT The relative weight of a molecule, assuming the weight of common carbon to be 12. It is the sum of the atomic weights (relative weights) of the atoms composing the molecule; also, the weight of a mole of a chemical substance in grams.

MOLECULE The smallest unit of a compound.

MONOCOTYLEDON A member of the class of flowering plants which have a single cotyledon (seed leaf).

MONOECIOUS Referring to a plant which has both staminate flowers (or pollen cones) and pistillate flowers (or seed cones) located on the same individual plant.

MORPHOGENESIS The processes resulting in the physiological and morphological differentiation of an organism.

MORPHOLOGY The branch of biology concerned with the study of the form and structure of organisms and their development.

MOSAIC Referring to the condition in a leaf which appears mottled due to unequal chlorophyll formation in the various parts of the leaf. This condition is usually produced by a viral infection.

MOTILE Having the ability to move.

MULTIPLE GENES Two or more separate gene pairs acting together in an organism to produce a combined effect on a single phenotypic characteristic.

MUTATION A change in the chemical composition of a gene which is reflected as a change in some inheritable characteristic.

MUTUALISM The situation in which two or more organisms live together, the association being of benefit to all involved.

MYCELIUM The mass of hyphae of which a fungal body is composed.

MYCOLOGY The branch of biology concerned with the study of fungi.

NATURAL SELECTION A process by which environmental parameters tend to eliminate those members of a natural population which are not as well adapted as others.

NEUTRON A particle found in the nucleus of an atom; its mass is about the same as a proton but uncharged.

NITRIFICATION Conversion of amonia to nitrate.

NITROGEN CYCLE Scheme by which nitrogen is recycled; involves fixation of nitrogen, usually by biological processes, incorporation into organic molecules, and eventual return of nitrogen gas to the atmosphere.

NITROGEN FIXATION   Conversion of nitrogen gas into a form available to plants; the process usually carried out by microorganisms.

NODE   Region where one or more leaves joins a plant stem.

NODULES   Enlarged areas on the roots of certain plants containing nitrogen-fixing microorganisms.

NUCLEAR ENVELOPE   Membrane surrounding the nucleus of a cell.

NUCLEOLUS   Small body found in the nucleus, which carries on ribosome production.

NUCLEUS   (1) Cellular organelle which contains the hereditary units of a cell.
(2) Central part of an atom.

OOGONIUM   A female sex organ that contains an egg.

ORGAN   Tissues organized into a structure serving a particular function.

ORGANELLE   Structural unit within the cytoplasm of a cell (e.g., chloroplast, mitochondria, nucleus, and endoplasmic reticulum).

OSMOSIS   Movement of water through a membrane in response to a concentration gradient.

OVARY   Enlarged part of the pistil which develops into the fruit.

OVULE   Structure containing the female gametophyte (including egg) in seed plants.

PALISADE LAYER   Tissue of the leaf mesophyll consisting of columnar-shaped parenchyma cells which contain chloroplasts. The palisade cells are oriented with their long axes at right angles to the leaf surface.

PARALLEL VENATION   A venation pattern in the leaf blade where the major veins are arranged parallel to one another along the longitudinal axis of the leaf; characterizes monocots.

PARENCHYMA   A type of cell which is living at maturity, thin-walled, and unspecialized, retaining the capacity for renewed cell division. The cells are usually isodiametric, though sometimes elongated in shape, having photosynthetic and storage functions; may also refer to the tissue composed of such cells.

PATHOGEN   Any organism having the ability to cause disease in another organism.

PEPTIDE   Two or more amino acids linked together via peptide bonds. The size division distinguishing between peptides and proteins is arbitrary.

PEPTIDE BOND   A bond formed between the acid group (COOH) of one amino acid and the basic amino group (NH_2) of another amino acid. When a peptide bond is formed, water (H_2O) is lost.

PERENNIAL PLANT A plant that remains living for several generations, as opposed to an annual plant.

PERICYCLE A layer of parenchyma cells located between the endodermis and the phloem in roots and some stems; area from which secondary roots arise.

PERMEABLE Describing materials through which diffusion may occur, usually refers to membranes.

PETAL A single unit of the corolla of a flower, usually strikingly colored.

PETIOLE The stalk which supports the leaf blade.

PHENOTYPE Appearance of an organism or a physiological condition which results from the inheritance and interaction of a particular genotype.

PHLOEM A tissue made up primarily of sieve cells and companion cells. The former functions in the long-distance transport of materials throughout the plant.

PHOTOAUTOTROPHIC Referring to organisms which use light energy to manufacture their own food (e.g., green plants).

PHOTOPERIODISM Time-measuring in organisms; the response of plants or animals to the day or night length (e.g., flowering).

PHOTOPHOSPHORYLATION The process in which chlorophyll-absorbed light energy is used in the production of high-energy, organic phosphate compounds.

PHOTOSYNTHESIS Process by which light energy is converted to chemical energy; the light-requiring production of organic molecules from CO_2 and water with the release of O_2.

PHOTOTROPISM The growth of a plant toward a source of light.

PHYLOGENY The evolutionary history of development of a group of organisms, especially in relation to that of other groups.

PHYSIOLOGY The branch of biology dealing with the functions and processes of living organisms.

PHYTOCHROME A plant pigment which exists in two reversible forms: P_R and P_{FR}. The latter form is thought to regulate many aspects of plant growth and development.

PIGMENT A complex organic compound, appearing colored, which has the ability to absorb certain wavelengths of light (e.g., chlorophyll).

PISTIL Centrally located flower organ composed of the stigma, style, and ovary (the latter made up of one or more carpels enclosing the ovules).

PIT A thin place or cavity in the plant cell wall; no secondary cell wall forms at a pit.

PITH The tissue found in the center of a stem interior to the vascular tissue, consisting primarily of parenchyma cells with other cell types sometimes occurring.

PLANKTON Aquatic organisms, usually microscopic in size, floating freely or having a weak ability to swim.

PLANT PATHOLOGY The branch of botany dealing with plant diseases.

PLASMA MEMBRANE Membrane surrounding and in immediate contact with the cytoplasm of the cell.

PLASMODESMATA Very small strands of cytoplasm which penetrate through pores in the cell wall serving to connect the living protoplasts of adjacent cells.

PLASMOLYSIS A situation where the cytoplasm of a cell has shrunk away from the cell wall as a result of water loss.

PLASTID A cell organelle bounded by a double membrane, located in the cytoplasm, and involved in food manufacture and storage.

POLAR NUCLEI Centrally located nuclei in the embryo sac. Fusion of the two polar nuclei with a male nucleus results in the formation of the endosperm.

POLLEN Male gametophytes of seed plants.

POLLEN MOTHER CELL In seed plants, a cell which undergoes meiosis giving rise to four microspores.

POLLEN TUBE Tubelike structure formed as the pollen grain germinates and carries the male sex cells to the ovule.

POLLINATION Transfer of pollen from the anther to a receptive female flower part, usually the stigma.

POLYMER A large molecule made of repeating or similar subunits; e.g., a protein is a polymer of amino acid units.

POLYPLOID A term which indicates that the cell, tissue, or plant has more than the diploid ($2n$) number of chromosomes.

POLYSACCHARIDE Polymers consisting of repeating sugar units.

PRIMARY GROWTH Growth which originates from root or shoot apical meristems, as opposed to secondary growth which originates from a cambium.

PRIMARY MERISTEM Meristematic tissue which is derived from the apical meristem; includes three kinds: protoderm, procambium, and ground meristem.

PRIMARY TISSUES Tissues originating from cell divisions within the apical meristems of the root and shoot.

PRIMARY WALL Wall layer (primarily cellulosic) deposited initially in growing plant cells; in some tissues displaced outward by the production of a chemically distinct secondary wall after the cell reaches its final size.

PROCARYOTIC Organisms which do not undergo mitosis and which lack membranes around various cell organelles (e.g., bacteria and blue-green algae).

PROTOPLASM The living material in the cells of all organisms.

PRODUCTIVITY The amount of energy stored as biomass by an entire ecosystem or one of its components.

PROPHASE A stage in mitosis in which the chromosomes become distinct and the nuclear envelope disappears.

PROTHALLUS The relatively independent gametophyte of ferns and certain other lower vascular plants. Originating from the germination of a spore and carrying the sex organs.

PROTON Positively charged particle found within the nucleus of an atom.

PSEUDO Prefix indicating something which is false.

PYRIMIDINE A nitrogenous base (e.g., cytosine, thymine, or uracil) which is an important constituent of the nucleic acids.

RADIAL SECTION A slice or section cut along a plane which passes longitudinally through the center of a stem or root.

RADICLE The embryonic root in a seed.

RADIOACTIVE Emitting radiation of alpha, beta, or gamma particles.

RADIOISOTOPE An isotope which emits radiation.

RAY (1) Narrow beam of radiation.

(2) A distinctive, narrow, flat set of cells arranged along the radial axis of vascular tissue; same as vascular ray.

(3) The flat, generally strap-shaped corolla of the marginal flowers in sunflowers, daisies, and their relatives.

RECESSIVE The trait or gene which is not expressed in a heterozygote, generally indicated by a lowercase letter. Compare with dominant.

REDUCTION DIVISION (1) The nuclear division which produces daughter nuclei with half the number of chromosomes as the parent nucleus, generally the first division of meiosis.

(2) In a heterozygote, the division of meiosis which separates the contrasting alleles.

REPLICATION Copying or reproducing; in biology, most commonly used with reference to molecules such as DNA.

RESPIRATION Consumption of free O_2 and food with the release of CO_2, H_2O, and energy. In plants this is strictly an intracellular biochemical process.

RHIZOID A filamentous or hairlike appendage of tissue which penetrates the substrate from the underside of some plants, such as mosses and liverworts; functions in anchorage and absorption.

RHIZOME An underground stem, usually horizontal and enlarged with storage tissue, such as potato.

RNA Ribonucleic acid, a polymer comprised of alternating sugar (ribose) and phosphate, with one of the bases adenine, cytosine, guanine, or uracil attached laterally to each sugar.

ROOT Most often the part of a plant which grows downward into the soil, has no pith, and serves in anchorage and absorption. Exception: Some stems have no pith; some stems grow underground; some roots grow above ground; roots never bear leaves. Compare with Stem.

ROOT CAP   The dome-shaped, loose protective covering of cells at the tips of many roots.

ROOT PRESSURE   The pressure created in the roots of some plants which aids in transporting sap upwards toward the stems and leaves.

RUNNER   A stem which grows horizontally along the surface of the ground and often roots at the nodes. *See also* Stolon.

SAP   Plant juice or fluid, generally that which circulates through the vascular tissue or is contained within the cell vacuole.

SAPROPHYTE   A heterotrophic plant which derives its food from dead organic matter. Compare with Parasite.

SAPWOOD   The light-colored outer layer of xylem which is most active in transport. Compare with Heartwood.

SCALARIFORM   Ladderlike, as the transverse secondary thickenings on some xylem elements.

SCHIZOCARP   A dry fruit which splits at maturity into carpels usually containing one seed each.

SCION   In grafting, the detached twig which is affixed to another stem with roots. *See also* Stock.

SCLERENCHYMA   Strengthening or protective tissue made of thick-walled cells. Long thin sclerenchyma cells are fibers, round-shaped ones are stone cells.

SECONDARY ROOT   Root which arises from another root. Compare with Primary root.

SECONDARY TISSUE   Tissue made up of cells which were produced by a cambium, not the apical meristem. Compare with Primary tissue.

SECONDARY WALL   The final layer of thickening of a plant cell wall, laid down by the protoplast upon the primary wall.

SEED   The matured ovule, consisting of the following parts: An outer covering called the seed coat or integument, a nutritive tissue made up of endosperm or cotyledons, and an embryo plant.

SEED SCARIFICATION   Scratching or rupturing the seed coat, presumably allowing entry of water or oxygen required for germination.

SEGREGATION   The separation of the two paired chromosomes and also their genes during meiosis, so that gametes carry one each.

SELECTION   The preferential survival and reproduction of some genotypes at the expense of others.

SELF-POLLINATION   Transfer of pollen from anther to a stigma on the same plant.

SELF-STERILITY   The inability to produce viable seed by self-pollination.

SEMIPERMEABLE Property of some membranes which allows free passage of some molecules while restricting others.

SEPALS In most complete flowers, the outermost appendages, usually green, leaflike in texture, and enclosing the bud.

SEXUAL In biology, refers to a type of reproduction involving the union of gametes. Compare with Asexual.

SHOOT Stem and leaves.

SHRUB A woody plant with several main stems from the base.

SIEVE PLATE The perforated end-wall between two adjacent cells in a sieve tube of phloem tissue.

SIEVE TUBE Tube in phloem formed by many cells arranged end to end and separated by sieve plates.

SOFTWOOD Wood lacking fibers; gymnosperm wood.

SOLUTE Substance dissolved in liquid medium. Compare with Solvent.

SOLUTION In biology, a combination of a liquid, usually water, with any of various sugars, salts, or other materials which disperse as molecules or ions in the liquid medium.

SOLVENT In biology, the liquid medium, usually water, in which other substances are dispersed and dissolved. Compare with Solute.

SORUS A cluster of sporangia on the underside of a fern leaf.

SPECIES The basic unit of classification; a kind of organism; a group of individuals which do, or are potentially capable of, interbreeding freely in nature.

SPERM A male gamete or sex cell, usually smaller and motile as compared with the larger, stationary egg.

SPINDLE A football-shaped set of protein microfibrils in a cell which functions in chromosome movement during mitosis and meiosis.

SPONGY TISSUE The layer of loosely packed, irregularly shaped cells within many leaves.

SPORANGIUM A special structure or case within which spores are produced.

SPORE A asexual reproductive cell.

SPORE MOTHER CELL A cell which produces spores by cell division; usually refers to the $2n$ cell which by meiosis produces four $1n$ cells.

SPOROPHYTE The diploid or $2n$ plant.

SPORT See Mutation.

SPRINGWOOD In an annual growth ring, the light-colored part comprised of large, thin-walled cells.

STAMEN The floral part which produces pollen, usually consisting of a long slender filament with an anther at its tip.

STARCH A food-storage carbohydrate in plants; a polymer of sugars, of the generalized formula $(C_6H_{12}O_5)_n$.

STELE The vascular tissue in a stem or root, comprised of xylem, phloem, endodermis, pericycle, and pith.

STEM Most often the part of a plant which grows upward and bears leaves. *See also* Root.

STIGMA The uppermost tip of the pistil where pollen is deposited during pollination.

STIPULE Pair of appendages at the base of the leaf on many plants; may be almost leaflike to minute in size.

STOCK (1) Garden ornamentals in the genus *Matthiola* in the mustard family.

(2) Any race or strain of similar organisms.

(3) In grafting, the rooted part to which a scion is grafted.

STOLON Same as runner; a horizontal stem growing along the surface of the ground and rooting at the nodes.

STOMATE The pore in epidermis, mostly in leaves, through which gases pass, controlled by guard cells; also called stoma.

STONE CELLS Very thick-walled, more or less isodiametric, strengthening and protective cells in sclerenchyma.

STROBILUS A compact set of overlapping scales borne on a central axis and bearing reproductive organs.

STRAIN A group of genetically very similar organisms of common descent; a race, breed, stock.

STYLE In a pistil, the often thin and elongate appendage between the ovary and stigma.

SUBERIN Waxy waterproof substance in the walls of some plant cells, particularly of cork.

SUCCESSION The sequence of kinds of vegetation occurring on a disturbed site or beginning with bare ground or rock and leading to a stable vegetation type.

SUCROSE Common sugar found in many plants; a disaccharide $C_{12}H_{22}O_{11}$.

SUMMER WOOD In an annual growth ring, the dark-colored part comprised of small, compact, thick-walled cells.

SUSPENSION A dispersion of tiny solid particles of water.

SYMBIOSIS The close association of two unrelated organisms with little or no ill effects to either.

SYNAPSIS The intimate pairing of chromosomes during meiosis.

SYNERGIDS The two nuclei which flank the egg nucleus in a mature embryo sac.

SYNGAMY The fusion of gametes in fertilization.

TANGENTIAL SECTION A slice or section cut along a plane which passes longitudinally through a stem or root off the center (i.e., at right angles to the radius).

TAP-ROOT SYSTEM A root system comprised of one main primary root together with the secondary roots.

TAXONOMY The study of variation in organisms, including their classification, identification, and nomenclature.

TELOPHASE The final stage of mitosis or meiosis during which the anaphase movement ceases, the chromosomes become diffuse, and nuclear envelopes form around the daughter nuclei.

TENDRIL A thin, elongated extension from a plant stem or leaf which generally coils around other suitable objects for support.

TETRAD Group of four; in botany, the group of four spores produced by meiosis. Sometimes, as in zoology, the group of four chromatids in a bivalent during meiosis.

TETRAPLOID Having the 4X number of chromosomes. Generally n and $2n$ refer to the gametophytic and sporophytic chromosome numbers, respectively, whereas X refers to the basic chromosome number for the group. Thus, 4X is the $2n$ number of a tetraploid, 2X the n number.

THALLOPHYTE A plant lacking vascular tissue throughout its life history (e.g., mosses, liverworts, algae, fungi, and bacteria).

THALLUS A plant body which lacks vascular tissue, such as in thallophytes.

TISSUE A group of cells with a common primary function.

TRACHEID An elongated tapered xylem cell with pitted walls; functions in support and water transport.

TRANSPIRATION The loss of water from stems and leaves, chiefly through the stomates.

TRANSVERSE SECTION A slice or section cut along a plane at right angles to the longitudinal axis, as of a stem or root.

TREE A woody plant having a single main stem from the base.

TRIPLOID Having the 3X number of chromosomes. *See also* Tetraploid.

TROPISM The bending movement of a plant in response to a simulus such as light or gravity.

TRUNK The main stem of a tree.

TUBE NUCLEUS The nucleus in a pollen grain or tube which leads the others as the pollen tube grows down the style.

TUBER A fleshy underground stem, such as a potato.

TUNDRA A low, treeless vegetation type of high elevations and the far north, growing over permanently frozen subsoil.

TURGOR PRESSURE The actual pressure within a plant cell resulting from water uptake.

VACUOLAR MEMBRANE Membrane separating the vacuole from the cytoplasm; also called tonoplast.

VACUOLE A large organelle (in mature plant cells) containing a watery solution of stored materials. Its function(s) is not entirely clear.

VARIATION Differences among individuals and populations.

VASCULAR BUNDLE General term referring to strands of xylem and phloem.

VASCULAR CAMBIUM A meristematic tissue which produces both secondary xylem and secondary phloem.

VASCULAR RAY Cells produced by the vascular cambium which extend in flat vertical plates along the radial axis into the secondary xylem and phloem.

VASCULAR TISSUE The xylem and phloem.

VEGETATIVE REPRODUCTION Reproduction via mitosis; reproduction without the fusion of gametes.

VEIN A term referring to the vascular bundle in a leaf.

VENATION The arrangement of veins with the plant, usually used in connection with a leaf.

VESSEL ELEMENT A particular cell type found in the xylem; characterized by a tubelike structure with thickened lateral walls and absence of end walls.

VIRUS A particle that can reproduce only within living cells; believed to represent the border between the living and nonliving.

VITAMIN Organic molecules divergent in structure which are essential in relatively small concentrations for normal growth and development. Green plants, but not animals, can synthesize their own vitamins.

WAVE LENGTH The distance between peaks of a wave.

WEED Any plant not valued for food or beauty and/or considered to compete with valued vegetation.

WHORL A circular arrangement of plant parts.

WOOD Xylem, usually secondary xylem.

WALL PRESSURE Pressure exerted by the cell wall on the protoplast.

WILTING Flaccid condition resulting from an abnormal loss of cellular water.

XEROPHYTE A plant which lives in very dry places.

XYLEM A complex plant tissue functioning primarily in water transport; consists of living storage cells, fibers, tracheids, and vessels.

ZEATIN A cytokinin isolated originally from corn.

ZOOSPORE A motile spore in algae and fungi.

ZYGOSPORE A thick-walled spore formed from the fusion of two isogametes.

ZYGOTE A diploid cell resulting from the fusion of two haploid gametes.

Index